Integrated Role of Nutrition and Physical Activity for Lifelong Health

Integrated Role of Nutrition and Physical Activity for Lifelong Health

Special Issue Editors

Karsten Koehler
Clemens Drenowatz

MDPI • Basel • Beijing • Wuhan • Barcelona • Belgrade

MDPI

Special Issue Editors
Karsten Koehler
Technical University of Munich
Germany

Clemens Drenowatz
University of Education Upper Austria
Austria

Editorial Office
MDPI
St. Alban-Anlage 66
4052 Basel, Switzerland

This is a reprint of articles from the Special Issue published online in the open access journal *Nutrients* (ISSN 2072-6643) from 2018 to 2019 (available at: https://www.mdpi.com/journal/nutrients/special_issues/Integrated_Nutrition_Physical_Health)

For citation purposes, cite each article independently as indicated on the article page online and as indicated below:

LastName, A.A.; LastName, B.B.; LastName, C.C. Article Title. *Journal Name* **Year**, *Article Number*, Page Range.

ISBN 978-3-03921-211-8 (Pbk)
ISBN 978-3-03921-212-5 (PDF)

Contents

About the Special Issue Editors

Karsten Koehler is an Assistant Professor of Exercise, Nutrition and Health at the Technical University of Munich and a Visiting Adjunct Assistant Professor at the University of Nebraska–Lincoln. After receiving formal training in Nutrition Sciences (MSc, University of Hohenheim, Germany) and Exercise Sciences (Ph.D., German Sports University Cologne, Germany), Dr. Koehler completed a postdoctoral fellowship at the Pennsylvania State University and was an Assistant Professor at the University of Nebraska–Lincoln, where he directed the Sports and Exercise Nutrition Lab. In 2019, Dr. Koehler moved to the Technical University of Munich, where his research focuses on the multiple pathways through which physical activity and exercise impact energy balance and food intake. Dr. Koehler has published more than 50 peer-reviewed papers and 5 book chapters. In March 2019, Dr. Koehler joined the editorial boards of Nutrients.

Clemens Drenowatz is a Professor in the Division of Physical Education at the University of Education Upper Austria, Austria. After receiving his Ph.D. in Kinesiology at Michigan State University, USA, he pursued postdoctoral studies at the University of Bath, UK, and Ulm University Medical Center, Germany, where he examined the association of correlates of energy balance (e.g., physical activity, diet) in children. As an Assistant Professor in the Arnold School of Public Health at the University of South Carolina, USA, Dr. Drenowatz continued his research on the regulation of energy balance, with a special emphasis on compensatory adaptations in response to exercise-based interventions. At his current position, his research focuses on the role of motor competence in the promotion of an active and healthy lifestyle. Dr. Drenowatz has published more than 70 peer-reviewed articles, 3 book chapters, and 2 books.

nutrients

MDPI

Editorial

Integrated Role of Nutrition and Physical Activity for Lifelong Health

Karsten Koehler [1,2,*] and Clemens Drenowatz [3]

1   Department of Sport and Health Sciences, Technical University of Munich, 80992 München, Germany
2   Department of Nutrition and Health Sciences, University of Nebraska-Lincoln, Lincoln, NE 68583, USA
3   Division of Physical Education, University of Education Upper Austria, 4020 Linz, Austria
*   Correspondence: karsten.koehler@tum.de; Tel.: +49-89-289-24488

Received: 19 June 2019; Accepted: 24 June 2019; Published: 26 June 2019

It is well established that healthy nutrition and physical activity (PA) are key lifestyle factors that modulate lifelong health through their ability to improve body composition, musculoskeletal health, and physical and cognitive performance, as well as to prevent metabolic diseases including obesity, diabetes mellitus, and cardiovascular disease across the lifespan. While the health benefits of nutrition and PA are often studied singularly, it has become more and more evident that the integration of nutrition and PA has the potential to produce greater benefits when compared to strategies focusing solely on one or the other. This Special Issue entitled "Integrated Role of Nutrition and Physical Activity for Lifelong Health" is devoted to manuscripts that highlight this integrational approach on various outcomes related to lifelong health. In response to our call, a total of 14 manuscripts were included. In addition to research focusing on the integrated benefits of nutrition and PA on various markers related to health and performance across a broad spectrum of life stages, several studies examining how PA has the potential to change food consumption were also included.

The featured article by Gustafson et al. reports that food choices are altered in the context of exercise. In this experiment, gym goers were asked to choose between a healthy and unhealthy snack to be consumed after completion of their exercise. Compared to when this choice was made prior to exercise, individuals after completing exercise were less likely to choose a healthy snack and instead chose the unhealthy option [1]. This possibly unhealthy impact of sport participation on food choices was extended in another study by Koenigstorfer, although participation was limited to passive viewership in this instance. Regardless, visits to sporting events were associated with an increased preference for unhealthy food items, although it was notable that a similar increase was also observed for some non-sport related venues, such as music concerts [2]. In a study focusing on the health effects of dietary choices in the context of an exercising population, Wirnizter et al. report that vegetarian and vegan athletes exhibited lower body weights and vegetarian athletes reported a lower prevalence of allergies when compared to their omnivore counterparts, while many other health-related outcomes were similar. These findings suggest that adherence to a vegetarian or even vegan diet is not detrimental to the health of endurance athletes [3].

Two studies [4,5] examine the role of diet and PA on health-related parameters in youth. Drenowatz et al. show that healthy dietary choices and sports participation are independently associated with motor competence, which is an important contributor to an active and healthy lifestyle [6,7]. Meng et al. report beneficial effects of an education-based obesity prevention program on dietary intake in adolescents. This study also highlights the importance of club sports participation, as PA levels were higher during the sport season compared to the off-season [5]. The promotion of leisure time PA outside-specific club settings thus remains a critical component for future research.

The importance of an active and healthy lifestyle is also shown by Van Elten et al. who examine the sustainability of a diet and PA intervention on cardiometabolic health in women. Even though a potential effect of snack intake on insulin resistance 3 to 8 years after the intervention was shown, the results emphasize the importance of current lifestyle choices for cardio-metabolic health. In order to achieve sustainable lifestyle changes, prolonged engagement in the intervention may be necessary [8]. A review by Balan et al. further emphasizes the importance of PA and diet to counteract age-related diseases by showing beneficial effects of fiber and unsaturated lipids on telomere health. While the authors acknowledge that more research is needed, they also suggest a protective effect of PA on telomere maintenance, which contributes to health in old age [9].

Given the importance of healthy lifestyle choices, a key question remains on how to promote healthy dietary choices and PA among a broad population. Electronic health (Ehealth) approaches could provide economically feasible opportunities—Doorn-van-Atten et al. examine the efficacy of an eHealth intervention in older adults, who are commonly less receptive to technology-based intervention strategies. Their results indicate beneficial effects of PA self-monitoring along with educational materials on lifestyle choices [10]. Educational materials were also shown to improve self-efficacy to engage in PA and consume a healthy diet in women as well as to provide growth-promoting animal protein to their stunted offspring [11].

In addition to the improvement of health outcomes, three articles address the role of nutrition and PA in modulating inflammation, which is associated with chronic disease and aging. When comparing elderly individuals across the PA spectrum, Ferrer et al. report that active individuals not only exhibited improved body composition but also improved blood profiles of inflammatory and anti-inflammatory markers [12]. These findings are strongly supported by Draganidis et al. who compared elderly individuals with low vs. high levels of systemic inflammation and found that low systemic inflammation is associated with greater levels of PA, particularly moderate-to-vigorous PA as well as increased antioxidant intake [13]. Although inflammation was not found to be significantly affected in a study assessing the impact of a purified vegan diet in exercising rats, this diet—especially when combined with an exercise regimen—was associated with improved body composition, metabolic markers, and physical performance [14].

Another topic in the area of exercise nutrition that is heavily debated involves protein requirements for exercisers. Isenmann et al. demonstrate the importance of adequate protein and carbohydrate intake from foodstuffs following an exercise bout for the facilitation of muscle regeneration while minimizing the inflammatory response [15]. Reckmann et al. present a novel method for quantifying exogenous protein oxidation using a breath test. While their findings failed to show alterations in whole-body metabolism in response to short-term fluctuations in protein intake, their data suggest that there is large inter-individual variability in response to protein-restricted diets. Accordingly, further research is needed to clarify the influence of dietary choices and nutrient intake on protein metabolism in active populations [16].

Taken together, the research presented in this Special Issue supports the previously emphasized role of integrating diet and PA on general health and well-being across the lifespan. A key issue for future research, therefore, will be the implementation of intervention strategies that promote an active and healthy lifestyle along with the exploration of the specific mechanisms that explain the individual and combined contribution of PA and nutrition to various health outcomes, including cardiovascular and metabolic diseases as well as orthopedic problems and depression.

Author Contributions: Conceptualization, Writing-Original Draft Preparation, and Writing-Review & Editing: K.K. and C.D.

Conflicts of Interest: The authors declare no conflict of interest.

References

1. Gustafson, C.R.; Rakhmatullaeva, N.; Beckford, S.E.; Ammachathram, A.; Cristobal, A.; Koehler, K. Exercise and the Timing of Snack Choice: Healthy Snack Choice is Reduced in the Post-Exercise State. *Nutrients* **2018**, *10*, 1941. [CrossRef] [PubMed]

2. Koenigstorfer, J. Childhood Experiences and Sporting Event Visitors' Preference for Unhealthy versus Healthy Foods: Priming the Route to Obesity? *Nutrients* **2018**, *10*, 1670. [CrossRef] [PubMed]
3. Wirnitzer, K.; Boldt, P.; Lechleitner, V.; Wirnitzer, G.; Leitzmann, C.; Rosemann, T.; Knechtle, B. Health Status of Female and Male Vegetarian and Vegan Endurance Runners Compared to Omnivores—Results from the NURMI Study (Step 2). *Nutrients* **2019**, *11*, 29. [CrossRef] [PubMed]
4. Drenowatz, C.; Greier, K. Association of Sports Participation and Diet with Motor Competence in Austrian Middle School Students. *Nutrients* **2018**, *10*, 1837. [CrossRef] [PubMed]
5. Meng, Y.; Manore, M.M.; Schuna, J.M.J.; Patton-Lopez, M.M.; Branscum, A.; Wong, S.S. Promoting Healthy Diet, Physical Activity, and Life-Skills in High School Athletes: Results from the WAVE Ripples for Change Childhood Obesity Prevention Two-Year Intervention. *Nutrients* **2018**, *10*, 947. [CrossRef] [PubMed]
6. Bremer, E.; Cairney, J. Fundamental movement skills and health-related outcomes: A narrative review of longitudinal and intervention studies targeting typically developing children. *Am. J. Lifestyle Med.* **2016**, *12*, 148–159. [CrossRef] [PubMed]
7. Robinson, L.E.; Stodden, D.F.; Barnett, L.M.; Lopes, V.P.; Logan, S.W.; Rodrigues, L.P.; D'Hondt, E.D. Motor competence and its effect on positive developmental trajectories of health. *Sports Med.* **2015**, *45*, 1273–1284. [CrossRef] [PubMed]
8. Van Elten, T.M.; Van Poppel, M.N.M.; Gemke, R.J.B.J.; Groen, H.; Hoek, A.; Mol, B.W.; Roseboom, T.J. Cardiometabolic Health in Relation to Lifestyle and Body Weight Changes 3–8 Years Earlier. *Nutrients* **2018**, *10*, 1953. [CrossRef] [PubMed]
9. Balan, E.; Decottignies, A.; Deldicque, L. Physical Activity and Nutrition: Two Promising Strategies for Telomere Maintenance? *Nutrients* **2018**, *10*, 1942. [CrossRef] [PubMed]
10. Van Doorn-van Atten, M.N.; De Groot, L.C.P.G.M.; De Vries, J.H.M.; Haveman-Nies, A. Determinants of Behaviour Change in a Multi-Component Telemonitoring Intervention for Community-Dwelling Older Adults. *Nutrients* **2018**, *10*, 1062. [CrossRef] [PubMed]
11. Mahmudiono, T.; Mamun, A.A.; Nindya, T.S.; Andrias, D.R.; Megatsari, H.; Rosenkranz, R.R. The Effectiveness of Nutrition Education for Overweight/Obese Mother with Stunted Children (NEO-MOM) in Reducing the Double Burden of Malnutrition. *Nutrients* **2018**, *10*, 1910. [CrossRef] [PubMed]
12. Ferrer, M.D.; Capó, X.; Martorell, M.; Busquets-Cortés, C.; Bouzas, C.; Carreres, S.; Mateos, D.; Sureda, A.; Tur, J.A.; Pons, A. Regular Practice of Moderate Physical Activity by Older Adults Ameliorates Their Anti-Inflammatory Status. *Nutrients* **2018**, *10*, 1780. [CrossRef] [PubMed]
13. Draganidis, D.; Jamurtas, A.Z.; Stampoulis, T.; Laschou, V.C.; Deli, C.K.; Georgakouli, K.; Papanikolaou, K.; Chatzinikolaou, A.; Michalopoulou, M.; Papadopoulos, C.; et al. Disparate Habitual Physical Activity and Dietary Intake Profiles of Elderly Men with Low and Elevated Systemic Inflammation. *Nutrients* **2018**, *10*, 566. [CrossRef] [PubMed]
14. Bloomer, R.J.; Schriefer, J.H.M.; Gunnels, T.A.; Lee, S.R.; Sable, H.J.; Van der Merwe, M.; Buddington, R.K.; Buddington, K.K. Nutrient Intake and Physical Exercise Significantly Impact Physical Performance, Body Composition, Blood Lipids, Oxidative Stress, and Inflammation in Male Rats. *Nutrients* **2018**, *10*, 1109. [CrossRef] [PubMed]
15. Isenmann, E.; Blume, F.; Bizjak, D.A.; Hundsdörfer, V.; Pagano, S.; Schibrowski, S.; Simon, W.; Schmandra, L.; Diel, P. Comparison of Pro-Regenerative Effects of Carbohydrates and Protein Administrated by Shake and Non-Macro-Nutrient Matched Food Items on the Skeletal Muscle after Acute Endurance Exercise. *Nutrients* **2019**, *11*, 744. [CrossRef] [PubMed]
16. Reckman, G.A.R.; Navis, G.J.; Krijnen, W.P.; Van der Schans, C.P.; Vonk, R.J.; Jager-Wittenaar, H. Whole Body Protein Oxidation Unaffected after a Protein Restricted Diet in Healthy Young Males. *Nutrients* **2019**, *11*, 115. [CrossRef] [PubMed]

nutrients

MDPI

Article

Comparison of Pro-Regenerative Effects of Carbohydrates and Protein Administrated by Shake and Non-Macro-Nutrient Matched Food Items on the Skeletal Muscle after Acute Endurance Exercise

Eduard Isenmann [1,2], Franziska Blume [1], Daniel A. Bizjak [1], Vera Hundsdörfer [1], Sarah Pagano [1], Sebastian Schibrowski [3], Werner Simon [3], Lukas Schmandra [1] and Patrick Diel [1,*]

[1]   Institute for Cardiovascular Research and Sports Medicine, Department of Molecular and Cellular Sports Medicine, German Sports University, 50333 Cologne, Germany; e.isenmann@dshs-koeln.de (E.I.); franzi.blume@web.de (F.B.); D.Bizjak@dshs-koeln.de (D.A.B.); verahundsdoerfer@gmail.com (V.H.); sara.pagano@gmx.de (S.P.); lukas.schmandra@gmail.com (L.S.)
[2]   Department of Fitness and Health, IST-University of Applied Sciences, 40233 Dusseldorf, Germany
[3]   Rheinische Fachhochschule Cologne, 50676 Cologne, Germany; sebastian.schibrowski@googlemail.com (S.S.); w.simon.314@web.de (W.S.)
*   Correspondence: Diel@dshs-koeln.de; Tel.: +49-221-4982-5860

Received: 20 February 2019; Accepted: 27 March 2019; Published: 30 March 2019

Abstract: Physical performance and regeneration after exercise is enhanced by the ingestion of proteins and carbohydrates. These nutrients are generally consumed by athletes via whey protein and glucose-based shakes. In this study, effects of protein and carbohydrate on skeletal muscle regeneration, given either by shake or by a meal, were compared. 35 subjects performed a 10 km run. After exercise, they ingested nothing (control), a protein/glucose shake (shake) or a combination of white bread and sour milk cheese (food) in a randomized cross over design. Serum glucose ($n = 35$), serum insulin ($n = 35$), serum creatine kinase ($n = 15$) and myoglobin ($n = 15$), hematologic parameters, cortisol ($n = 35$), inflammation markers ($n = 27$) and leg strength ($n = 15$) as a functional marker were measured. Insulin secretion was significantly stimulated by shake and food. In contrast, only shake resulted in an increase of blood glucose. Food resulted in a decrease of pro, and stimulation of anti-inflammatory serum markers. The exercise induced skeletal muscle damage, indicated by serum creatine kinase and myoglobin, and exercise induced loss of leg strength was decreased by shake and food. Our data indicate that uptake of protein and carbohydrate by shake or food reduces exercise induced skeletal muscle damage and has pro-regenerative effects.

Keywords: endurance exercise; skeletal muscle damage; inflammation; protein; carbohydrates; protein shake; food

1. Introduction

Consumption of proteins and carbohydrates after exercise via whey protein and glucose shakes is a common strategy for the enhancement of regeneration and physical performance after training [1]. There are reports that isolated amino acids, mainly leucine, can increase strength after exercise and result in a stimulation of the recovery of skeletal muscle after exercise [2]. Damage of muscle fibers by physical activity modulates protein synthesis, but simultaneously protein degradation [3]. Disturbance of the balance of muscle fiber degradation and synthesis leads to fiber degeneration and muscle atrophy, reduction of strength, and increase in muscle soreness [4,5].

Molecular mechanisms involved in muscle recovery include a modulation of protein synthesis and protein breakdown [6], an inhibition of the inflammatory response, and the activation of satellite cells [7]. Modulation of the balance of protein breakdown and protein synthesis is an important mechanism in skeletal muscle recovery and adaptation after exercise. It has been demonstrated that muscle protein breakdown rates are elevated in the days following resistance exercise [8]. In mouse knock out models, the inhibition of muscle protein breakdown impairs muscle recovery [9]. Some authors claim that modulation of protein breakdown is an important mechanism in muscle recovery in the period immediately after muscle damage [10]. In contrast, the stimulation of protein synthesis seems to be important for long term regeneration and hypertrophy. Therefore, all strategies affecting the balance between protein breakdown and synthesis in the period after damage will directly influence skeletal muscle recovery.

Protein metabolism in skeletal muscle has been shown to be stimulated by an ingestion of dietary proteins [11,12]. Whey protein supplementation results in a high availability of amino acids within the blood [13]. However, there are conflicting data about the effects of protein supplementation on physiological markers related to muscle recovery, like muscle strength, muscle soreness and serum creatine kinase (CK) concentrations after exercise [14,15]. These conflicting data are mainly caused by variables in the different studies. These variables include different amounts of nutritional feeding, timing, habitual food intake and type of exercise, volume of exercise, recovery measurements, and the timing of recovery measurement [16].

Apart from protein shakes, carbohydrate shakes are also frequently consumed to stimulate recovery after physical activity, containing protein/carbohydrates combinations. These drinks are recommended by the manufacturing companies to be ingested immediately after exercise, in order to have the most effective pro-regenerative effect. Research on this subject shows highly conflicting results. Some studies demonstrate that adding protein to carbohydrates improves the process of the recovery of endurance performance [17–21]. Others, focusing on the modulating of protein breakdown and synthesis, provide no evidence that carbohydrates increase exercise-induced protein secretion versus protein alone [22].

Mechanistically, it is believed that, beside the compensation of carbohydrate loss during exercise, carbohydrates also activate molecular mechanisms related to pro-regenerative effects. Here, insulin is claimed to be an important factor. It has been shown that a protein/carbohydrate combination leads to a higher increase in blood sugar and insulin concentration than just carbohydrate intake [23]. The molecular mechanism as to how proteins stimulate insulin secretion is unknown so far [24]. It also has been demonstrated that the amount of glycogen storage in skeletal muscle is higher with protein/carbohydrate combinations than with carbohydrates alone [23]. It is discussed that any uptake of carbohydrates immediately after exercise results in a strong increase of serum insulin, followed by a strong decrease of blood glucose. Binding of insulin to IGF-1 receptors should result in the activation of skeletal muscle specific signal transduction pathways, and via mTOR (Rapamycin), it should also result in a stimulation of protein synthesis in skeletal muscle [25,26]. Moreover, increasing insulin secretion also should stimulate amino acid uptake in skeletal muscle cells [26]. However, other investigations demonstrate that insulin does not stimulate muscle protein synthesis under physiological conditions in humans. It is very likely that the experimental design will have a huge impact on the results of the studies.

Beside ingestion of isolated amino acids, effects of protein/carbohydrate combinations on skeletal muscle could also be observed after the uptake of food containing protein and carbohydrates in a ratio of 70% to 30% [27]. In previous studies, we could observe pro-regenerative effects on the skeletal muscle after exercise, by eating combinations of dairy products and white bread [28]. Based on these observations, we concluded that eating food containing suitable concentrations of protein and carbohydrates may be an alternative strategy to promote skeletal muscle regeneration after exercise. Therefore, the aim of this follow up study was to compare the pro-regenerative effects of any protein/carbohydrate shake on the recovery of skeletal muscle after endurance exercise, with the ingestion of similar amounts of protein/carbohydrate combinations by eating a nearly iso-caloric meal.

Our hypothesis was that providing proteins/carbohydrate combinations will result in pro-regenerative effects, independent as to whether it was given by a shake or by eating a meal. Therefore, 35 nonspecifically trained subjects consumed either a whey protein/glucose shake, or a meal of white bread and sour milk cheese directly after endurance exercise. The following mechanistic read outs were determined: Serum glucose and serum insulin, serum creatine kinase (CK) and myoglobin (Myo) as muscle damage markers, hematologic parameters, cortisol, serum levels of the inflammation markers interleukin 6 (IL 6) and 10 (IL 10), as well as the macrophage migration inhibitory factor (MIF). In addition, leg strength as a functional marker for skeletal muscle regeneration was measured.

2. Methods

2.1. Participants

The study protocol has been approved by the local ethics committee (German Sports University, Cologne) and is in accord with the Declaration of Helsinki. It is registered in the German Clinical Trials Register under the title NUPROMU, and under the registration number DRKS-ID: DRKS00013359.

All participants provided written informed consent prior to their participation. The study excluded subjects who were currently taking any dietary supplements, sports drinks or functional food intended to enhance performance, or had taken any of these in the previous month. Moreover, subjects with known hypersensitivity to any of the constituents of the products under study (milk protein or lactose), were excluded. Throughout the study, subjects maintained their usual training routines and diets. Based on a statistical power analysis a total of 35 male participants (age: 23.2 ± 2.3 years; height: 181.5 ± 6 cm; weight: 77.9 ± 8.4 kg; mean ± standard deviation) were recruited for the study. Power was calculated with the program G *power (University of Düsseldorf, (Heinrich-Heine-Universität Düsseldorf (**HHU**)), Germany) based on effect sizes measured for the biological endpoints creatine kinase (CK) IL6 and IL10 in our pilot study [28]. Power analysis revealed that a number of 15 participants would be sufficient. Nevertheless, the group size was doubled, because additional parameters not measured in the pilot study were included. All participants were healthy and free of injury in the time period preceding the study. They were not specifically endurance trained, but they were physically active sports students. Anthropomorphic characteristics of the participants are indicated in Table 1.

Table 1. Compositions and calories of the standardized breakfast.

	Carbohydrates (g)	Protein (g)	Fat (g)	Calories (kcal)
30 g Cornflakes	25.2	2.1	0.3	111.9
250 mL milk (1.5% fat)	12.0	8.5	3.8	116.2
One banana (100 g)	27.4	1.3	0.4	118.4
Sum	64.6	11.9	4.5	346.5

2.2. Experimental Procedure

The aim of this study was to investigate the beneficial effects of a co-ingestion of protein and carbohydrate from a traditional food source in amateur sportsmen. For this purpose, 35 non-specific endurance trained male subjects performed a 10 km run with an intensity at 80% of their individual anaerobic threshold (IAT). Intensity was chosen based on previous investigations, demonstrating that 80% IAT and a distance of 10 km is a sufficient load for nonspecific endurance trained subjects to increase their serum CK [29]. The IAT was determined using an incremental field test with lactate determination according to a standard protocol [30]. Participants started with a speed of 2.5 m/s and distance of 800 m. After each successful termination of load level, the speed was increased by 0.5 m/s. The load duration was always between 5–6 min. Between the speed levels, 5 µL of capillary blood was taken from all participants. If a load level could not be successfully passed, the test was terminated.

The anaerobic threshold was used to calculate the meters per second and the resulting lap time in a track and field stadium (400 m lap length) for the 10 km run. Between the field test and the first investigation a wash out period of three weeks was used. Before running, the individuals were randomly divided into three intervention conditions—control, shake or food. Through a crossover design, each subject participated in each condition. However, the order was given after the first random assignment—control, shake, food.

Between the respective interventions there was a minimum wash out period of three weeks. The experimental design of the procedure at the intervention days is shown in Figure 1.

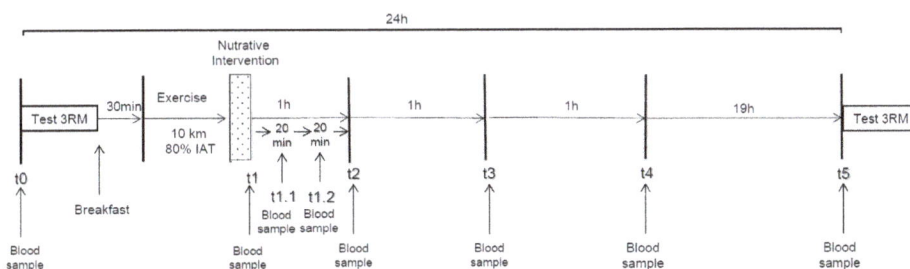

Figure 1. Experimental design of the study. IAT = individual anaerobic threshold. 3RM = leg strength.

Blood samples were determined for different time points (Figure 1). In the morning, blood sample t0 was collected from the overnight fasted participants. Leg strength was tested by maximum three repetitions in back squat (3-RM BS) followed by s a defined small breakfast (Table 1).

30 min after breakfast, all participants started with the 10 km run with 80% of IAT. Immediately after exercise, subjects ingested either nothing (control), a combination of carbohydrates by eating white bread (35.3 g) and 100 g of a sour milk cheese high in protein (36.1 g) but very low in fat (3.5 g)(Loose GmbH, Leppersdorf, Germany), or by drinking a whey protein (52 g)/glucose (45 g) shake. The composition of the shake used was taken from a published study [21] where effects regarding skeletal muscle regeneration have already been demonstrated. It serves as a standard. The nutritional values of the foodstuffs used in each intervention are shown in Table 2. As a consequence of using whole foods, it is not possible to match the macronutrient content of the shake and foods exactly. Whole foods, like dairy products, usually contain fat. Nevertheless, we have opted for low-fat protein and carbohydrate sources as indicated in Table 2. In addition we have tried as accurately as possible to be isocaloric. For practical reasons our participants got the food as a sandwich composed of two slices of white bread, and 100g of sour milk cheese. So, we differed from the shake with regard to the calories around 17% (Table 2).

Table 2. Compositions and calories of protein/carbohydrate interventions.

		Carbohydrates (g)	Protein (g)	Fat (g)	Calories (kcal)
Food	76 g white bread and 100 g sour milk cheese	35.3	36.1	3.5	321
Shake	45 g Glucose 52 g Whey protein	45	52	0	386

Immediately after ingestion, blood samples were determined at t1 (directly after exercise), followed by t.1.1 (+20 min after exercise), t1.2. (+40 min), t2 (+60 min), t3 (+120 h), t4 (+180 h) and t5 (+22 h). During the testing period, starting 12 h before exercise and 24 h after exercise, all participants were kept under a standardized nutrition to exclude additional effects by nutrition. All food for this 36 h period was provided to them.

12 h before exercise participants had a standardized dinner—spaghetti with tomato sauce (377 kcal, 15 g protein, 2.5 g fat, 72 g carbohydrates), and fasted until the standardized breakfast the next day. On the intervention day, participants were not allowed to eat until 3 h after exercise (t4; +180 min) except the provided nutritional compositions. After t4 they were allowed to eat the food provided in a period of 5 h after exercise until 8 pm. After 8 pm on the exercise day, participants fasted until the next day, when they were given the standardized breakfast. The standardized 24 h nutrition at the intervention day was calculated to fulfill all daily requirements regarding macro- and micronutrients, and provide sufficient calories.

Including the nutritive intervention, participants consumed a total of 3000 kcal a day. Daily macronutrient content was 146 g protein, 77 g fat and 411 g carbohydrates. Participants were allowed to drink water ad libitum.

3. Measurements

3.1. Determination of Serum Glucose and Serum Insulin

Samples were analyzed for glucose by oxidase method (COBAS Mira Plus; Roche Diagnostic Systems, Rotkreuz, Switzerland) and insulin by EIA (ALPCO diagnostics 1–2–3 Human Insulin EIA, Windham, NH, USA). Both serum concentrations were measured at all time points.

3.2. Determination of Serum Cortisol

Serum Cortisol concentrations were determined using the COBAS Mira Plus system (Roche Diagnostic Systems by ElectroChemiLumineszenz ImmunoAssay, ECLIA Rotkreuz, Switzerland). The serum cortisol concentration was measured at the time points t0, t1 (directly after exercise), t2 (+60 min), t3 (+120 min), t4 (+180 min) and t5 (+22 h).

3.3. Hematology

Blood samples were analyzed for routine CBC including total WBC count and differential for WBC sub fractions using a BC2300 hematology analyzer (Mindray Medical International Systems, Shenzhen, Peoples' Republic of China). The Hematology parameters were determined at t0, t4 (+180 min) and t5 (+22 h).

3.4. Skeletal Muscle Creatine Kinase (CK mm) and Myoglobin (Myo)

Skeletal muscle specific creatine kinase activity (CKmm) and myoglobin (Myo) concentrations in the serum were determined using the COBAS h 232 Point-of-Care-System (Roche Diagnostic Systems, Rotkreuz, Switzerland) at t0 and t5 (+22 h).

3.5. Serum Cytokine Levels

IL-6 concentrations of serum samples were analyzed using the Human IL-6 ELISA Kit High. Sensitivity (Abcam, Cambridge, UK). IL10 serum concentrations were analyzed using a human IL-10. ELISA Kit (Abcam, Cambridge, UK). MIF serum concentrations were analyzed using a human MIF. ELISA Kit (Abcam, Cambridge, UK). Cytokine concentrations were determined at t0 and t5.

3.6. Leg Strength—3-RM Back Squat

The strength test was performed t0 and t5 (+22 h). The strength protocol is based on the guidelines of NSCA [31]. All subjects performed three warm-up sets and started at 50% of their planned maximum power with 10 repetitions. In the following warm-up sets, the weight was increased by about 10–20% and the repetition number reduced to five and three repetitions. This was followed by four 3-RM tests, starting with about 90% of the planned 3-RM. Between the sets a four-minute break was taken. The subsequent increases were made individually. If a load level was not successfully mastered twice in succession, the test was stopped.

3.7. Statistical Analyses

Quantitative variables were presented as mean values and standard deviations (SD). All measurement parameters were tested for the Normal Distribution. As a result, the Wilcoxon sign rank test (repeated measurement) and the Kruskal-Wallis test were used for Myo, CK, IL6, IL10, glucose and insulin. For MIF and 3-RM BS, a 2-way Analysis of Variance (ANOVA) with time and condition effects, and a dependent (Student's) t-test, were used. The current version of SPSS (IBM SPSS Statistics 25.0, Ehningen, Germany) was used for statistical analysis, and $p < 0.05$ was taken as the level of statistical significance for all procedures. The images were created using GraphPad PRISM software (GraphPad Software, Inc. La Jolla, CA, USA).

4. Results

4.1. Effects of Exercise and Protein/Carbohydrate on Blood Glucose and Insulin Concentrations

It has been shown that the consumption of protein/carbohydrate combinations after exercise influences the blood sugar and insulin response, which is discussed to improve regeneration.

In Figure 2A, showing mean serum glucose concentrations of all participants ($n = 35$), a significant increase of serum glucose compared to t0 is detectable in the control and the shake condition at t1.1 (+20 min). A significant decrease compared to t0 is detectable in the shake condition and the food condition at t2 (+60 min).

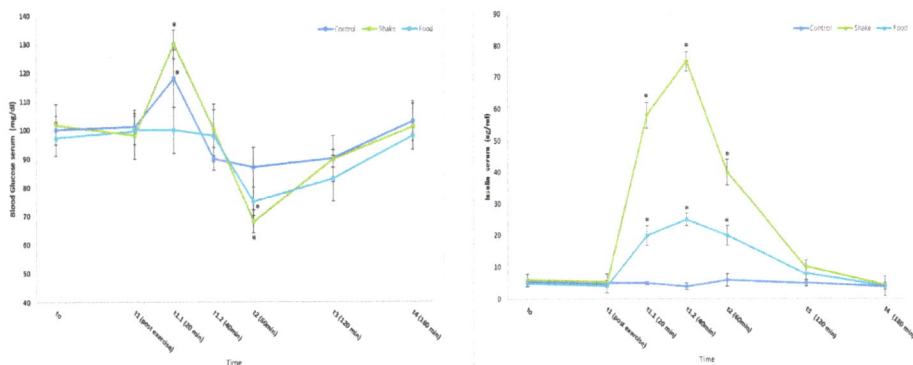

Figure 2. Effects of protein/carbohydrate uptake via food and shake, immediately after exercise, on serum glucose and serum insulin serum concentrations. Left: Mean serum blood glucose concentrations of all participants. All data are expressed as mean ± SD. * = $p \leq 0.05$. Statistically significant differences between marked time points and respective t0 value are observed. Right: Serum insulin concentrations of all participants (Mean ± SD). * = $p \leq 0.01$ show statistically significant differences between marked time point and respective t0 value in each condition.

In Figure 2B, mean serum insulin concentrations of all participants are shown. A significant increase of serum insulin compared to t0 is detectable in the food and the shake condition at t1.1 (+20 min), t1.2 (+40 min), t2 (+60 min) and t3 (+120 min).

4.2. Effects of Exercise and Protein/Carbohydrate on Hematopoietic Parameters

In agreement with published data [32] a significant decrease in the total leucocyte number in conditions could be observed 3 h (t4; +180 min) after exercise. This effect was not affected by any nutritive intervention (data not shown). After 24h (t5; +22 h), the total leucocyte number was again on the baseline in each condition. All other investigate hematopoietic parameters remained also unaffected by exercises and the nutritive intervention (data not shown).

4.3. Effects of Exercise and Protein/Carbohydrate on Blood Cortisol Levels

Serum Cortisol levels in athletes have been demonstrated to be influenced by physical activity and nutrition [33]. Figure 3 shows mean cortisol serum concentrations from all participants. The well described circadian rhythm of cortisol could be observed in all intervention conditions. Neither physical activity nor the nutritive interventions resulted in significant effects.

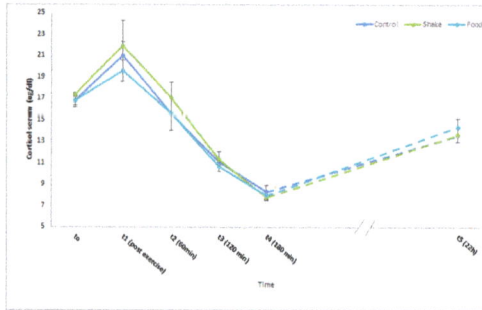

Figure 3. Effects of protein/carbohydrate uptake via food or shake, immediately after exercise, on serum cortisol concentrations. Mean serum cortisol concentrations of all participants. Mean ± SD. $p \leq 0.05$ show statistically significant differences between marked group and respective t0 value.

4.4. Effects of Exercise and Protein/Carbohydrate on Markers of Inflammation

Skeletal muscle damage results in an induction of inflammation. Interleukin 6 (IL 6), Interleukin 10 (IL 10), and Macrophage migration inhibitory factor (MIF) in the serum at the time point's t0 and t4 (+180 min) were measured after exercise as markers for inflammation.

MIF serum levels ($n = 18$) were significantly increased in all conditions, however, in the food and shake conditions, the increase was significantly lower, compared to the control. (Figure 4A) IL 6 serum levels ($n = 27$) were significantly increased by exercise (Figure 4B). No significant differences could be observed between the shake and the food condition. Il10 serum levels ($n = 28$) were significantly increased in all conditions after exercise; however, the increase in the shake and food condition was significantly higher (Figure 4C).

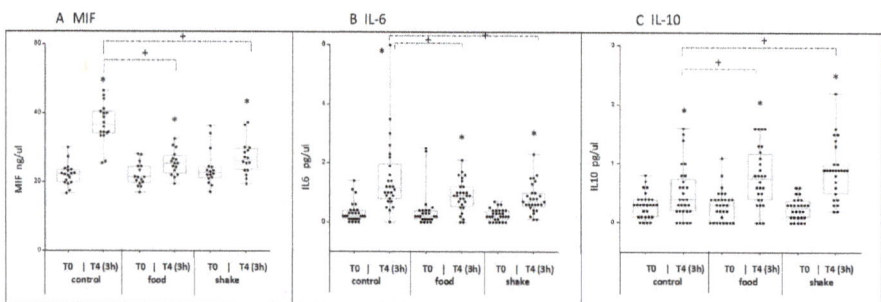

Figure 4. Effects of protein/carbohydrate uptake via food and shake, immediately after exercise, on serum levels of MIF (**A**) IL 6 (**B**) IL 10 (**C**) and 3h after exercise (t4; +180 min)). Shown are mean individual serum concentrations of the respective cytokines (A, B, C). Shown is mean ± SD. * = $p \leq 0.05$, statistically significant differences between marked time point and respective t0 value. + = $p \leq 0.05$ show statistically significant differences between marked groups.

4.5. Effects of Exercise and Protein/Carbohydrate on Serum Markers for Skeletal Muscle Damage

Skeletal muscle creatine kinase (CKmm) and myoglobin (Myo) are markers for skeletal muscle damage. Figure 5A shows the mean absolute Myo serum concentrations (*n* = 15) and Figure 5B shows the mean absolute of CKmm (*n* = 15).

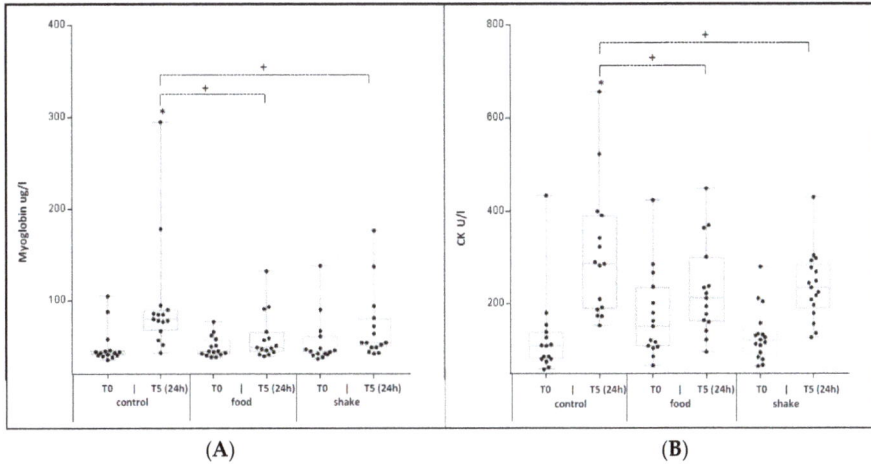

(A)   (B)

Figure 5. Effects of protein/carbohydrate uptake via food and shake on serum creatine kinase (CK) and myoglobin (Myo) concentrations, immediately after exercise, and 24 h after exercise. Shown are mean individual serum concentrations Myo (**A**) and CK (**B**). Mean of all individuals $\pm$ SD. * = $p \leq 0.05$ shows statistically significant differences between marked time point and respective t0 value. + = $p \leq$ 0.05 show statistically significant differences between marked groups.

Exercise in the control condition resulted in a significant increase of CKmm and Myo in the blood at t5 (+22 h). In the shake but also in the food condition exercise induced increase of skeletal muscle damage markers and was significantly lower compared to the control.

4.6. Effects of Exercise and Protein/Carbohydrate on Leg Strength as a Functional Marker for Skeletal Muscle Regeneration

We postulated that a better regeneration after endurance exercise will result in a lower loss of muscle strength (*n* = 15). As a functional approach for leg strength the 3-repetition maximum back squat was used. As seen in Figure 6, the uptake of protein/carbohydrate resulted in no significant loss of leg strength after endurance exercise. Only in the control condition did endurance exercise resulte in significantly lower leg strength at t5 (+22 h) after exercise.

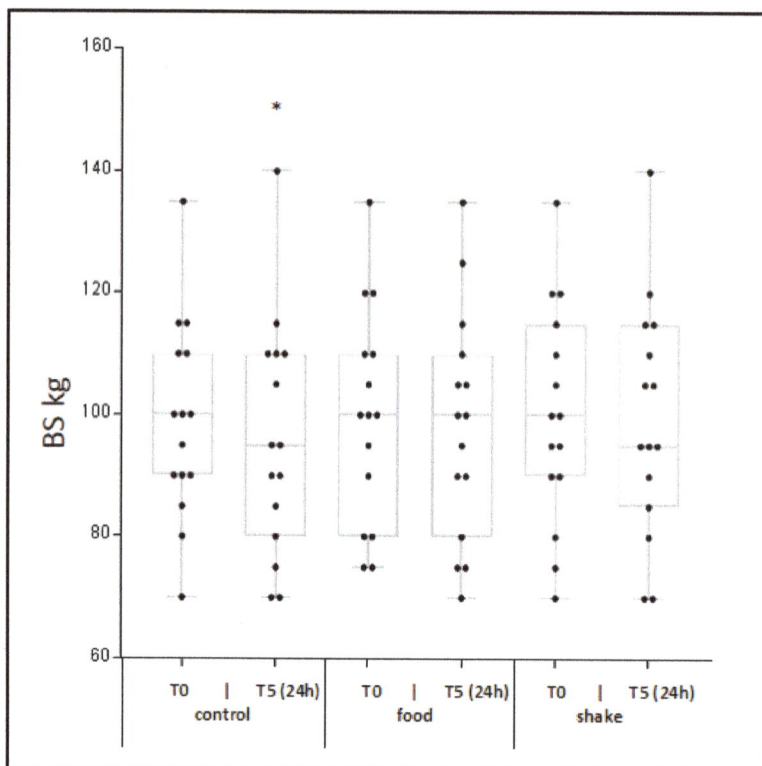

Figure 6. Effects of protein/carbohydrate uptake via food and shake on leg strength measured by the 3-repetition maximum back squat. In Figure 6 individual leg strength is shown Mean ± SD. * = $p \leq$ 0.01 show statistically significant differences between t5 and respective t0 value.

5. Discussion

In the present study, we compared pro-regenerative effects of protein and carbohydrates administration on the skeletal muscle regeneration after exercise, given either by shake or by a standardized meal. Pro-regenerative effects of protein/carbohydrate combinations after endurance exercise have been described and mechanistically linked to a stimulation of insulin secretion [21]. Therefore, in this investigation blood glucose levels and insulin levels were measured at different time points after exercise and subsequent protein/carbohydrate ingestion.

It was clearly visible that exercise results in an increase of blood glucose concentrations 20 min after exercise termination. This is in agreement to our previous investigations [28] and confirms observations that moderate exercise increases blood glucose concentrations in healthy non-diabetic persons when a required exercise volume is achieved [34]. As expected, blood glucose concentration after administration of the shake extended a maximum 40 min after ingestion. In contrast, after food administration, no significant increase of blood glucose concentration could be observed, although a comparable amount of carbohydrates was ingested (Table 2). Both shake and also food ingestion resulted in a significant increase of insulin serum concentration, starting 20 min after exercise/ingestion, and reaching a maximum after 40 min. The increase of insulin in the food condition is lower, compared to a shake, which could be explained by the lower content of total carbohydrates (35 g compared to 45 g in the shake condition). Nevertheless, the kinetics of insulin in both conditions is absolutely comparable.

A possible explanation for the missing increase of blood sugar levels in the food condition may be the direct uptake of the resorbed sugars in the skeletal muscle and the liver. In the shake condition, same mechanisms may be relevant, but the faster and higher load of glucose could not be compensated completely. In general, our observation is supported by data demonstrating that, combining protein and carbohydrate increases insulin levels, but does not improve glucose response [35]. Our observation is also in line with published data, showing that the uptake of specific amounts of carbohydrates results in a much stronger stimulation of insulin secretion when they are combined with proteins [17,21].

Hematopoietic factors and cortisol serum concentrations were not affected in our study by any ingestion of carbohydrates and proteins, but the response of pro- and anti-inflammatory serum markers to exercise was strongly influenced. Here it has to be mentioned that the role of inflammation in the skeletal muscle's adaptation to exercise is complex. Acute inflammatory response to exercise seems to promote skeletal muscle training adaptation and regeneration. In contrast, persistent, low-grade inflammation, as seen in a multitude of chronic diseases, is obviously detrimental [36]. Acute exercise has been described to induce an inflammatory response in the skeletal muscle. This can be detected by changes in the serum concentrations of inflammation-related cytokines [37,38]. Strenuous exercise results in changed serum concentrations of pro-inflammatory and anti-inflammatory cytokines like TNF alpha, IL-1, IL-6, IL-1 receptor antagonist, TNF receptors, IL-10, IL-8 and macrophage inflammatory protein-1 [38]. Increase of pro- and anti-inflammatory cytokines has been also described in animal models were skeletal muscle damage is induced by notoxin [39]. For these reasons the circulating levels of the pro-inflammatory cytokines IL6, and MIF and the anti-inflammatory cytokine IL10, were investigated 3 h after exercise in this study. We observed a significant increase of the serum concentrations of the pro-inflammatory cytokines IL6 and MIF after exercise in the control condition, which was significantly lower in both protein/carbohydrate conditions. In contrast, the serum concentrations of the anti-inflammatory cytokine IL10 were even more increased after exercise in both protein/carbohydrate conditions compared to the control condition. Remarkably, the effects after ingestion of the proteins by shake and by food are quantitatively and qualitatively comparable. These regulation patterns demonstrate that administration of protein/carbohydrate after exercise reduces the excretion of pro-inflammatory cytokines, but increases the excretion of anti-inflammatory proteins in the serum, which can be interpreted as an anti-inflammatory effect. Indeed, in an animal model where muscle damage was induced by administration of notoxin, accelerated skeletal muscle recovery correlated with increased IL10 expression and decreased TNF-alpha expression in the respective skeleton muscles [39]. As our study demonstrated, while a complete suppression of inflammation after exercise, e.g., after administration of glucocorticoids [40], promotes catabolic effects in the skeletal muscle, a modulation of the inflammatory response can be interpreted as an indication for pro-regenerative effects.

An important physiological endpoint investigated in this study was skeletal muscle damage and post-exercise recovery. Some studies show no effects of proteins/carbohydrates ingestion on skeletal muscle recovery [11] after exercise, while other investigations, however, have demonstrated a reduced post-exercise muscle soreness [42] and lower plasma concentrations of Myo [43] and CK [44] by administration of proteins/carbohydrates before, after, or during endurance exercise. Therefore, in our investigation CKmm and Myo concentrations were measured time-dependently after exercise, and a significant increase of the mean serum CK and Myo concentrations was detectable in the control condition 22h after training. The increase was not detectable in both protein/carbohydrate conditions. Our observation might be interpreted as an indication that skeletal muscle damage, induced by exercise, is lower after the uptake of protein/carbohydrate combinations. This is in line with the observation of Diel et al. [28] and previous described results [14]. Our data also indicate that administration of protein/carbohydrate combinations by food was as effective as administration by shake.

This hypothesis is further supported by our functional regeneration marker, the measurement of leg strength. The reduction of muscle strength especially after endurance training is well reported [45]. As molecular mechanisms for this reduction, reduced numbers of connective structures between actin and myosin filaments, and a reduced sensitivity for calcium, are discussed [46–48]. We postulate that a better regeneration after endurance exercise will result in a lower loss of muscle strength. As a functional approach, we have chosen the 3-repetition maximum back squat assay, which is believed to be one of the most effective training approaches based on the interaction of several muscle groups in the leg [49]. We found that the loss of leg strength after endurance is significantly lower in the nutritive intervention conditions, compared to the control condition. This is an obvious functional indication that uptake of protein/carbohydrate results in pro-regenerative effects. Again, food and shake are of comparable effectiveness.

Our study has several limitations. One limitation is that the food and the shake in our study do not match perfectly for macronutrient content and calories. However, for calories both groups differ only by 17%. With respect to macronutrient content, a perfect match is not possible because protein/carbohydrate shakes do not contain fat. Moreover, in our study, although there is no perfect match, food is lower in calories, protein and glucose compared to shake, but nevertheless we could observe comparable effects. Another limitation of our study is that important parameters related to skeletal muscle recovery, protein synthesis and protein breakdown, were not investigated. These are of course important endpoints which should be addressed in future studies and in a similar design.

Our study also has strength. As biological markers for skeletal muscle damage, we did not only focus on CK. CK is highly debated as a suitable parameter [50]. Therefore we have also analyzed Myo and used two independent markers for skeletal muscle damage in comparison. Looking for effects on the immune response, we have investigated a panel of different cytokines, as well pro- and anti-inflammatory ones. This also strengthens the interpretation of our results.

In summary, our results demonstrate that ingestion of protein and carbohydrate combinations by a shake, but also by a nearly isocaloric meal immediately after endurance exercise, affects a variety of physiological endpoints. In our study, uptake of protein and carbohydrate combinations by shake and meal resulted in a strong increase of insulin serum concentrations of the participants. Moreover, a modulation of any inflammatory response of the skeletal muscle towards exercise, indicated by reduced concentrations of pro-inflammatory markers, and an increase in anti-inflammatory markers, could be observed. The endurance training induced an increase of serum CKmm and Myo, markers of skeletal muscle damage, and as a functional marker for regeneration, the loss of leg strength was observed. All this could be antagonized by protein/carbohydrate combinations, given either by a shake or a meal, respectively. Mechanistically our data provide evidence that a combined uptake of protein and carbohydrates appears to reduce skeletal muscle damage after endurance exercise via a modulation of the immune response of the skeletal muscle. Also, insulin seems to be involved in initiation and mediation of the pro-regenerative effects. The most important part our observation is that all these beneficial effects can be achieved, either by the ingestion of food containing sufficient concentrations of carbohydrates and protein, or with the same efficiency as consuming a shake. This finding demonstrates again that it is possible to develop concepts to support training by a suitable diet without the need to consume nutrition supplements or isolated proteins.

Author Contributions: E.I. = main investigator; F.B., D.A.B., V.H., S.P., L.S. = supporting investigators; S.S., W.S. = statistics, P.D. = study design.

Funding: This research received no external funding.

Conflicts of Interest: The authors declare no conflicts of interest.

Abbreviations

ANS	anaerobic threshold
CK	Creatine kinase
CKmm	Creatine kinase skeletal muscle
Myo	Myoglobin
IL 10	Interleukin 10
IL 6	Interleukin 6

MIF	Macrophage migration inhibitor factor
SD	standard deviations
SEM	standard error of the mean
TNF alpha	Tumor necrosis receptor alpha
km	kilometer
IGF 1	insulin like growth factor 1
mTOR	mechanistic Target of Rapamycin

References

1. Maughan, R.J.; Depiesse, F.; Geyer, H. The use of dietary supplements by athletes international association of athletics federations. *J. Sports Sci.* **2007**, *25*, 103–113. [CrossRef]
2. Reidy, P.T.; Rasmussen, B.B. Role of Ingested amino acids and protein in the promotion of resistance exercise-induced muscle protein anabolism. *J. Nutr.* **2016**, *146*, 155–183. [CrossRef]
3. Bonaldo, P.; Sandri, M. Cellular and molecular mechanisms of muscle atrophy. *Dis. Models Mech.* **2013**, *6*, 25–39. [CrossRef]
4. Armstrong, R.B.; Warren, G.L.; Warren, J.A. Mechanisms of exercise-induced muscle fibre injury. *Sports Med.* **1991**, *12*, 184–207. [CrossRef]
5. Trappe, T.A.; White, F.; Lambert, C.P.; Cesar, D.; Hellerstein, M.; Evans, W.J. Effect of ibuprofen and acetaminophen on postexercise muscle protein synthesis. *Am. J. Physiol. Endocrinol. Metab.* **2002**, *282*, 551–556. [CrossRef]
6. Koopman, R.; Wagenmakers, A.J.; Manders, R.J.; Zorenc, A.H.; Senden, J.M.; Gorselink, M.; Keizer, H.A.; van Loon, L.J. Combined ingestion of protein and free leucine with carbohydrate increases postexercise muscle protein synthesis in vivo in male subjects. *Am. J. Physiol. Endocrinol. Metab.* **2005**, *288*, 645–653. [CrossRef]
7. Mackley, A.L.; Rasmussen, L.K.; Kadi, F.; Schjerling, P.; Helmark, I.C.; Ponsot, E.; Aagaard, P.; Durigan, J.L.; Kjaer, M. Activation of satellite cells and the regeneration of human skeletal muscle are expedited by ingestion of nonsteroidal anti-inflammatory medication. *FASEB J.* **2016**, *30*, 2266–2281. [CrossRef]
8. Phillips, S.M.; Tipton, K.D.; Aarsland, A.; Wolf, S.E.; Wolfe, R.R. Mixed muscle protein synthesis and breakdown after resistance exercise in humans. *Am. J. Physiol.* **1997**, *273*, 9–107. [CrossRef]
9. Kitajima, Y.; Tashiro, Y.; Suzuki, N.; Warita, H.; Kato, M.; Tateyama, M.; Ando, R.; Izumi, R.; Yamazaki, M.; Abe, M.; et al. Proteasome dysfunction induces muscle growth defects and protein aggregation. *J. Cell Sci.* **2014**, *127*, 5204–5217. [CrossRef]
10. Sandri, M. Signaling in muscle atrophy and hypertrophy. *Physiology* **2008**, *23*, 160–170. [CrossRef]
11. Tang, J.E.; Moore, D.R.; Kujbida, G.W.; Tarnopolsky, M.A.; Phillips, S.M. Ingestion of whey hydrolysate, casein, or soy protein isolate: Effects on mixed muscle protein synthesis at rest and following resistance exercise in young men. *J. Appl. Physiol.* **2009**, *107*, 987–992. [CrossRef]
12. Symons, T.B.; Sheffield-Moore, M.; Wolfe, R.R.; Paddon-Jones, D. A moderate serving of high-quality protein maximally stimulates skeletal muscle protein synthesis in young and elderly subjects. *J. Am. Diet. Assoc.* **2009**, *109*, 1582–1586. [CrossRef]
13. Markus, C.R.; Olivier, B.; de Haan, E.H. Whey protein rich in α-lactalbumin increases the ratio of plasma tryptophan to the sum of the other large neutral amino acids and improves cognitive performance in stress-vulnerable subjects. *Am. J. Clin. Nutr.* **2002**, *75*, 1051–1056. [CrossRef]
14. White, J.P.; Wilson, J.M.; Austin, K.G.; Greer, B.K.; St John, N.; Panton, L.B. Effect of carbohydrate-protein supplement timing on acute exercise-induced muscle damage. *J. Int. Soc. Sports Nutr.* **2008**, *5*, 5. [CrossRef]
15. Buckley, J.D.; Thomson, R.L.; Coates, A.M.; Howe, P.R.; Denichilo, M.O.; Rowney, M.K. Supplementation with a whey protein hydrolysate enhances recovery of muscle force-generating capacity following eccentric exercise. *J. Sci. Med. Sport* **2010**, *13*, 178–181. [CrossRef]
16. Trommelen, J.; Betz, M.W.; van Loon, L.J.C. The muscle protein synthetic response to meal ingestion following resistance-type exercise. *Sports Med.* **2019**, *49*, 185–197. [CrossRef]
17. Manninen, A.H. Hyperinsulinaemia, hyperaminoacidaemia and postexercise exercise: Influence on performance and recovery. *Int. J. Sport Nutr. Exerc. Metab.* **2006**, *17*, 87–103.

18. Saunders, M.J.; Kane, M.D.; Todd, M.K. Effects of a carbohydrate-protein beverage on cycling endurance and muscle damage. *Med. Sci. Sports Exerc.* **2004**, *36*, 1233–1238. [CrossRef]
19. Kerksick, C.; Harvey, T.; Stout, J.; Campbell, B.; Wilborn, C.; Kreider, R.; Antonio, J. International society of sports nutrition position stand: Nutrient timing. *J. Int. Soc. Sports Nutr.* **2008**, *5*, 17. [CrossRef]
20. Valentine, R.J.; Saunders, M.J.; Todd, M.K.; Laurent, T.G. Influence of carbohydrate-protein beverage on cycling endurance and indices of muscle disruption. *Int. J. Sport Nutr. Exerc. Metab.* **2008**, *18*, 363–378. [CrossRef]
21. Hill, K.M.; Stathis, C.G.; Grinfeld, E.; Hayes, A.; McAinch, A.J. Coingestion of carbohydrate and whey protein isolates enhance PGC-1α mRNA expression: A randomised, single blind, cross over study. *J. Int. Soc. Sports Nutr.* **2013**, *10*, 8. [CrossRef]
22. Staples, A.W.; Burd, N.A.; West, D.W.; Currie, K.D.; Atherton, P.J.; Moore, D.R.; Rennie, M.J.; Macdonald, M.J.; Baker, S.K.; Phillips, S.M. Carbohydrate does not augment exercise-induced protein accretion versus protein alone. *Med. Sci. Sports Exerc.* **2011**, *43*, 1154–1161. [CrossRef] [PubMed]
23. van Loon, L.J.; Saris, W.H.; Verhagen, H.; Wagenmakers, A.J. Plasma insulin responses after ingestion of different amino acid or protein mixtures with carbohydrate. *Am. J. Clin. Nutr.* **2000**, *72*, 96–105. [CrossRef] [PubMed]
24. Behrends, C.; Sowa, M.E.; Gygi, S.P.; Harper, J.W. Network organization of the human autophagy system. *Nature* **2010**, *466*, 68–76. [CrossRef] [PubMed]
25. Anthony, J.C.; Lang, C.H.; Crozier, S.J.; Anthony, T.G.; MacLean, D.A.; Kimball, S.R.; Jefferson, L.S. Contribution of insulin to the translational control of protein synthesis in skeletal muscle by leucine. *Am. J. Physiol. Endocrinol. Metab.* **2002**, *282*, 1092–1101. [CrossRef]
26. Fujita, S.; Rasmussen, B.B.; Cadenas, J.G.; Grady, J.J.; Volpi, E. Effect of insulin on human skeletal muscle protein synthesis is modulated by insulin-induced changes in muscle blood flow and amino acid availability. *Am. J. Physiol. Endocrinol. Metab.* **2006**, *291*, 745–754. [CrossRef] [PubMed]
27. Ivy, J.L.; Goforth, H.W., Jr.; Damon, B.M.; McCauley, T.R.; Parsons, E.C.; Price, T.B. Early postexercise muscle glycogen recovery is enhanced with a carbohydrate-protein supplement. *J. Appl. Physiol.* **2002**, *93*, 1337–1344. [CrossRef] [PubMed]
28. Diel, P.; Le Viet, D.; Humm, J.; Huss, J.; Oderkerk, T.; Simon, W.; Geisler, S. Effects of a nutritive administration of carbohydrates and protein by foodstuffs on skeletal muscle inflammation and damage after acute endurance exercise. *J. Nutr. Health Food Sci.* **2017**, *5*, 17. [CrossRef]
29. Mader, A.; Liesen, H.; Heck, H.; Phillipi, H.; Rost, R.; Schürch, P. Zur beurteilung der sportartspezifischen ausdauerleistungsfähigkeit im labor. *Sportarzt Sportmed.* **1976**, *27*, 80–112.
30. Nicholson, R.M.; Seivert, G.G. Indices of lactate threshold and their relationship with 10-km running velocity. *Med. Sci. Sports Exerc.* **2001**, *33*, 339–342. [CrossRef] [PubMed]
31. Todd, M.; Haywood, K.M.; Roberton, M.A.; Getchell, N. *National Strength & Conditioning Association, SA's Guide to Tests and Assessments*; Human Kinetics: Champaign, IL, USA, 2012.
32. Johannsen, N.M.; Swift, D.L.; Johnson, W.D.; Dixit, V.D.; Earnest, C.P.; Blair, S.N.; Church, T.S. Effect of different doses of aerobic exercise on total white blood cell (WBC) and WBC subfraction number in postmenopausal women: Results from DREW. *PLoS ONE* **2012**, *7*, E31319. [CrossRef]
33. Lima-Silva, A.E.; Pires, F.O.; Lira, F.S.; Casarini, D.; Kiss, M.A. Low carbohydrate diet affects the oxygen uptake on-kinetics and rating of perceived exertion in high intensity exercise. *Psychophysiology* **2011**, *48*, 277–284. [CrossRef] [PubMed]
34. Adams, P. The impact of brief high-intensity exercise on blood glucose levels. *Diabetes Metab. Syndr. Obes.* **2013**, *6*, 113–122. [CrossRef] [PubMed]
35. Ang, M.; Müller, A.S.; Wagenlehner, F.; Pilatz, A.; Linn, T. Combining protein and carbohydrate increases postprandial insulin levels but does not improve glucose response in patients with type 2 diabetes. *Metabolism* **2012**, *61*, 1696–1702. [CrossRef]
36. Beiter, T.; Hoene, M.; Prenzler, F.; Mooren, F.C.; Steinacker, J.M.; Weigert, C.; Nieß, A.M.; Munz, B. Exercise, skeletal muscle and inflammation: ARE-binding proteins as key regulators in inflammatory and adaptive networks. *Exerc. Immunol. Rev.* **2015**, *21*, 42–57.
37. Walsh, N.P.; Gleeson, M.; Pyne, D.B.; Nieman, D.C.; Dhabhar, F.S.; Shephard, R.J.; Oliver, S.J.; Bermon, S.; Kajeniene, A. Position statement part two: Maintaining immune health. *Exerc. Immunol. Rev.* **2011**, *17*, 64–103. [PubMed]

38. Petersen, A.M.; Pedersen, B.K. The role of IL-6 in mediating the anti-inflammatory effects of exercise. *J. Physiol. Pharmacol.* **2006**, *57*, 43–51.
39. Velders, M.; Schleipen, B.; Fritzemeier, K.H.; Zierau, O.; Diel, P. Selective estrogen receptor-β activation stimulates skeletal muscle growth and regeneration. *FASEB J.* **2012**, *26*, 1909–1920. [CrossRef] [PubMed]
40. Klein, G.L. The effect of glucocorticoids on bone and muscle. *Osteoporos. Sarcopenia.* **2015**, *1*, 39–45. [CrossRef] [PubMed]
41. Burnley, E.C.D.; Olson, A.N.; Sharp, R.L.; Baier, S.M.; Alekel, D.L. Impact of protein supplements on muscle recovery after exercis—induced muscle soreness. *J. Exerc. Sci. Fit.* **2010**, *8*, 89–96. [CrossRef]
42. Outlaw, J.J.; Wilborn, C.D.; Smith-Ryan, A.E.; Hayward, S.E.; Urbina, S.L.; Taylor, L.W. Effects of a pre-and post-workout protein-carbohydrate supplement in trained crossfit individuals. *SpringerPlus* **2014**, *3*, 369. [CrossRef] [PubMed]
43. Millard-Stafford, M.; Warren, G.L.; Thomas, L.M.; Doyle, J.A.; Snow, T.; Hitchcock, K. Recovery from run training: Efficacy of a carbohydrate-protein beverage? *Int. J. Sport Nutr. Exerc. Metab.* **2005**, *15*, 610–624. [CrossRef]
44. Seifert, J.G.; Kipp, R.W.; Amann, M.; Gazal, O. Muscle damage, fluid ingestion, and energy supplementation during recreational alpine skiing. *Int. J. Sport Nutr. Exerc. Metab.* **2005**, *15*, 528–536. [CrossRef] [PubMed]
45. Saunders, M.J. Coingestion of carbohydrate-protein during endurance muscle anabolism: The search for the optimal recovery drink. *Br. J. Sports Med.* **2007**, *40*, 900–905.
46. Westerblad, H.; Allen, D.G.; Bruton, J.D.; Andrade, F.H.; Lännergren, J. Mechanisms underlying the reduction of isometric force in skeletal muscle fatigue. *Acta Physiol. Scand.* **1998**, *162*, 253–260. [CrossRef] [PubMed]
47. Nocella, M.; Colombini, B.; Benelli, G.; Cecchi, G.; Bagni, M.A.; Bruton, J. Force decline during fatigue is due to both a decrease in the force per individual cross-bridge and the number of cross-bridges. *J. Physiol.* **2011**, *589*, 3371–3381. [CrossRef] [PubMed]
48. Lamb, G.D.; Westerblad, H. Acute effects of reactive oxygen and nitrogen species on the contractile function of skeletal muscle. *J. Physiol.* **2011**, *598*, 2119–2127. [CrossRef] [PubMed]
49. Meyer, D.G.; Kushner, M.A.; Brent, L.J.; Schoenfeld, J.B.; Hugentobler, J.; Lloyd, S.R.; Vermeil, A.; Chu, A.D.; Harbin, J.; McGill, M.S. The back squat: A proposed assessment of functional deficits and technical factors that limit performance. *Strength Cond. J.* **2014**, *36*, 4–27. [CrossRef] [PubMed]
50. Baird, M.F.; Graham, S.M.; Baker, J.S.; Bickerstaff, G.F. Creatine-kinase- and exercise-related muscle damage implications for muscle performance and recovery. *J. Nutr Metab.* **2012**, *2012*, 960363. [CrossRef]

nutrients

MDPI

Article

Whole Body Protein Oxidation Unaffected after a Protein Restricted Diet in Healthy Young Males

Gerlof A.R. Reckman [1,2], Gerjan J. Navis [3], Wim P. Krijnen [2], Cees P. van der Schans [2,4], Roel J. Vonk [5] and Harriët Jager-Wittenaar [2,6,*]

[1]   Department of Internal Medicine, Division of Nephrology, University of Groningen, University Medical Center Groningen, AA53, PO Box 30.001, 9700 RB Groningen, The Netherlands; g.a.r.reckman@pl.hanze.nl
[2]   Research Group Healthy Ageing, Allied Health Care and Nursing, Centre of Expertise Healthy Ageing, Hanze University of Applied Sciences, Petrus Driessenstraat 3, 9714 CA Groningen, The Netherlands; w.p.krijnen@pl.hanze.nl (W.P.K.); c.p.van.der.schans@pl.hanze.nl (C.P.v.d.S.)
[3]   Department of Internal Medicine, Division of Nephrology, University of Groningen, University Medical Center Groningen, AA53, PO Box 30.001, 9700 RB Groningen, The Netherlands; g.j.navis@umcg.nl
[4]   Department of Rehabilitation and Health Psychology, University of Groningen, University Medical Center Groningen, CD44, PO Box 30.001, 9700 RB Groningen, The Netherlands
[5]   Department of Cell Biology, University of Groningen, University Medical Center Groningen, FB33, PO Box 30.001, 9700 RB Groningen, The Netherlands; r.j.vonk@umcg.nl
[6]   Department of Maxillofacial Surgery, University of Groningen, University Medical Center Groningen, BB70, PO Box 30.001, 9700 RB Groningen, The Netherlands
*   Correspondence: ha.jager@pl.hanze.nl; Tel.: +31-623-668-897

Received: 26 November 2018; Accepted: 29 December 2018; Published: 8 January 2019

Abstract: Protein oxidation may play a role in the balance between anabolism and catabolism. We assessed the effect of a protein restricted diet on protein oxidation as a possible reflection of whole body protein metabolism. Sixteen healthy males (23 ± 3 years) were instructed to use a 4-day isocaloric protein restricted diet (0.25 g protein/kg body weight/day). Their habitual dietary intake was assessed by a 4-day food diary. After an overnight fast, a 30 g ^{13}C-milk protein test drink was administered, followed by 330 min breath sample collection. Protein oxidation was measured by Isotope Ratio Mass Spectrometry. To assess actual change in protein intake from 24-h urea excretion, 24-h urine was collected. During the 4-day protein restricted diet, the urinary urea:creatinine ratio decreased by $56 \pm 9\%$, which is comparable to a protein intake of ~0.65 g protein/kg body weight/day. After the protein restricted diet, $30.5 \pm 7.3\%$ of the 30 g ^{13}C-milk protein was oxidized over 330 min, compared to $31.5 \pm 6.4\%$ (NS) after the subject's habitual diet (1.3 $\pm$ 0.3 g protein/kg body weight/day). A large range in the effect of the diet on protein oxidation (-43.2% vs. $+44.0\%$) was observed. The residual standard deviation of the measurements was very small (0.601 ± 0.167). This suggests that in healthy males, protein oxidation is unaffected after a protein restricted diet. It is uncertain how important the role of fluctuations in short-term protein oxidation is within whole body protein metabolism.

Keywords: Protein; oxidation; anabolic competence; breath test; naturally enriched ^{13}C-milk proteins

1. Introduction

Adequate protein intake and subsequent utilization of protein is of great importance for health. The recommended daily protein intake for healthy adults is 0.8 g protein/kg body weight/day, and is suited for maintaining normal body composition and meeting metabolic demand [1]. Patients with disease-related malnutrition (DRM) have an absolute or relative deficiency and inadequate utilization of energy, protein, and other nutrients caused by a concomitant disease. Compromised outcomes, such as impaired clinical outcome from disease, and diminished physical and mental function have been described in relation to DRM [2–4]. Nutrition, exercise, and the hormonal milieu are essential to reach a state which optimally supports protein synthesis and lean body mass (LBM), global aspects of muscle and organ function, and the immune response, a paradigm also known as "anabolic competence" [5].

Prevention and treatment of LBM loss could benefit from direct measurement and monitoring of disturbed protein metabolism. Current methods to measure protein metabolism focus on protein synthesis, which requires blood sampling, muscle biopsies, and/or the use of expensive synthetic labelled amino acids [6–8]. Therefore, these methods are not suitable for the clinical setting. A non-invasive bedside method to measure protein metabolism would be more suitable, as direct measurements of the metabolic state could lead to more insight in optimal protein intake and optimal physical activity, which then enables tailored improved treatment for each patient, resulting in improved outcomes of disease.

Measuring protein oxidation is a feasible and non-invasive technique and can be performed with naturally labelled ^{13}C-protein, which is relatively inexpensive. All oxidized ^{13}C-protein will be exhaled as $^{13}CO_2$ [9]. However, it is unknown to what extent variations in protein oxidation occur under various physiological conditions, such as changes in protein intake. Generally, after the ingestion of protein, the protein derived amino acids will be incorporated into new proteins until protein synthesis requirements are met. The lack of protein storage leads to the oxidation of surplus amino acids [10,11]. Accordingly, an altered protein intake could modify protein oxidation under normal conditions. Thus, we hypothesized that restriction in protein intake in healthy subjects leads to decreased activity of the oxidation pathway, as assessed by the $^{13}CO_2$ breath test.

To test this hypothesis, in the current study, we aimed to measure the effect of a four-day protein restricted diet, compared to their habitual diet, on protein oxidation, as assessed by the $^{13}CO_2$ breath test in healthy subjects.

2. Materials and Methods

2.1. Subjects

Healthy young males were included as being a representative group for healthy subjects. The decision to recruit young subjects versus older subjects was based on logistics, as the pool of young healthy subjects is more easily accessible for study. Women were excluded to rule out possible effects of the menstrual cycle on protein metabolism, and to exclude possible effects of differences in body composition between women and men. Furthermore, subjects having a disease and/or undergoing or starting medical treatment were excluded. Sixteen healthy young male subjects were recruited via local advertising. To obtain a homogeneous group of subjects, reducing the possible influence of covariates, the following inclusion criteria were applied: Age between 18–30 years, body mass index (BMI) between 20–25 kg/m^2, and being able to fast overnight. Exclusion criteria were: Having a disease and/or being medically treated, milk protein allergy or intolerance, smoking, use of drugs, drinking on average more than 2 glasses of alcohol per day, waist circumference larger than 102 cm, and using a vegetarian diet. This design was chosen to minimize possible confounding effects of subject characteristics over the protein restricted diet intervention.

The study was approved by the local Medical Ethical Committee at the University Medical Center Groningen (NL56982.042.16, METc 2016.144), conducted in accordance with the Helsinki Declaration of 2013, and registered in the Dutch Trial Register under the registration number, NTR6101. After receiving an information letter about the purpose and practical procedures of the study, and an informative meeting with the researcher, every subject gave his written informed consent prior to participation.

2.2. Study Protocol

In each subject, age (year), height (cm), waist circumference (cm), bodyweight (kg), BMI (kg/m^2), and LBM (kg) were measured. LBM was measured by bioelectrical impedance analysis (Quadscan 4000, Bodystat Ltd., Isle of Man, British Isles). After these measurements, subjects were instructed to keep a four-day food diary with respect to their habitual food intake to calculate the average daily intake of energy (kcal), protein (g), protein, en%, animal protein (g), plant protein (g), carbohydrates (g), carbohydrates, en%, fat (g), and fat, en%. The calculations on dietary intake were performed with Evry (Evry BV), which uses the NEVO 2013, RIVM database [12].

On each subject, at two separate days, two breath tests were performed; one after the subject's habitual diet and one after an isocaloric protein restricted diet (0.25 g protein/kg body weight/day). Between the breath tests, there was a washout period of at least a week to return to baseline $^{13}CO_2$ levels. On the evening before the breath test, subjects were instructed to start fasting overnight (only consumption of water and tea, or coffee without milk and sugar was allowed) from 22:00 p.m. onwards to arrive sober the next morning at 08:45 a.m. During each test, 3 basal breath samples were collected and averaged to establish the subject's baseline $^{13}CO_2$:$^{12}CO_2$ ratio. At 09:15 a.m., 30 g naturally enriched ^{13}C-milk protein dissolved in 500 mL water was consumed within 5 minutes. The isotope, ^{13}C, is a stable isotope. From 09:25 a.m. until 14:45 p.m. (5.5 h), a breath sample was collected every 10 minutes. During this period, subjects were instructed to remain seated in upright position and not to eat or drink during the remainder of the breath test. Subjects were allowed to work on a laptop, to read, and to write.

The four-day protein restricted diet was given as a food menu, which described in detail what and when to eat, to facilitate energy and protein intake as prescribed. The subjects were instructed to use the food menus. The food menu was tailored to each subject's habitual energy intake, which was calculated from the four-day food diaries. Therefore, the created protein restricted diet was isocaloric to the habitual diet. Consequently, by both reducing protein intake and keeping the protein restricted diet isocaloric, the macronutrient composition of the diet changed, as the energy lost from protein intake was replaced by mainly an increase in carbohydrates and to a lesser extent with an increase in fat, as food which contains fat also contains protein. Each subject underwent the breath tests in the same order: Starting the first after habitual diet and second after the protein restricted diet. The four-day protein restricted diet was tailored to each subject.

During five days, 24-h urine was collected, on the fourth day of the subject's habitual diet and, next, every day during the four-day isocaloric protein restricted diet. From each 24-h urine sample, urea and creatinine concentrations were measured to calculate the subject's actual protein intake [13]. The urea:creatinine ratio was calculated to assess the compliance to the isocaloric protein restricted diet. A change to a protein restricted diet will reduce 24-h urinary urea production, while 24-h urinary creatinine production, which depends almost exclusively on muscle mass, will remain steady, and therefore the urea:creatinine ratio will decrease.

2.3. Calculations and Statistical Analysis

Disturbances in short-term protein oxidation can be measured by pulse-labelling with the use of naturally enriched ^{13}C-proteins. After ingestion of ^{13}C-proteins, the protein oxidation can be quantified by measuring the $^{13}CO_2$:$^{12}CO_2$ ratio in exhaled breath over time. To calculate the amount of CO_2 produced on each timepoint, the CO_2 production in rest was calculated by the following regression formula: 300 mmol CO_2/hour × body surface area (BSA), which was calculated with the Haycock formula [14]:

$$BSA\ (m^2) = weight\ (kg)^{0.5378} \times height\ (cm)^{0.3964} \times 0.024265 \quad (1)$$

The breath samples were measured for their $^{13}CO_2$:$^{12}CO_2$ ratio with an isotope ratio mass spectrometer (IRMS) and compared to a high ^{13}C-enriched international standard, Pee Dee Belemnite

(PDB), which has an accepted absolute $^{13}C/^{12}C$ ratio of 0.0112372. The differences (delta, δ) between the breath samples and the standard is expressed in parts per 1000 (:‰) as [15] follows:

$$\delta \text{ 13C sample} = ((13C/12C\text{sample})/(13C/12C\text{standard}) - 1) \times 1000 \qquad (2)$$

The PDB standard $^{13}C/^{12}C$ ratio is defined as 0:‰. To calculate the $^{13}C/^{12}C$ ratio from the IRMS delta values, the following inversion formula was used [10]:

$$13C/12C\text{ratio} = ((\text{deltavalue}/1000) + 1) \times 0.0112372 \qquad (3)$$

Next, the $^{13}C/^{12}C$ ratio of each breath sample is used to calculate the %^{13}C [16] by:

$$\%13C = ((13C/12C\text{ratio})/(13C/12C\text{ratio} + 1)) \times 100 \qquad (4)$$

The baseline ($t = 0$) breath sample $^{13}C/^{12}C$ ratio was subtracted from each subsequent breath sample to acquire the change from the subject's baseline. The estimated CO_2 production, together with the delta value on each timepoint, and the enrichment of the ^{13}C-milk protein was used to calculate the protein oxidation rate of the subject at each timepoint with 10 minutes in between.

The change in the $^{13}CO_2$:$^{12}CO_2$ ratio over time has been described [17,18] by the general concentration model as follows:

$$y(t) = a \times t^b \times e^{(-kt)} + \varepsilon \text{ (normally distributed)} \qquad (5)$$

The model function was fitted to the measurement data for each subject over time, with t ranging from zero to 330 minutes. That is, the parameters, a, b, and k, were determined by fitting the oxidation rate curves to the measurement data per subject per day, with the goal of finding optimal parameters per subject per day. Total protein oxidation was calculated as the integral over the curve representing the area under the curve (AUC). The term, ε, is normally distributed with a zero mean, and its variance is estimated by the residual error variance. Each resulting curve starts at the natural amount, $y = 0$, as no ^{13}C-milk protein has been ingested at timepoint, $t = 0$. After ingestion, the stomach releases the ^{13}C-milk protein into the digestive track where the proteins are digested and taken up by the gut. The digested proteins are circulated and become available to the cells for protein synthesis or oxidation. The process of protein oxidation is reflected by the oxidation curve. From $t = 0$ onwards, the oxidation curve ascends, reaching its maximum and thereafter, the oxidation rate of ^{13}C-amino acids descends towards the subject's baseline over multiple hours.

Per breath test, the values of the fitted parameters, a, b, and k, differ over persons and type of diet. For larger values of a and b, the ascending slope becomes steeper, whereas a higher value for the constant, k, leads to a steeper descending slope after the maximum was reached.

The difference in the average AUC after a habitual diet and protein restricted diet within subjects was tested with the paired t-test. Two other important characteristics directly calculated from the parameters of each fitted curve per subject were the timepoint (t_{max}) in minutes at which the maximum oxidation rate was reached, as well as the corresponding maximum oxidation rate, $y(t_{max})$, where $t_{max} = b/k$. The latter was expected to be similar within subjects in both experimental conditions, as t_{max} is mainly determined by the rate of stomach emptying [19], which is dependent on the test drink. The latter was identical for both breath tests. The standard error of the t_{max} per person is computed by the delta method [20]. Both the maximum oxidation rate, $y(t_{max})$, and the total oxidation (AUC) were expected to be lower after the protein restricted diet compared to the habitual diet, as a deficit of amino acids in the body is hypothesized to lead to less oxidation of the 30 g ingested milk protein. From each concentration model, the timepoint at which 1% oxidation/hour was reached was calculated using the above formula and a numerical intersection method.

All statistical analyses were performed by the statistical programming language, R (R Core Team, 2017), with the package, "car" [21], specifically using the non-linear least squares function [22] to fit the concentration curve to the measurements over time and the delta method to determine the standard error of t_{max} [20]. The difference in the urea:creatinine ratio between the habitual diet and day four of the protein restricted diet was investigated with the paired Student's *t*-test. The associations between total protein oxidation and the demographic characteristics, i.e., age, bodyweight, BMI, LBM, habitual protein intake, habitual energy intake, and baseline urea:creatinine ratio, were investigated with the Pearson correlation coefficient, *r*. All data are represented as mean $\pm$ standard deviation (SD).

3. Results

Baseline characteristics of the 16 male subjects are presented in Table 1.

Table 1. Baseline characteristics of the subjects (*n* = 16).

	Mean	SD
Age (years)	23.0	3.1
Height (cm)	185.4	8.6
Body weight (kg)	77.1	9.5
Body Mass Index (kg/m^2)	22.3	1.1
Lean Body Mass (%)	88.3	2.7
Habitual diet		
Protein intake (g protein/kg body weight/day)	1.3	0.3
Protein intake (g protein/day)	102	25
En% protein (%)	17	4
En% carbohydrates (%)	47	5
En% mono- and disaccharides (%)	20	8
En% fat (%)	35	6
En% saturated fat (%)	13	4
En% unsaturated fat (%)	19	7
Protein restricted diet		
En% protein (%)	3	1
En% carbohydrates (%)	73	7
En% mono- and disaccharides (%)	53	8
En% fat (%)	22	7
En% saturated fat (%)	9	5
En% unsaturated fat (%)	12	6
Baseline breath $^{13}CO_2$ enrichment (delta value)	−26.18	0.50

During the four-day protein restricted diet, the urea:creatinine ratio in 24-h urine decreased with an average of $56 \pm 9\%$, as compared to the habitual diet from day 0 to day 4 (Figure 1). The mean difference in the urea:creatinine ratio between the habitual diet day 0 and protein restricted diet day 4 was statistically significant ($p < 0.001$, $t = 12.837$, $df = 15$). Based on the change in the urea:creatinine ratio, the protein intake decreased to 0.65 g protein/kg body weight/day, compared to the habitual protein intake of 1.3 g protein/kg body weight/day.

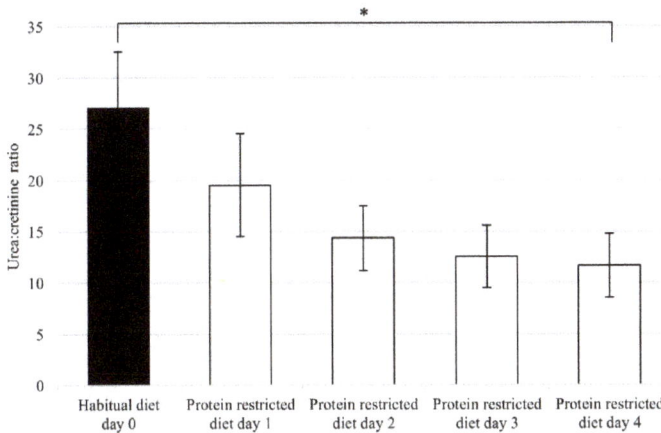

Figure 1. Urea:creatinine ratio calculated from 24-h urine collections ($n = 16$). The symbol "*" denotes the statistically significant change from day 0 to day 4 with $p < 0.001$. Protein intake during the habitual diet was 1.3 g protein/kg body weight/day $\pm$ 0.3 g; the prescribed four-day protein restricted diet was 0.25 g protein/kg body weight/day

The average protein oxidation kinetics of all subjects during the 330 minute breath tests, separated by the habitual diet and the protein restricted diet, are shown in Figure 2. Total oxidation (AUC) after the habitual diet and the protein restricted diet was 31.5 $\pm$ 6.4% and 30.5 $\pm$ 7.3%, respectively. The difference in the mean total oxidation between the habitual and protein restricted diet was not statistically significant ($p = 0.530$, $t = 0.643$, $df = 15$). The mean total protein oxidation of ~30%, corresponds to ~10 g oxidized. Time to t_{max}, after the habitual and protein restricted diet, was 137 minutes $\pm$ 24 and 138 minutes $\pm$ 18, respectively ($p = 0.854$, $t = -0.188$, $df = 15$). The maximum %oxidation rate per hour, after the habitual diet and the protein restricted diet, was 8.05 $\pm$ 1.27% and 8.11 $\pm$ 1.62% ($p = 0.868$, $t = -0.170$, $df = 15$).

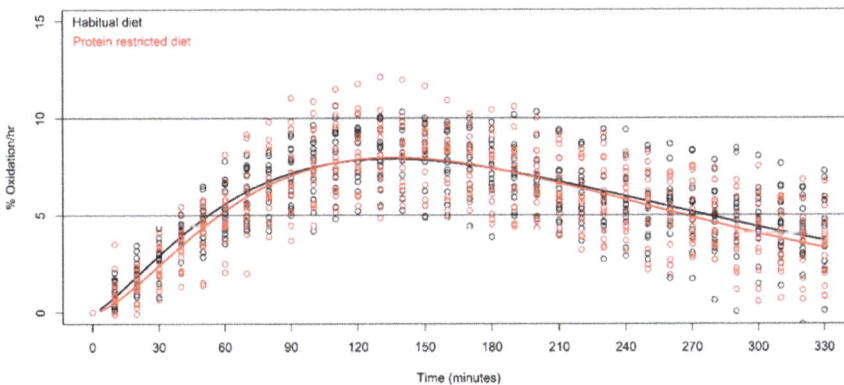

Figure 2. Protein oxidation kinetics after the habitual diet (black) and protein restricted diet (red) ($n = 16$).

The following means and confidence intervals were calculated with all 32 breath tests (16 subjects $\times$ 2 breath tests). From the curve fitting with the function, $y(t) = a \times t^b \times e^{(-kt)}$, the mean of constant, a, was 0.037 $\pm$ 0.041. For the constant, b, the mean was 1.636 $\pm$ 0.458. For the constant, k, the mean was 0.012 $\pm$ 0.004. The concentration curve fitted well with the breath test measurements with a mean R^2 of 0.930 $\pm$ 0.033. The timepoint of maximum oxidation (t_{max}) was obtained with a mean 137 ± 21 minutes. The mean proportionate decrease from the maximum oxidation rate (mean 8.1% oxidation/hour) to the final timepoint (mean 3.5% oxidation/hour) over all 32 breath tests was $-57 \pm 16\%$. The mean time to reach 1% oxidation/hour was 502 ± 119 minutes. The total oxidation was positively correlated with the maximal oxidation rate (0.95). Residual standard deviation of the measurements to the fitted curve was 0.601 ± 0.167. The associations between the total protein oxidation and demographic characteristics, i.e., age, body weight, BMI, LBM, habitual protein intake, habitual energy intake, and baseline urea:creatinine ratio, were fair to poor ($r < 0.4$) [23].

Differences in the total protein oxidation after each subject's habitual diet (1.3 ± 0.3 g protein/kg body weight/day), and after a prescribed isocaloric protein restricted diet (0.25 g/kg body weight/day) are shown in Figure 3. The results of the subjects are ordered based on the strongest relative decrease in protein oxidation from their habitual diet to the protein restricted diet, towards the strongest relative increase. A large range in the effect on the total protein oxidation (relative change -43.2% vs. $+44.0\%$) was observed, however, the large range of the increase and decrease in the total protein oxidation canceled each other out, as shown in Figure 2.

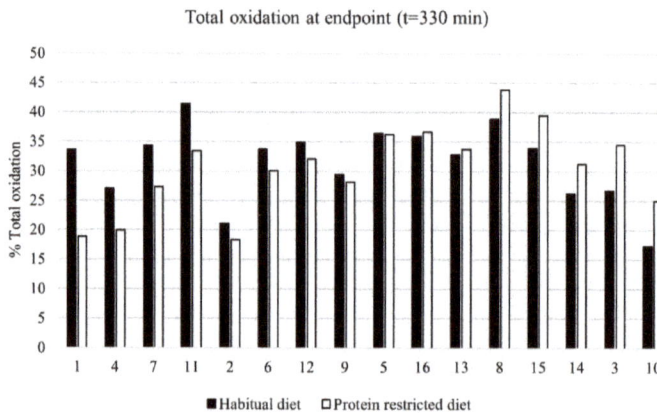

Figure 3. Total protein oxidation (% of given 30 g dose) measured with the breath test after a habitual diet (black bars) versus the protein restricted diet (0.25 g/kg body weight/day)(white bars) ($n = 16$). Subjects are ordered from left to right, based on the strongest relative reduction in protein oxidation from their habitual diet to the protein restricted diet, towards the strongest relative increase.

4. Discussion

In the current study, we assessed the effect of a protein restricted diet on protein oxidation, as assessed by the $^{13}CO_2$ breath test in healthy young males, as a possible reflection of whole body protein metabolism. On the group level, the total protein oxidation was not affected by the two-fold reduction in mean protein intake, which decreased from 1.3 to 0.65 g protein/kg body weight/day. We found a large variation in the total protein oxidation response after the protein restricted diet compared to the habitual diet, which ranged from a decrease of 43.2% to an increase of 44.0%. Combined with the precision of the breath test, this implies that the range in effects on the oxidation rate may have a biological background.

This is the second study that has assessed the overall protein oxidation with naturally ^{13}C-enriched milk protein during different states of protein intake. An explorative investigation was performed with a three-day protein restricted isocaloric diet, in which the protein intake was reduced to ~10 g of protein per day, which corresponds to ~0.15 g protein/kg body weight/day [9]. In that study, the total protein oxidation after the protein restricted diet had a lower mean than after the habitual diet, but the difference in the mean was not significant ($p = 0.142$). Further comparison of our results with those of other studies, which either measured the effect of several conditions or the effect of different exercise regimens combined with protein ingestion on the protein synthetic response, is difficult. However, our finding that ingestion of 30 g of ^{13}C-milk protein was associated with an overall oxidation of ~30%, corresponding to ~10 g oxidation over 5.5 hours, seems in line with the study by Moore et al. [8]. In that study, in which protein muscle synthesis was measured with primed ^{13}C-leucine infusion over 4 hours, muscle protein synthesis after resistance exercise was maximally stimulated with 20 g of ingested whole egg protein and dietary protein ingested in excess of 20 g stimulated ^{13}C-leucine oxidation. Therefore, Moore et al. imply that they would have found ~10 g of whole egg protein oxidized if they had tested a 30 g dose of whole egg protein, which is comparable to the amount of ^{13}C-milk protein oxidized in our study. Our findings did not confirm our hypothesis that a change in protein intake would be associated with a change in overall protein oxidation. This could mean that a change in protein intake in the range investigated here does indeed not affect protein oxidation in healthy males, or, alternatively, that there were methodological limitations in detecting such an alleged change in the current methodological design of our study.

The first limitation that needs to be taken into account is that there could be a pre-meal effect of the non-standardized evening meal, prior to the breath test, on the utilization and oxidation subsequent to the 30 g protein test drink. However, no relationship between the evening meal energy intake, en% carbohydrates, and en% protein intake on the subsequent protein oxidation was found, which could in part be related to the small sample size. Second, during the protein restricted diet, the subjects did not reach the intended target of 0.25 g protein/kg body weight/day intake, as the mean level of protein intake decreased to only 0.65 g protein/kg body weight/day as measured by the urea:creatinine ratio. This level of intake is only slightly lower than the general protein intake recommendation of 0.80 g/kg body weight/day for healthy adults set by the World Health Organization (WHO) [1]. It would seem that in this study, the subjects were probably still within the range of adequate protein intake to maintain protein homeostasis. On the other hand, whether the subjects in this study were in steady state is uncertain, as they acutely adjusted their protein intake from habitual to the four-day protein restricted diet. The use of 24-hour urinary urea for the estimation of dietary protein intake is most reliable in subjects who are in a steady state [24,25]. Therefore, it could be argued that in these experimental circumstances, the change in the urea:creatinine ratio does not accurately reflect protein homeostasis and also the actual protein intake. However, it does at least underline a clear reduction in protein intake. Third, the replacement of protein with carbohydrates to obtain an isocaloric protein restricted diet changed the macronutrient ratios within the diet, and might have resulted in a decreased uptake of amino acids into tissue and, consequently, an increased amino acid oxidation in the splanchnic area due to a possible altered insulin response [26,27]. Fourth, the current protocol includes an overnight fast, to forgo breakfast, and after consumption of the test drink, the subjects did not eat or drink for 5.5 hours. This was in order to minimize the influence of differences in starting conditions, such as the stomach emptying rate, between the subjects. These requirements enable better interpretable measurements and are unlikely to harm healthy subjects. However, fasting can further deteriorate the condition of clinical populations. For future studies in clinical populations, adaptations to the protocol should be made to minimize the burden. Potential targets to reduce the burden are a standardized breakfast and reduction in the collection of breath samples over time.

The major strengths of this study were the well-controlled design, with subjects as their own control, and the precise protein oxidation measurements. First, in the current tests, natural enriched ^{13}C-milk protein was used, which implies that all amino acids are enriched with ^{13}C and therefore the exhaled $^{13}CO_2$ represents the oxidation of all amino acids, which reflects the total body protein oxidation [10]. Oxidation studies with specific amino acids, like ^{13}C-leucine, most likely do not reflect overall amino acid oxidation, as all the amino acids have various biological functions, aside from being building blocks for synthesizing protein [28]. Second, the breath test is reliable as it measures the protein oxidation process well, with the concentration function fitting well to all breath test measurements, as demonstrated by a mean R^2 of 0.930 ± 0.033. The formula of each curve provided estimated parameters per subject. These parameters, such as the timepoint of the maximum, and the oxidation rate at the timepoint of the maximum, had small standard errors. Moreover, the residual standard deviation of the oxidation rate was very small (0.601 ± 0.167). As the breath test is reliable, the personal parameters found have a biological basis. Finding potentially important biological factors involved in protein oxidation and protein utilization are a next step in understanding whole body protein metabolism.

In conclusion, this study has shown that on the group level, the total protein oxidation was not affected by a short-term reduction in protein intake in healthy subjects. This suggests that over the range of protein intake investigated here, the overall protein metabolism is robust against challenges. However, due to large variations found on the individual level with respect to the change in total protein oxidation between the habitual and protein restricted diet, and the poor to fair associations of total protein oxidation with demographic characteristics, it is uncertain how important the role of fluctuations in short-term protein oxidation is within whole body protein metabolism.

Author Contributions: Conceptualization, G.A.R.R., G.J.N., C.P.v.d.S., R.J.V. and H.J.-W.; methodology, G.A.R.R., G.J.N., W.P.K., R.J.V. and H.J.-W.; formal analysis, G.A.R.R. and W.P.K.; writing—original draft preparation, G.A.R.R., G.J.N., W.P.K., C.P.v.d.S., R.J.V. and H.J.-W.; writing—review and editing, G.A.R.R., G.J.N., W.P.K., C.P.v.d.S., R.J.V. and H.J.-W.; supervision, G.J.N., C.P.v.d.S., R.J.V. and H.J.-W.

Funding: This research received no external funding.

Conflicts of Interest: The naturally enriched ^{13}C-milk protein has been made available as an in kind contribution by Hanze Nutrition B.V., which is owned by co-author R.J. Vonk.

References

1. WHO. Protein and Amino Acid Requirements in Human Nutrition. In *Report of a Joint WHO/FAO/UNU Expert Consultation*; WHO Technical Report Series; WHO: London, UK, 2007.
2. Barker, L.A.; Gout, B.S.; Crowe, T.C. Hospital malnutrition: Prevalence, identification and impact on patients and the healthcare system. *Int. J. Environ. Res. Public Health* **2011**, *8*, 514–527. [CrossRef] [PubMed]
3. Bell, C.L.; Lee, A.S.W.; Tamura, B.K. Malnutrition in the nursing home. *Curr. Opin. Clin. Nutr. Metab. Care* **2015**, *18*, 17–23. [CrossRef]
4. Cederholm, T.; Barazzoni, R.; Austin, P.; Ballmer, P.; Biolo, G.; Bischoff, S.C.; et al. ESPEN guidelines on definitions and terminology of clinical nutrition. *Clin. Nutr.* **2017**, *36*, 49–64. [CrossRef] [PubMed]
5. Langer, C.J.; Hoffman, J.P.; Ottery, F.D. Clinical Significance of weight loss in cancer patients: Rationale for the use of anabolic agents in the treatment of cancer-related cachexia. *Nutrition* **2001**, *17*, S1–S21. [CrossRef]
6. Groen, B.B.L.; Horstman, A.M.; Hamer, H.M.; De Haan, M.; Van Kranenburg, J.; Bierau, J.; et al. Post-prandial protein handling: You are what you just ate. *PLoS ONE* **2015**, *10*, 1–22. [CrossRef] [PubMed]
7. Elango, R.; Chapman, K.; Rafii, M.; Ball, R.O.; Pencharz, P.B. Determination of the tolerable upper intake level of leucine in acute dietary studies in young men. *Am. J. Clin. Nutr.* **2012**, *96*, 759–767. [CrossRef] [PubMed]
8. Moore, D.R.; Robinson, M.J.; Fry, J.L.; Tang, J.E.; Glover, E.I.; Wilkinson, S.B.; Prior, T.; Tarnopolsky, M.A.; Phillips, S.M. Ingested protein dose response of muscle and albumin protein synthesis after resistance exercise in young men. *Am. J. Clin. Nutr.* **2009**, *89*, 161–168. [CrossRef] [PubMed]
9. Reckman, G.A.R.; Koehorst, M.; Priebe, M.; Schierbeek, H.; Vonk, R.J. 13C Protein Oxidation in Breath: Is It Relevant for the Whole Body Protein Status? *J. Biomed. Sci. Eng.* **2016**, *9*, 160–169. [CrossRef]
10. Nolles, J.A.; Verreijen, A.M.; Koopmanschap, R.E.; Verstegen, M.W.A.; Schreurs, V.V.A.M. Postprandial oxidative losses of free and protein-bound amino acids in the diet: Interactions and adaptation. *J. Anim. Physiol. Anim. Nutr.* **2009**, *93*, 431–438. [CrossRef]
11. Millward, D.J. Knowledge Gained from Studies of Leucine Consumption in Animals and Humans. *J. Nutr.* **2012**, *142*, 2212S–2219S. [CrossRef]
12. RIVM. *NEVO Online Version 2013/4.0*, RIVM: Bilthoven, The Netherlands, 2013.
13. Maroni, B.; Steinman, T.; Mitch, W. A method for estimating nitrogen intake of patients with chronic renal failure. *Kidney Int.* **1985**, *27*, 58–65. [CrossRef] [PubMed]
14. Haycock, G.B.; Schwartz, G.J.; Wisotsky, D.H. Geometric method for measuring body surface area: A height-weight formula validated in infants, children, and adults. *J. Pediatr.* **1978**, *93*, 62–66. [CrossRef]

15. Lefebvre, P.; Mosora, F.; Lacroix, M.; Luyckx, A.; Lopez-Habib, G.; Duchesne, J. Naturally labeled 13C-glucose. Metabolic studies in human diabetes and obesity. *Diabetes* **1975**, *24*, 185–189. [CrossRef] [PubMed]

16. Evenepoel, P.; Geypens, B.; Luypaerts, A.; Hiele, M.; Ghoos, Y.; Rutgeerts, P. Digestibility of cooked and raw egg protein in humans as assessed by stable isotope techniques. *J. Nutr.* **1998**, *128*, 1716–1722. [CrossRef]

17. Ghoos, Y.F.; Maes, B.D.; Geypens, B.J.; Mys, G.; Hiele, M.I.; Rutgeerts, P.J.; et al. Measurement of gastric emptying rate of solids by means of a carbon-labeled octanoic acid breath test. *Gastroenterology* **1993**, *104*, 1640–1647. [CrossRef]

18. Sanaka, M.; Yamamoto, T.; Anjiki, H.; Osaki, Y.; Kuyama, Y. Is the pattern of solid-phase gastric emptying different between genders? *Eur. J. Clin Investig.* **2006**, *36*, 574–579. [CrossRef] [PubMed]

19. Sanaka, M.; Nakada, K.; Nosaka, C.; Kuyama, Y. The Wagner-Nelson method makes the [13C]-breath test comparable to radioscintigraphy in measuring gastric emptying of a solid/liquid mixed meal in humans. *Clin. Exp. Pharmacol. Physiol.* **2007**, *34*, 641–644. [CrossRef] [PubMed]

20. Weisberg, S. *Applied Linear Regression*, 4th ed.; Section 6.1.2; Wiley: Hoboken, NJ, USA, 2014.

21. Fox, J.; Weisberg, S. *An R Companion to Applied Regression*; SAGE: Newcastle upon Tyne, UK, 2011.

22. Bates, D.; Watts, D. *Nonlinear Regression Analysis and Its Applications*, 2nd ed.; Wiley: Hoboken, NJ, USA, 1988.

23. Chan, Y.H. Biostatistics 104: Correlation analysis. *Singapore Med. J.* **2003**, *44*, 614–619. [PubMed]

24. Bingham, S.A.; Cummings, J.H. Urine nitrogen as an independent validatory measure of dietary intake: A study of nitrogen balance in individuals consuming their normal diet. *Am. J. Clin. Nutr.* **1985**, *42*, 1276–1289. [CrossRef] [PubMed]

25. Fouillet, H.; Juillet, B.; Bos, C.; Mariotti, F.; Gaudichon, C.; Benamouzig, R.; et al. Urea-nitrogen production and salvage are modulated by protein intake in fed humans: Results of an oral stable-isotope-tracer protocol and compartmental modeling. *Am. J. Clin. Nutr.* **2008**, *87*, 1702–1714. [CrossRef] [PubMed]

26. Horst, K.W.; Schene, M.R.; Holman, R.; Romijn, J.A.; Serlie, M.J. Effect of fructose consumption on insulin sensitivity in nondiabetic subjects: A systematic review and meta-analysis of diet intervention trials. *Am. J. Clin. Nutr.* **2016**, *104*, 1562–1576. [CrossRef] [PubMed]

27. Biolo, G.; Wolfe, R.R. Insulin action on protein metabolism. *Baillieres Clin. Endocrinol. Metab.* **1993**, *7*, 989–1005. [CrossRef]

28. Wu, G. Amino acids: Metabolism, functions, and nutrition. *Amino Acids* **2009**, *37*, 1–17. [CrossRef] [PubMed]

nutrients

MDPI

Article

Health Status of Female and Male Vegetarian and Vegan Endurance Runners Compared to Omnivores—Results from the NURMI Study (Step 2)

Katharina Wirnitzer [1,2,*], Patrick Boldt [3], Christoph Lechleitner [4], Gerold Wirnitzer [5], Claus Leitzmann [6], Thomas Rosemann [7] and Beat Knechtle [7]

[1]   Center for Research and Knowledge Management, Pedagogical University Tyrol, 6020 Innsbruck, Austria
[2]   Department of Sport Science, University of Innsbruck, 6020 Innsbruck, Austria
[3]   Faculty of Medicine, University of Gießen, 35390 Gießen, Germany; paboldt@aol.com
[4]   ITEG, 6020 Innsbruck, Austria; christoph.lechleitner@iteg.at
[5]   AdventureV & change2V, 6135 Stans, Austria; gerold@wirnitzer.at
[6]   Institute of Nutrition, University of Gießen, 35390 Gießen, Germany; claus@leitzmann-giessen.de
[7]   Institute of Primary Care, University of Zurich, 8091 Zurich, Switzerland; thomas.rosemann@usz.ch (T.R.); beat.knechtle@hispeed.ch (B.K.)
*   Correspondence: info@nurmi-study.com; Tel.: +43-(0)650-590-17-94

Received: 15 November 2018; Accepted: 20 December 2018; Published: 22 December 2018

Abstract: Health effects of vegetarian and vegan diets are well known. However, data is sparse in terms of their appropriateness for the special nutritional demands of endurance runners. Therefore, the aim of this study was to investigate the health status of vegetarian (VER) and vegan endurance runners (VGR) and compare it to omnivorous endurance runners (OR). A total of 245 female and male recreational runners completed an online survey. Health status was assessed by measuring health-related indicators (body weight, mental health, chronic diseases, and hypersensitivity reactions, medication intake) and health-related behavior (smoking habits, supplement intake, food choice, healthcare utilization). Data analysis was performed by using non-parametric ANOVA and MANOVA. There were 109 OR, 45 VER and 91 VGR. Significant differences ($p < 0.05$) were determined for the following findings: (i) body weight for VER and VGR was less than for OR, (ii) VGR had highest *food choice* scores, and (iii) VGR reported the lowest prevalences of allergies. There was no association ($p > 0.05$) between diet and mental health, medication intake, smoking habits, supplement intake, and healthcare utilization. These findings support the notion that adhering to vegetarian kinds of diet, in particular to a vegan diet, is associated with a good health status and, thus, at least an equal alternative to an omnivorous diet for endurance runners.

Keywords: vegetarian; vegan; half-marathon; marathon; running; health conscious; recreational athlete

---

1. Introduction

During an endurance event, such as a marathon running, body and mind are challenged to an extremely high degree. Athletes are exposed to several physiological and psychological challenges, in particular with regard to energy metabolism, body temperature and fluid balance [1–3]. A study by Hausswirth and Lehénaff highlighted the importance of fat metabolism, since an increase in free fatty acids and glycerol at the end of long-distance races crucially affects running economy and, thus, performance when the athlete is almost at the finish line [3]. Further important parameters with regard to running economy are maximal oxygen consumption, lactate-threshold, and metabolic efficacy [2]. Moreover, completing a long-distance race is a psychological challenge which requires favorable character traits, such as inhibitory control, the ability not only to inhibit motor response, but also to suppress processing of irrelevant information, and the ability to protect cognitive performance so that it is less influenced by emotional stimuli [1]. In order to meet all these requirements, a good health status and a strong mind are necessary and will contribute to good exercise performance [1,2].

An essential requirement for a good health status is the choice of an appropriate, healthy, and sustainable diet [4,5]. As endurance running is known as a kind of sport with high energy expenditure and, thus, consumption, an endurance athlete's need for vitamins, trace elements and other valuable food ingredients besides macronutrient requirements is very high [4]. Therefore, a well-balanced energy turnover is crucial [4], resulting in the creation of a well-planned and reasonable nutrition strategy [5]. Current evidence suggests that one strategy could be adhering to a meatless diet rich in vegetables and fruits, such as a vegetarian kind of diet [6–8] (pp. 419–437). Vegetarian kind of diet is an umbrella term which subsumes four main dietary patterns: lacto-ovo-vegetarian, lacto-vegetarian, ovo-vegetarian, and vegan. Lacto-ovo vegetarians consume dairy products and eggs but no meat, poultry or seafood. Lacto-vegetarians eat dairy products but avoid eggs, meat, poultry, and seafood. Ovo-vegetarians eat eggs, but no dairy products, meat, poultry, or seafood. A vegan diet is characterized by the rejection of all products from animal sources, such as meat, fish/shellfish, milk and dairy products, eggs, and honey. A dietary pattern without any restriction is referred to as an omnivorous kind of diet [7].

Healthy vegetarian kinds of diet usually include complex carbohydrates, fiber, fruits, vegetables, and antioxidants [9]. Although potentially lower in some nutrients, such as zinc and vitamin B12 [9], carefully planned vegetarian kinds of diet meet or even exceed the nutritional requirements of athletes, in particular with regard to the intake of proteins, fatty acids and iron [6–8] (pp. 419–437). More than this, vegetarian kinds of diet are known to have further beneficial effects on health than just energy intake, in particular in terms of body weight control [10,11], the prevention of diabetes mellitus type 2 [12,13], ischemic heart disease [11,14], and protection against depression [15]. In addition, a vegetarian diet has been found to reduce the risk for some types of cancer, such as colon and prostate cancer [9,16]. Despite immediate health-related effects due to the consumption of healthy foods, being a vegetarian or vegan is often associated with a healthy lifestyle characterized by the avoidance of adverse health behavior, such as smoking and alcohol consumption, a high level of physical activity, and time for relaxation [8] (p. 393).

To date, little is known about the health status and health-related behavior of vegetarian and vegan endurance runners [17,18]. Most researchers have not classified their subjects by dietary subgroup [19]. Beyond that, these studies usually dealt with athletes in general, so that data in terms of endurance runners is sparse. A well-founded comparison of health characteristics between vegetarian, vegan and omnivorous endurance runners is lacking. Specific knowledge about the interconnectedness of diet choice and health could provide a better basis for athletes and their coaches, physicians, and nutritionists/dietitians, in order to optimize training and treatment strategies.

The aim of the study, therefore, was to investigate the health status of endurance runners and to compare athletes who adhere to a vegetarian or vegan diet to those who follow an omnivorous diet. Since a good state of health of non-active vegetarians and vegans is sound and compares favorably to that of omnivores [8] (p. 411), it was hypothesized that vegetarian and vegan endurance runners would have a better health status than omnivorous endurance runners.

2. Materials and Methods

2.1. Study Protocol and Ethics Approval

The study protocol [20] was approved by the ethics board of St. Gallen, Switzerland on 6 May 2015 (EKSG 14/145). The trial registration number is ISRCTN73074080.

2.2. Participants

The NURMI (Nutrition and Running High Mileage) Study was conducted in three steps following a cross-sectional design. Endurance runners, mainly from German-speaking countries including

Germany, Austria, and Switzerland, were recruited. In addition, people from around the world were addressed. Participants were contacted mainly via social media, websites of the organizers of marathon events, online running communities, email lists, and runners' magazines, as well as via magazines for health, vegetarian, and/or vegan nutrition and lifestyle, sports fairs, trade fairs on vegetarian and vegan nutrition and lifestyle, and through personal contacts. The characteristics of the subjects are presented in Table 1.

Table 1. Anthropometric and demographic characteristics of the subjects displayed by diet group.

		Omnivorous	Vegetarian	Vegan
Number of Subjects		100% (109)	100% (45)	100% (91)
Sex	Female	47% (51)	58% (26)	70% (64)
	Male	53% (58)	42% (19)	30% (27)
Age (years) (median)		43 (IQR 18)	39 (IQR 16)	37 (IQR 15)
Body Weight (kg) (median)		68 (IQR 16.7)	62 (IQR 11.3)	64 (IQR 10)
BMI_{CALC} (kg/m^2)	≤18.49	4% (4)	7% (3)	9% (8)
	18.50–24.99	80% (87)	87% (39)	82% (75)
	≥25–29.99	17% (18)	7% (3)	9% (8)
Race Distance	10 km	34% (37)	33% (15)	43% (39)
	Half-marathon	36% (39)	44% (20)	33% (30)
	Marathon/Ultramarathon	30% (33)	22% (10)	24% (22)
Academic Qualification	No Qualification	0% (0)	0% (0)	1% (1)
	Upper Secondary Education/Technical Qualification/GCSE or Equivalent	38% (41)	38% (17)	27% (25)
	A Levels or Equivalent	24% (26)	16% (7)	22% (20)
	University Degree/Higher Degree (i.e., doctorate)	30% (33)	38% (17)	36% (33)
	No Answer	8% (9)	9% (4)	13% (12)
Marital Status	Divorced/Separated	3% (3)	4% (2)	11% (10)
	Married/Living with Partner	75% (82)	58% (26)	62% (56)
	Single	22% (24)	38% (17)	27% (25)
Country of Residence	Austria	21% (23)	18% (8)	14% (13)
	Germany	70% (76)	76% (34)	74% (67)
	Switzerland	7% (8)	4% (2)	3% (3)
	Other	2% (2)	2% (1)	9% (8)
Motive for Diet Choice	Health, wellbeing	81% (21)	85% (28)	90% (79)
	Sporting performance	54% (14)	33% (11)	59% (52)
	Food scandals	15% (4)	55% (18)	32% (28)
	Animal welfare	46% (12)	79% (26)	90% (79)
	Ecological aspects	50% (13)	76% (25)	83% (73)
	Social aspects	35% (9)	55% (18)	57% (50)
	Economic aspects	8% (2)	12% (4)	22% (19)
	Religion/spirituality	0% (0)	12% (4)	7% (6)
	Custom/tradition	15% (4)	0% (0)	2% (2)
	Taste/enjoyment	42% (11)	33% (11)	44% (39)
	No specific reason	4% (1)	0% (0)	0% (0)

10 km = 10-km control group. BMI_{CALC} = Body Mass Index (calculated). IQR = interquartile range.

2.3. Procedures

2.3.1. Experimental Approach

Participants completed an online survey within the NURMI Study Step 2, which was available in German and English at www.nurmi-study.com from 1 February 2015 to 31 December 2015. Prior to completing the questionnaires on physical and psychological health, participants were provided with a written description of the procedures and gave their informed consent to take part in the study.

For successful participation in the study, the following inclusion criteria were required: (1) written informed consent, (2) at least 18 years of age, (3) completed questionnaire, (4) successful participation in a running event of at least half-marathon distance in the past two years. Participants were classified into three dietary subgroups (Scheme 1): omnivorous (commonly known as Western diet, no dietary restrictions) diet; vegetarian (no meat); and vegan (no products from animal sources) [7]. In addition, they were categorized according to race distance: 10-km, half-marathon, and marathon/ultramarathon. Marathoners and ultramarathoners were pooled together since the marathon distance is included in an ultramarathon. A total of 91 highly-motivated runners provided accurate and useful answers with plenty of high-quality data. However, they had not successfully participated in either a half-marathon or marathon, but rather in a 10-km race. In order to avoid an irreversible loss of these valuable datasets, those who met all inclusion criteria but named a 10-km race as their running event were kept as the control group.

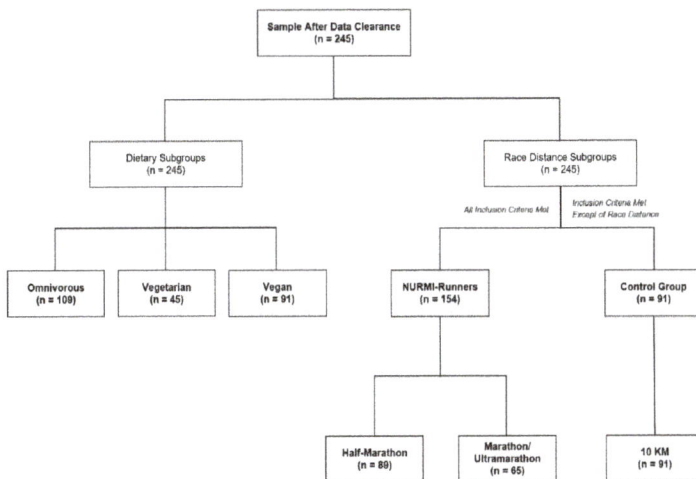

Scheme 1. Categorization of participants.

According to the WHO [21,22] the goal for individuals should be to maintain a BMI in the range 18.5–24.9 kg/m^2 (BMI$_{NORM}$) in order to achieve optimum health. They point to an increased risk of co-morbidities for a BMI 25.0–29.9 kg/m^2, and moderate to severe risk of co-morbidities for a BMI > 30 kg/m^2 [21,22]. Therefore, the calculated Body Mass Index (BMI$_{CALC}$) was classified into three categories of body weight-to-height ratio (kg/m^2): $\leq 18.49 <$ BMI$_{NORM}$: 18.50–24.99 kg/m^2 ≥ 25. Since the BMI of active runners could be below BMI$_{NORM}$ [23], but in addition people with a higher BMI might start running in order to achieve and maintain a stable, healthy body weight, participants with BMI < 30 were included. BMI has been shown to be a significant performance-determining parameter for speed improvement in running over various distances, with a continuous increase in BMI from 19.57 (1.29) kg/m^2 in marathoners to 23.3 (1.67) kg/m^2 over the 100 m distance [24]. An optimal BMI for high running pace, reported for the best performers over 10 km and marathon distance, was found to be between 19–20 kg/m^2.

2.3.2. Data Clearance

In order to control for measures of (1) diet and (2) running, two groups of control questions were included, each within different sections of the survey. In order to control for a minimal status of health linked to a minimum level of fitness and to further enhance the reliability of datasets, the BMI approach following the WHO [21,22] was used. With a BMI ≥ 30 other health-protecting and/or weight loss strategies than running would be necessary to safely reduce body weight. Therefore, three participants with a BMI ≥ 30 were excluded from the data analysis.

A total of 317 endurance runners completed the survey. Incomplete, inconsistent, and conflicting datasets were excluded from the data analysis. After data clearance a total of 245 runners with complete datasets were included for descriptive statistical analysis (Scheme 2).

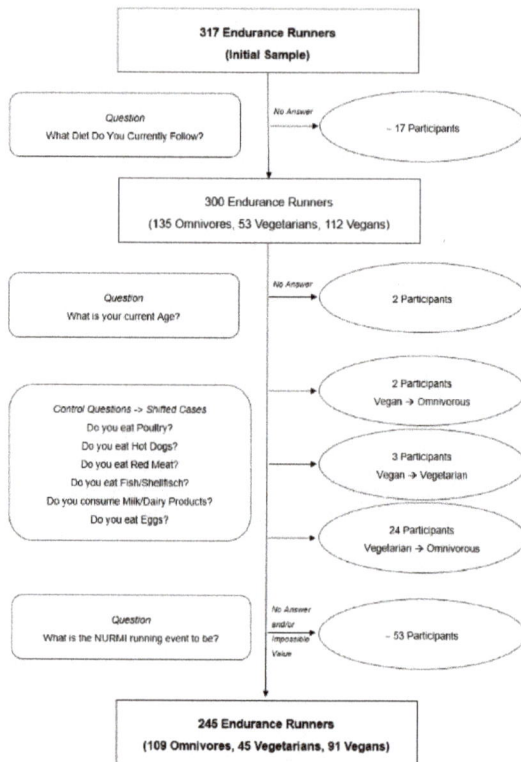

317 Endurance Runners
(Initial Sample)

Question
What Diet Do You Currently Follow?

No Answer → – 17 Participants

300 Endurance Runners
(135 Omnivores, 53 Vegetarians, 112 Vegans)

Question
What is your current Age?

No Answer → 2 Participants

Control Questions -> Shifted Cases
Do you eat Poultry?
Do you eat Hot Dogs?
Do you eat Red Meat?
Do you eat Fish/Shellfisch?
Do you consume Milk/Dairy Products?
Do you eat Eggs?

2 Participants
Vegan → Omnivorous

3 Participants
Vegan → Vegetarian

24 Participants
Vegetarian → Omnivorous

Question
What is the NURMI running event to be?

No Answer and/or Impossible Value → – 53 Participants

245 Endurance Runners
(109 Omnivores, 45 Vegetarians, 91 Vegans)

Scheme 2. Flow chart of participants' enrollment.

2.4. Measures

Health status (latent variable) was derived by using both the two clusters 'Health-related Indicators' and 'Health-related Behavior'. Each cluster pooled four dimensions, with each defined by specific items based on manifest measures. An overview of the variables is presented in Table 2.

The following health-related indicators described health outcomes: (1) body weight/BMI, (2) mental health (stress perception), (3) chronic diseases and hypersensitivity reactions: prevalence of chronic diseases (heart disease, state after heart attack, cancer), prevalence of metabolic diseases (diabetes mellitus 1, diabetes mellitus 2, hyperthyroidism, hypothyroidism), prevalence of hypersensitivity reactions (allergies, intolerances), and (4) medication intake (for thyroid disease, for hypertension, for cholesterol level, for contraception).

Table 2. Overview of the variables in order to derive health status of endurance runners.

Cluster	Dimension	Indicator	Item	Measure
Health-related Indicators	Body weight/BMI	BMI_{calc} [1]	Your current body weight (kg)? Your height (m)?	Body weight (kg) Height (m)
	Mental health	Stress perception	Are you under pressure and/or are you suffering from stress?	Yes No
	Chronic diseases and Hypersensitivity reactions	Cardiovascular diseases and Cancer	Are you currently suffering from the following chronic diseases or their direct consequences?	Heart disease requiring treatment Heart attack Cancer (now or in the past)?
		Metabolic diseases	Are you currently suffering from one of the following metabolic disease(s)?	Diabetes mellitus type 1 Diabetes mellitus type 2 Hyperthyroidism Hypothyroidism
		Hypersensitivity reactions	Are you currently suffering from ...?	Allergies Intolerances
	Medication intake		Do you take medicaments regularly (every day), for example, ...?	Thyroid High blood pressure Cholesterol and/or other blood serum lipid values contraceptive pill
Health-related Behavior	Smoking habits	Current consumption of cigarettes	Do you currently smoke?	Yes No
		Former consumption of cigarettes	Have you ever smoked?	Yes No
	Supplement intake	Substance intake for medical reasons	Do you take supplements prescribed by a doctor regularly (everyday)?	Yes No
		Intake of Performance enhancement substance	Do you take anything to boost your performance in your daily life, at work or while doing sport (e.g., energy drinks)?	Yes, regularly every day Yes, occasionally No
		Substance intake for stress coping	Do you take anything to help you cope with stress in your daily life, at work or while doing sport?	Yes, regularly every day Yes, occasionally No
	Food choice	Motivation for food choice	Do you choose ingredients and food on the basis Of the following (e.g., in view of the disease mentioned above or other illnesses)?	Healthy (e.g., if you are ill) Health-promoting (e.g., to prevent ill-health) Good for maintaining health (e.g., wholefoods)

33

Table 2. *Cont.*

Cluster	Dimension	Indicator	Item	Measure
		Avoided ingredients	Do you choose food in order to avoid particular ingredients or nutrients (e.g., in view of the diseases mentioned above or other illnesses or effects on health)?	Refined sugar Sweetener Fat in general Saturated fats Cholesterol Products made with white flour Sweet things (e.g., Jelly beans, chocolate drops, cream cakes) Nibbles (e.g., crisps, salted peanuts) Alcohol Caffeine or other stimulants (e.g., in coffee or energy drinks)
		Desired ingredients	Do you choose food because of particular valuable ingredients or nutrients (e.g., to prevent the diseases mentioned above or other illnesses or effects on health)?	Vitamins Minerals/trace elements Antioxidants Phytochemicals Fiber Other
	Healthcare utilization	Frequency of doctor consultations	How often have you seen a doctor in the last 12 months (except dentist and for routine check-ups)?	Never Once a month Every 2 months Every 3 months (four times a year) Every 6 months (twice a year) Once a year
		Utilization of regular health check-ups	Do you go for regular check-ups or routine health checks?	Yes No

[1] BMI_{CALC} = Body Mass Index calculated.

The following variables of health-related behavior described health outcomes: (1) smoking habits (current and former smoking), (2) supplement intake (supplements prescribed by a doctor, supplements for performance enhancement, supplements to cope with stress), (3) food choice (motivation, desired ingredients, avoided ingredients), and (4) healthcare utilization (regular check-ups). Resulting from this, eight domain scores (body weight/BMI, mental health, chronic diseases, and hypersensitivity reactions, medication intake, smoking, supplement intake, food choice, healthcare utilization) were derived, which generated scores between 0 and 1. Low scores indicate detrimental health effects, while higher scores indicate beneficial health effects (given as mean scores plus standard deviation, and percentage (%)).

2.5. Statistical Analysis

The statistical software R version 3.5.0 Core Team 2018 (R Foundation for Statistical Computing, Vienna, Austria) performed all statistical analyses. Exploratory analysis was performed by descriptive statistics (median and interquartile range (IQR)). Significant differences between dietary subgroups and domain scores to describe health status were calculated by using a non-parametric ANOVA. Chi-square test and Kruskal-Wallis test were used to examine the association between dietary subgroups and domain scores with nominal scale variables, and Wilcoxon test and Kruskal-Wallis test (ordinal and metric scale) approximated by using the F distributions.

Statistical modeling. State of health as the latent variable was derived by manifest variables (e.g., body weight, cancer, smoking, etc.). In order to scale the health status displayed by measures, items and dimensions, a heuristic index between 0 and 1 was defined (equivalence in all items). To test the statistical hypothesis considering significant differences between dietary subgroups, race distance and sex for each dimension a MANOVA was performed to define health status. The assumptions of the ANOVA were verified by residual analysis.

The level of statistical significance was set at $p \leq 0.05$.

3. Results

A total of 317 endurance runners completed the survey, of whom 245 (141 women and 104 men) remained after data clearance with a mean age of 39 (IQR 17) years, from Germany ($n = 177$), Switzerland ($n = 13$), Austria ($n = 44$) and from other countries ($n = 11$; Belgium, Brazil, Canada, Italy, Luxemburg, Netherlands, Poland, Spain, UK).

A total of 109 participants followed an omnivorous diet, 45 reported to adhere to a vegetarian diet, and 91 to a vegan diet. In addition, there were a total of 91 10-km runners, 89 half-marathoners, and 65 marathoners/ultramarathoners.

3.1. Cluster 'Health-Related Indicators'

3.1.1. Dimension of Body Weight/BMI

There was a significant difference in body weight between dietary subgroups ($F_{(2, 242)} = 6.86$, $p = 0.001$), with vegetarians and vegans showing lower body weight than omnivores. However, there was no difference in the health-related item BMI between dietary subgroups ($\chi^2_{(4)} = 6.08$, $p = 0.193$) (Table 3). Moreover, vegans had the highest counts for the health-related indicator *body weight/BMI* (0.69 (0.40), $F_{(2, 242)} = 0.41$, $p = 0.662$) (Figure 1a).

3.1.2. Dimension of Mental Health

There was no significant association between diet group and stress perception ($\chi^2_{(2)} = 1.78$, $p = 0.412$) (Table 3). However, vegans had the highest score with regard to *mental health* (0.66 (0.48), $F_{(2, 219)} = 0.88$, $p = 0.415$) (Figure 1a).

Table 3. Descriptive results and ANOVA of the 'Health-Related Indicators' cluster.

Dimension	Omnivorous	Vegetarian	Vegan	Statistics
Body Weight/BMI				
Body Weight (kg) (median)	68.00 (IQR 16.70)	62.00 (IQR 11.30)	64.00 (IQR 10.00)	$F_{(2, 242)} = 6.86, p = 0.001$
BMI$_{CALC}$				$x^2_{(4)} = 6.08, p = 0.193$
≤18.49	4% (4)	7% (3)	9% (8)	
18.50–24.99	80% (87)	87% (39)	82% (75)	
≥25–29.99	17% (18)	7% (3)	9% (8)	
Mental Health				$x^2_{(2)} = 1.78, p = 0.412$
Stress Perception				
Yes	36% (35)	46% (18)	34% (29)	
No	64% (63)	54% (21)	66% (56)	
Chronic Diseases and Hypersensitivity Reactions				
Prevalence of Chronic Diseases				$x^2_{(4)} = 2.88, p = 0.578$
Heart Disease	1% (1)			
Heart Attack				
Cancer			1% (1)	
No Diseases	99% (97)	100% (39)	99% (84)	
Prevalence of Metabolic Diseases				$x^2_{(10)} = 7.14, p = 0.713$
Diabetes Mellitus 1	1% (1)	3% (1)		
Diabetes Mellitus 2	2% (2)			
Hyperthyroidism	1% (1)	3% (1)	1% (1)	
Hypothyroidism	5% (5)	3% (1)	8% (7)	
Other Diseases	1% (1)		1% (1)	
No Diseases	90% (88)	92% (36)	89% (76)	
Prevalence of Hypersensitivity Reactions				$x^2_{(4)} = 12.87, p = 0.012$
Allergies	32% (31)	36% (14)	20% (17)	
Intolerances	1% (1)	10% (4)	12% (10)	
No Reactions	67% (66)	54% (21)	68% (58)	
Medication Intake (regularly)				$x^2_{(6)} = 7.58, p = 0.271$
Thyroid Disease	6% (6)	8% (3)	11% (9)	
Hypertension	5% (5)	3% (1)		
Cholesterol Level				
Other Medication	5% (5)		5% (4)	
No Medication	84% (82)	90% (35)	85% (72)	
Contraceptives	12% (12)	10% (4)	15% (13)	$x^2_{(2)} = 0.70, p = 0.704$

BMI$_{CALC}$ = Body Mass Index (calculated). IQR = interquartile range.

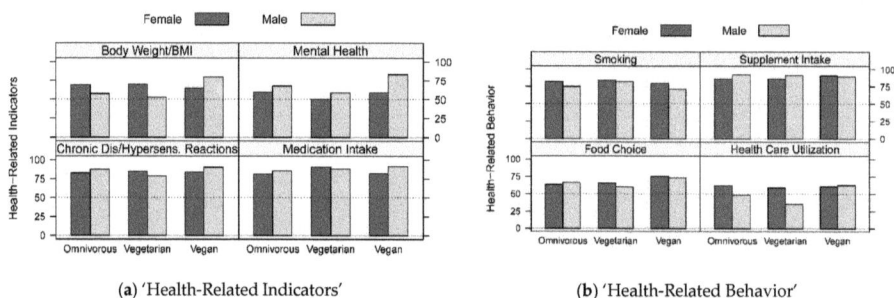

(a) 'Health-Related Indicators'

(b) 'Health-Related Behavior'

Figure 1. Indices of both clusters 'Health-Related Indicators' and 'Health-Related Behavior' of female and male endurance runners, displayed by dietary subgroups (as percentage, %). Low scores indicate detrimental health effects, high scores indicate beneficial health effects.

3.1.3. Dimension of Chronic Diseases and Hypersensitivity Reactions

There was no significant association between diet and the prevalence of cardiovascular diseases and cancer ($x^2_{(4)} = 2.88$, $p = 0.578$), and even between diet and prevalence of metabolic diseases ($x^2_{(10)} = 7.14$, $p = 0.713$). However, there was a significant difference between the prevalence of hypersensitivity reactions and diet ($x^2_{(4)} = 12.87$, $p = 0.012$), where vegan endurance runners stated least often that they had at least one allergy. In addition, omnivores reported having a food intolerance least often (Table 3). Omnivorous, vegan, and vegetarian runners scored similarly with regard to the health-related indicator *chronic diseases and hypersensitivity reactions* (respectively, 0.85 (0.20), 0.82 (0.20), and 0.85 (0.18), $F_{(2, 219)} = 0.58$, $p = 0.562$) (Figure 1a).

3.1.4. Dimension of Medication Intake

There was no significant association between medication intake and dietary subgroup ($\chi^2_{(6)} = 7.58$, $p = 0.271$) (Table 3). Furthermore, there was no significant effect diet on the use of contraceptives ($\chi^2_{(2)} = 0.70$, $p = 0.704$). However, vegetarians had the highest scores with regard to medication intake, even though all dietary subgroups had similar scores (respectively, 0.84 (0.37), 0.90 (0.31), and 0.85 (0.36), $F_{(2, 219)} = 0.41$, $p = 0.663$) (Figure 1a).

3.2. Cluster 'Health-Related Behavior'

3.2.1. Dimension of Smoking Habits

Diet and current or former smoking were not significantly associated ($\chi^2_{(4)} = 8.96$, $p = 0.062$) (Table 4). Vegetarians showed the best health-related behavior with regard to *smoking habits* (0.83 (0.29), $F_{(2, 219)} = 1.30$, $p = 0.275$) (Figure 1b).

Table 4. Descriptive results and ANOVA of the 'Health-Related Behavior' cluster.

Dimension	Omnivorous	Vegetarian	Vegan	Statistics
Smoking Habits				$\chi^2_{(4)} = 8.96$, $p = 0.062$
Non-Smoker	58% (57)	72% (28)	54% (46)	
Ex-Smoker	40% (39)	23% (9)	46% (39)	
Smoker	2% (2)	5% (2)		
Supplement Intake				
prescribed by doctor	8% (8)	10% (4)	6% (5)	$\chi^2_{(2)} = 0.79$, $p = 0.675$
to boost your performance (occasionally)	10% (10)	21% (8)	11% (9)	
to boost your performance (regularly)	3% (3)	0 % (0)	2% (2)	$\chi^2_{(4)} = 4.09$, $p = 0.394$
to cope with stress (occasionally)	5% (5)	5% (2)	8% (7)	
to cope with stress (regularly)	2% (2)		2% (2)	$\chi^2_{(4)} = 1.79$, $p = 0.774$
Food Choice				
Motivation				
because it is healthy	67% (66)	74% (29)	75% (64)	$\chi^2_{(2)} = 1.59$, $p = 0.452$
because it is health-promoting	81% (79)	79% (31)	88% (75)	$\chi^2_{(2)} = 2.41$, $p = 0.300$
because it is good for maintaining health	85% (83)	92% (36)	95% (81)	$\chi^2_{(2)} = 5.99$, $p = 0.050$
Avoided Ingredients				
Refined Sugar	62% (61)	56% (22)	73% (62)	$\chi^2_{(2)} = 3.95$, $p = 0.138$
Sweetener	74% (73)	59% (23)	80% (68)	$\chi^2_{(2)} = 6.16$, $p = 0.046$
Fat in General	39% (38)	46% (18)	49% (42)	$\chi^2_{(2)} = 2.17$, $p = 0.339$
Saturated Fats	53% (52)	46% (18)	72% (61)	$\chi^2_{(2)} = 9.82$, $p = 0.007$
Cholesterol	34% (33)	31% (12)	65% (55)	$\chi^2_{(2)} = 21.60$, $p < 0.001$
White Flour	64% (63)	59% (23)	74% (63)	$\chi^2_{(2)} = 3.42$, $p = 0.181$
Sweets	58% (57)	62% (24)	69% (59)	$\chi^2_{(2)} = 2.52$, $p = 0.284$
Nibbles	62% (61)	59% (23)	62% (53)	$\chi^2_{(2)} = 0.15$, $p = 0.928$
Alcohol	55% (54)	51% (20)	56% (48)	$\chi^2_{(2)} = 0.29$, $p = 0.864$
Caffeine	26% (25)	36% (14)	46% (39)	$\chi^2_{(2)} = 8.30$, $p = 0.016$
Desired Ingredients				
Vitamins	81% (79)	72% (28)	86% (73)	$\chi^2_{(2)} = 3.48$, $p = 0.175$
Minerals/Trace Elements	70% (69)	72% (28)	75% (64)	$\chi^2_{(2)} = 0.56$, $p = 0.757$
Antioxidants	47% (46)	44% (17)	60% (51)	$\chi^2_{(2)} = 4.25$, $p = 0.119$
Phytochemicals	42% (41)	31% (12)	59% (50)	$\chi^2_{(2)} = 9.93$, $p = 0.007$
Fiber	68% (67)	62% (24)	75% (64)	$\chi^2_{(2)} = 2.58$, $p = 0.276$
Health Care Utilization				
Regular check-ups or routine health checks	54% (53)	49% (19)	61% (52)	$\chi^2_{(2)} = 1.91$, $p = 0.385$

3.2.2. Dimension of Supplement Intake

There was no significant association between diet and supplement intake prescribed by a doctor ($\chi^2_{(2)} = 0.79$, $p = 0.675$), the consumption of performance-enhancing substances ($\chi^2_{(4)} = 4.09$, $p = 0.394$) or the intake of substances to cope with stress ($\chi^2_{(4)} = 1.79$, $p = 0.774$) (Table 4). Vegans showed the best health-related behavior with regard to *supplement intake* (0.91 (0.19), $F_{(2, 219)}$ = 0.35, p = 0.708) (Figure 1b).

3.2.3. Dimension of Food Choice

There was no significant association between diet and food choice (i) because it is healthy ($\chi^2_{(2)}$ = 1.59, p = 0.452) and health-promoting ($\chi^2_{(2)}$ = 2.41, p = 0.300); or (ii) in order to obtain vitamins ($\chi^2_{(2)}$ = 3.48, p = 0.175), minerals/trace elements ($\chi^2_{(2)}$ = 0.56, p = 0.757), antioxidants ($\chi^2_{(2)}$ = 4.25, p = 0.119) and fiber ($\chi^2_{(2)}$ = 2.58, p = 0.276) (Table 4). Moreover, there was no significant association between diet and the avoidance of the following ingredients (Table 4): refined sugar ($\chi^2_{(2)}$ = 3.95, p = 0.138), fat in general ($\chi^2_{(2)}$ = 2.17, p = 0.339), white flour ($\chi^2_{(2)}$ = 3.42, p = 0.181), sweets ($\chi^2_{(2)}$ = 2.52, p = 0.284), nibbles ($\chi^2_{(2)}$ = 0.15, p = 0.928), and alcohol ($\chi^2_{(2)}$ = 0.29, p = 0.864).

However, there was a significant effect of diet on *food choice*, both (i) because it is good for maintaining health ($\chi^2_{(2)}$ = 5.99, p = 0.050), with vegetarians and vegans reporting doing so more often; and (ii) in order to obtain phytochemicals ($\chi^2_{(2)}$ = 9.93, p = 0.007), with vegans reporting doing so more often. Moreover, there was a significant association between diet and the avoidance of the following ingredients (Table 4): sweetener ($\chi^2_{(2)}$ = 6.16, p = 0.046), saturated fats ($\chi^2_{(2)}$ = 9.82, p = 0.007), cholesterol ($\chi^2_{(2)}$ = 21.60, p < 0.001), and caffeine ($\chi^2_{(2)}$ = 8.30, p = 0.016). Vegans were more likely to report considering avoiding these ingredients in their food choice than vegetarians and omnivores.

Vegan athletes had the highest scores in *food choice* compared to the other dietary subgroups (0.75 (0.20), $F_{(2, 219)}$ = 6.76, p = 0.001) (Figure 1b).

3.2.4. Dimension of Healthcare Utilization

There was no significant association between the use of regular health check-ups and diet ($\chi^2_{(2)}$ = 1.91, p = 0.385) (Table 4). Vegan athletes had the highest scores with regard to *healthcare utilization* (0.61 (0.49), $F_{(2, 219)}$ = 0.95, p = 0.389) (Figure 1b).

3.3. Results of the MANOVA

The findings of the MANOVA considering state of health are presented in Table 5, indicating significant differences (p < 0.05) for the following results: (i) race distance (F = 3.39, Df = 2, p = 0.036) and sex (F = 4.06, Df = 1, p = 0.045) had an effect on *mental health*, (ii) race distance had an impact on *chronic diseases* and *hypersensitivity reactions* (F = 3.27, Df = 2, p = 0.040), (iii) an association between sex and *smoking habits* (F = 4.22, Df = 1, p = 0.041), and (iv) an association between *food choice* and diet (F = 6.10, Df = 2, p = 0.003), with vegans having the highest scores (0.75).

Table 5. Results of the MANOVA considering health status.

Cluster	Dimension	Subgroup	F	Df	p
Health-related Indicators	Body weight/BMI	Diet	0.75	2	0.475
		Race Distance	0.49	2	0.613
		Sex	0.62	1	0.432
	Mental health	Diet	0.91	2	0.402
		Race Distance	3.39	2	0.036
		Sex	4.06	1	0.045
	Chronic diseases and hypersensitivity reactions	Diet	0.49	2	0.611
		Race Distance	3.27	2	0.040
		Sex	0.77	1	0.381
	Medication intake	Diet	0.41	2	0.665
		Race Distance	0.15	2	0.859
		Sex	1.06	1	0.304

Table 5. *Cont.*

Cluster	Dimension	Subgroup	F	Df	p
	Smoking habits	Diet	0.80	2	0.451
		Race Distance	1.78	2	0.172
		Sex	4.22	1	0.041
	Supplement intake	Diet	0.14	2	0.866
		Race Distance	0.93	2	0.395
		Sex	1.91	1	0.168
	Food choice	Diet	6.10	2	0.003
		Race Distance	1.11	2	0.331
		Sex	0.08	1	0.779
	Healthcare utilization	Diet	0.96	2	0.385
		Race Distance	1.52	2	0.222
		Sex	2.14	1	0.145

(Cluster column: Health-related Behavior)

F = F-value. Df = Degrees of freedom. p = p-value for difference among groups.

However, the overall health status derived from all dimensions showed differences between race distances with statistical trend (F = 1.83, Df = 2, p = 0.71), but no significant differences were found for either diet or sex.

4. Discussion

This study intended to investigate the health status of vegetarian and vegan endurance runners and to compare it to omnivorous athletes, regarding potential differences in body weight, smoking habits, stress perception, the prevalence of chronic and metabolic diseases, the prevalence of allergies and food intolerances, medication and supplement intake, food choice, consumption of performance-enhancing substances, and healthcare utilization. In terms of assessing the state of health of endurance runners, it is generally accepted, that body weight, BMI and smoking behavior were known to affect running performance.

The main findings were: (i) vegetarians and vegans weighed significantly less than omnivores, (ii) vegans had the highest *food choice* scores, (iii) vegans reported choosing food because it is good for maintaining health more often, (iv) vegans reported avoiding sweeteners, saturated fats, cholesterol, and caffeine when choosing food more often, (v) vegans reported choosing food in order to obtain phytochemicals more often, and (vi) vegans reported the lowest prevalence of allergies.

4.1. Body Weight and BMI

A first important finding was that both vegetarians and vegans had lower body weight (62.00 kg (IQR 11.30) and 64.00 (IQR 10.00) kg, respectively) than omnivores (68.00 kg (IQR 16.70)). At the same time, the majority of all participants had a BMI which was within the normal range of 18.50–24.99 kg/m^2 (80 % in omnivores vs. 87 % in vegetarians vs. 82 % in vegans) [21,22,24], with vegans having the best *body weight/BMI* health scores.

BMI is a relevant parameter, since it is associated with an increased risk for diseases, such as cardiovascular diseases, if it is higher than BMI_{NORM}, and with a couple of other disorders, such as anorexia nervosa, if it is below BMI_{NORM} [21,22]. In addition, it is a key factor with regard to running performance [24]. However, careful use and interpretation of the BMI is required. For example, the BMI of active runners could be below the normal range without being pathological [23].

In the light of this, the findings of the present study were in line with previous literature, where vegetarians and vegans also had lower BMI than meat-eaters [25–27]. Spencer et al. [28] attributed these differences in body weight and BMI mainly to differences in macronutrient intake between vegetarians, vegans, and omnivores. High protein and low fiber intakes were the factors most strongly associated with increasing BMI. Considering the fact that running speed and endurance performance are significantly associated with body mass and BMI [24], vegetarian kinds of diet are known to be a good basis for body weight control strategies for endurance athletes [7,16,29]. Meanwhile, athletes, as well as their coaches, have to be particularly aware of unintended body weight loss [30], which is why regular monitoring of body weight is recommended [27]. Beyond athletic concerns, vegetarian, but in particular vegan, dietary patterns are known as to be useful for body weight control for people who suffer from obesity and diabetes mellitus type 2 and hypercholesterinemia [12,13].

4.2. Vegetarians' and Vegans' Attitudes Towards Food Choice

While only the dimension *food choice* showed significant differences between dietary subgroups, overall the vegan dietary subgroup displayed the highest health scores from all dimensions (except for *medication intake*) and contributed to runners' good state of health, ranging from 61%–91%.

A main result was that vegans showed the highest score (75%) in endurance runners in the dimension *food choice* to contribute most beneficially to the overall state of health. This means that they reported choosing food ingredients because they are good for maintaining health. This finding was consistent with available scientific literature.

Studies of vegetarians and vegans have identified a range of motivations for dietary choices [8] (p. 395), although personal health and animal welfare were predominant motives [31–33] (pp. 24–28). It has also been shown that vegetarians and vegans usually have healthier lifestyles than omnivores [8, 34] (p. 393). Their healthy lifestyle is characterized by the avoidance of adverse health behaviors, such as smoking and alcohol consumption, a high level of physical activity, and time for relaxation. Moreover, vegetarians and vegans are usually well-educated, have a certain degree of intellectual curiosity, and are open to new experiences [8] (p. 393). These findings match the results from the present study and support the characterization of vegetarians, but vegans in particular, as being health-conscious. However, all participants, meaning vegetarian, vegan, and omnivorous endurance runners, reported health-reasons as being important for food and ingredients choice. This supports the notion that athletes in general are health-conscious [35], but vegan athletes are supposed to be those who care most about this specific health-related strategy [8] (p. 393).

However, there was no significant major effect of dietary subgroups on whether food or ingredients had been chosen because they were healthy or health-promoting, even though there was a slight predominance of vegetarian and vegan runners. This was not entirely in line with current scientific evidence, as it has been shown that vegetarians and vegans are usually more health-conscious than omnivores [33,34,36]. Notwithstanding this, the contradiction might be explained by the composition of the sample. As all participants were endurance runners, who are known to be health-conscious compared to non-active people of the general population [35], the predominance of vegetarians and vegans might have been compensated for in this regard. Furthermore, the survey was based on self-reporting, so the definition of what is healthy or health-promoting in terms of food ingredients would depend on individual definitions based on personal suggestions and beliefs. Therefore, the results might have been biased to a certain degree. However, as the majority of all runners reported considering health aspects when choosing food, the findings support the characterization of the participants as being health-conscious.

A further main result was that vegan participants reported choosing food ingredients in order to avoid cholesterol, caffeine, sweetener and saturated fats more often. This finding was in line with the literature as well [37] and supports the fact that vegans in particular are supposed to be health-conscious.

Even though caffeine and cholesterol do not have detrimental health effects or may even have beneficial health implications if they are consumed conscientiously [38,39], cholesterol, in particular, is believed to be a crucial factor in the genesis of cardiovascular diseases [39]. Cholesterol is known to be an important risk factor for cardiovascular disease due to the induction of the elevation of LDL levels. It has also been found that HDL levels, which protect against cardiovascular diseases, increase after cholesterol consumption, so moderate consumption has been recommended in some studies [39].

However, to date the interactions between cholesterol intake and LDL and HDL blood levels have not been revealed completely [40]. With regard to caffeine, moderate consumption can increase physical and mental performance, while excessive intake can induce abuse or dependence [39]. Thus, it seems likely that both substances can be consumed moderately without any severe harm. However, being aware of potential detrimental side effects and therefore conscientious consumption is recommended.

Consumption of a high number of saturated fats is associated with cardiovascular diseases, such as stroke, myocardial infarction, and hypertension [8] (p. 414). Since vegan diets are characterized by a low percentage of saturated fats and a high percentage of omega-3 and omega-6 fatty acids [41], adhering to a plant-based diet can be a good way to improve cardiovascular health.

Health-effects of artificial sweeteners are controversial. While a couple of these products, such as aspartame, have previously received a generally recognized status as being safe from the United States Food and Drug Administration, there is also evidence for detrimental effects, such as the manifestation of glucose intolerance, weight gain and triggering of migraine in susceptible individuals [42]). Moreover, carcinogen effects could not be ruled out yet [43]. Overall, avoiding these agents appears to be advisable, so that the fact that the vegan endurance runners of our sample reported avoiding ingestion of sweeteners characterized them as being particularly health-conscious once again.

In addition to the avoidance of harmful substances, such as cholesterol and saturated fats, vegans reported choosing food in order to obtain phytochemicals. This finding supports the fact that vegan athletes are particularly health-conscious, since the consumption of phytochemical-rich foods is an important benefit of any plant-based diet in that it might help to mitigate the effects of excess inflammation and to promote recovery from training [41].

4.3. Allergies and Food Intolerances

There was a significant association between the prevalence of hypersensitivity reactions and diet, whereby vegan endurance runners reported least often that they had at least one allergy (20% in vegans vs. 32% in omnivores and 36% in vegetarians). Among those vegan endurance runners, 10-km runners had the lowest prevalence of allergies. At the same time, omnivores reported having a food intolerance least often (1% in omnivores vs. 10% in vegetarians and 12% in vegans).

Current evidence is sparse in this regard. One study has detected higher allergy rates among vegetarians [44], whereas others found a protective effect of a diet rich in fruits and vegetables on the occurrence of allergic asthma [45] and food allergies [46,47]. However, a relatively high incidence of allergies in a sample of endurance runners is not unexpected. It is well known that endurance athletes are more likely to have allergies (prevalence up to 13%) than people from the general population (prevalence 7% to 8%) [48]. This is usually attributed to the amount of time runners spend outdoors, which is supposed to be associated with a drying of the airways and an increased exposure to airborne allergens [49]. In the light of this, the finding that the vegan 10-km runners reported the lowest prevalence of allergies appears to be plausible because they usually have to cope with smaller training volumes (daily and weekly mileage) to successfully compete over shorter race distances. As a consequence, these runners do not spent as much time outdoors as long-distance runners, such as half-marathoners and (ultra-)marathoners.

Regarding food intolerances, the current literature does not provide clear data in this regard. One study indicated that a vegan diet might beneficially affect the intestinal flora, which seems to lower the risk of irritable bowel disease [47], whereas another study identified a long-term vegetarian diet as being the reason for the occurrence of irritable bowel disease [50]. However, endurance athletes, in general, are supposed to be more susceptible to symptoms of food sensitivities, which can be similar to those of irritable bowel disease. Constant training challenges the bowel to an extreme degree and endurance running, in particular, might cause gastrointestinal complaints. Thus, the ability to cope with additional gastrointestinal stress induced by food intolerances would be reduced [51].

4.4. Stress Perception

There was no significant difference found between vegetarians, vegans, and omnivores in reported stress and perceived pressure. *Mental health* scores were high, regardless of diet choice. However, vegan endurance runners had the highest scores for *mental health*. These findings were in line with previous studies, which showed that both endurance running [52,53] and adhering to a vegan dietary pattern caused good mood states [54]. Certain characteristics of vegans, such as a high degree of health-awareness [8] (p. 393), and the beneficial effects of endurance running, such as relaxation due to physical activity and an increase in stress resilience [52,53], appear to be the key factors in this regard.

In the light of this, finding the optimal dose of endurance running appears to be relevant, since too little exercise does not lead to a reduction in stress, whilst too much exercise might even increase stress levels [52]. According to the findings of the present study, half-marathon running appears to be a good way to cope with stress. These findings (unpublished data from our laboratory) are discussed in detail elsewhere [55]. Moreover, among the participants of the present study, there was a slight male predominance among those runners who reported as not suffering from stress. This was in line with previous research where it was reported that male endurance athletes possess a slightly higher degree of mental toughness than their female counterparts, allowing them to cope better with stress during exercise and in everyday life [56].

4.5. Chronic Diseases

There was no significant differences between the dietary subgroups when considering heart disease requiring treatment, state after heart attack, cancer, diabetes mellitus type 1 and 2, hypothyroidism and hyperthyroidism. In addition, there was a low overall incidence of these diseases among our participants. The only exceptions seemed to be apparently higher rates of cancer and hypothyroidism among vegetarians and vegans, which could be explained by a statistical bias.

There were five females who had suffered from breast cancer. Three of them had decided to change their dietary habits in favor of a vegetarian kind of diet after diagnosis of cancer, which skewed the results. The higher prevalence of hypothyroidism could be explained by the female predominance among the vegetarian and vegan subjects, as it is well known that eight times as many women suffer from thyroid diseases in general, and in particular from hypothyroidism, than men [57].

The fact that there was no association between diet and the prevalence rates of chronic diseases partially contradicts the body of evidence. Adhering to a vegetarian or vegan diet is usually associated with a lower incidence of diabetes mellitus type 2 [7,12,13], hypothyroidism (Tonstad et al. 2013 [58]), coronary artery disease [11,14], depression [54] and obesity [11] compared to an omnivorous diet. However, this effect might be compensated for by the fact that all our subjects were endurance athletes, who are usually supposed to be health-conscious, especially compared to non-active people of the general population [35]. Furthermore, the mean age of our participants was quite low (43.00 ± 18.00 in omnivores, 39.00 ± 16.00 in vegetarians, and 37.00 ± 15.00 in vegans), so that it can be assumed that the peak age for the manifestation of most diseases had not been reached yet. Furthermore, the fact that people who suffer from severe diseases usually do not become endurance runners might have led to a certain decrease in prevalence rates as well.

4.6. Medication Intake

There was no significant association found between the intake of medication with diet. All subgroups had similar *medication intake* scores. As there was a low prevalence of chronic diseases among our subjects, it was not surprising that there was also a low number of athletes who had to take any medication on a regular basis. The only exceptions were the intake of hormones and medication for the thyroid. The relatively high number of athletes who take hormones could be explained by the use of contraceptive pills or other interventions among the female runners. With regard to thyroid medication, the relatively high incidence rates of hypothyroidism among the female subjects (8%) explains the number of subjects who were taking thyroid medication.

4.7. Smoking Habits

There was no significant association between diet and current or former smoking. Yet, a low rate of smokers in vegetarian, vegan and omnivorous runners was observed. Vegetarians had the best scores considering *smoking habits*. These findings were in line with previous research, which also showed low numbers of smokers among vegetarians and vegans [59,60]. Although the low rates among endurance runners could be explained by undesired performance limitations due to smoking [59], vegetarians and vegans are often particularly health-conscious and therefore the number of smokers would be quite low among them [8,60] (p. 393). In addition, we found that women were more likely to be non-smokers compared to men, which was in line with previous research [61]. Nonetheless, in the past years, the number of female smokers has increased, which is particularly displayed in the prevalence of smoking associated diseases, such as lung cancer [61].

4.8. Supplement Intake

The finding that percentages of supplement intake were similar in all diet groups is consistent with current evidence. At the same time, vegans had the highest *supplement intake* scores. These findings are in line with previous research which showed that vegetarian kinds of diet are not lacking in critical micronutrients and macronutrients, per se, but rather that nutrient deficits can occur in any kind of diet [62]. Plant-based diets, such as a vegan dietary pattern, are not worse in terms of daily nutrient intake than omnivorous kinds of diet [63]. A recent study showed that an omnivorous diet does not meet the required amount of intake of six nutrients on average (calcium, folate, magnesium, iron, copper, vitamin E), whereas in vegetarian diets the amount of the daily intake of three nutrients is too low on average (calcium, zinc, vitamin B12) [9]. Another study has even revealed higher diet quality scores in vegetarian runners than in non-vegetarian runners [17]. Thus, supplementation of certain nutrients can be recommended for omnivores, vegetarians, and vegans alike [63]. More than this, these findings underpinned the fact that vegans are particularly health-conscious, which has been confirmed in other studies as well [8,60] (p. 393).

The most frequently taken supplement mentioned by the participants was vitamin D. Vitamin D deficiency is usually not associated with a vegetarian or vegan diet [64], but is a common problem in the general population [65] and in particular among endurance runners. It was found that there is a very large difference between necessary and real intake in athletes, regardless of whether they adhere to a vegetarian, vegan, or omnivorous diet [66]. Thus, all endurance athletes have to be aware of vitamin D levels, irrespective of their dietary patterns.

4.9. Enhancement Substances

There was no significant association between dietary subgroups and the consumption of enhancement substances or anything to cope with stress. Vegans reported the lowest use of enhancement substances. As there was a low overall number of subjects who reported using such substances (n = 32 for the consumption of enhancement substances, n = 18 for the consumption of substances to cope with stress) it could be expected that the number among the dietary subgroups would be quite low as well. It is noteworthy that these findings contradicted a previous study by Wilson [18] who found that 40% of male marathon finishers reported the recent use of performance-enhancing supplements. However, since our subjects did (almost) not report using such substances, they are probably aware of the detrimental effects of substances to increase performance and would, therefore, have avoided intake. This applies especially to the vegan participants, who are known to be particularly health-conscious [8] (p. 393).

4.10. Healthcare Utilization

Vegans had the highest scores in *healthcare utilization*, although scores were similar for all dietary subgroups. Scientific data is sparse in this regard. In one study a higher need for healthcare has been found among vegetarians [44]. However, since our results showed a good state of health in vegan, vegetarian and omnivorous endurance runners, there seems to be no need for frequent doctor consultations. Furthermore, physical activity, such as endurance running, prevents diseases which could require consulting a doctor more frequently [52]. However, about half of the participants (54% in omnivores, 49% in vegetarians, and 61% in vegans) reported making use of routine health checks. Considering that the mean age of our participants was around 40 years, this was an encouraging result, as most health checks for the early recognition and treatment of severe diseases in Europe are recommended for people who are aged 40 years and older [67].

4.11. Limitations and Implications for Future Research

Some limitations of the study should be noted. The survey was based on self-reporting. Thus, the reliability of the data depended on the conscientiousness of the participants. However, this effect was controlled for diet, participation in running events and race distance by using control questions, each separated from the respective main question and included in different sections of the questionnaire. In addition, the small sample size and the pre-selection of the participants due to the fact that mainly highly motivated runners took part led to a lack of statistical representativeness which might have affected the results. Nonetheless, the high intrinsic motivation of the participants would also have led to an increase in the accuracy of their answers and thus to a high quality of the generated data. Moreover, it was striking that most subjects came from Germany (*n* = 177). This imbalance in the composition of the sample may have several causes. First, Germany has a population of 82 million, making it the largest German-speaking country [68]. As the core area of the present study were German-speaking countries, this predominance is displayed in the sample of the present study. Second, Germany has large vegetarian and vegan populations [69]. Since a couple of subjects were addressed via trade fairs on vegetarian and vegan nutrition and lifestyle, it was likely that the number of German participants would increase. Third, some of the largest running events, such as the Berlin Marathon [70], take place in Germany. Together, this might have led to an increase in German participants.

Nevertheless, the data contributes to the growing scientific interest in and research on vegetarianism and veganism as it relates to sports and exercise and can be taken as one step towards creating a broad body of evidence in this regard. Future studies should be performed on large randomized samples in order to improve statistical representativeness. Furthermore, measurement of the health status could be elaborated by including additional parameters, such as energy metabolism and fluid balance regulation. Thereby, the data generated from the participants' self-report could be specified.

5. Conclusions

In summary, the findings revealed that all endurance runners had a good health status, regardless of the diet choice. At the same time, vegan athletes appeared to be extraordinarily health-conscious, in particular due to their food choice habits. These findings support the notion that adhering to vegetarian kinds of diet, but in particular to a vegan dietary pattern, is compatible with ambitious endurance running and can be an appropriate, at least equal and healthy alternative to an omnivorous diet for athletes.

Author Contributions: Conceptualization: B.K., C.L. (Claus Leitzmann), and K.W.; data curation: C.L. (Christoph Lechleitner), G.W., and K.W.; formal analysis: C.L. (Christoph Lechleitner), G.W., P.B., and K.W.; funding acquisition: K.W.; Investigation, K.W.; methodology: K.W.; project administration: K.W.; resources: C.L. (Christoph Lechleitner), G.W., and K.W.; software: C.L. (Christoph Lechleitner) and G.W.; supervision: B.K., C.L. (Claus Leitzmann), and K.W.; validation: K.W.; writing—original draft: P.B., T.R., and K.W.; writing—review and editing: B.K. and K.W.

Funding: This research received no external funding.

Acknowledgments: There are no professional relationships with companies or manufacturers who will benefit from the results of the present study. Beyond that, this research did not receive any specific grant from funding agencies in the public, commercial, or not-for-profit sectors. The results of the present study do not constitute endorsement of the product by the authors.

Conflicts of Interest: The authors declare no conflict of interest.

References

1. Cona, G.; Cavazzana, A.; Paoli, A.; Marcolin, G.; Grainer, A.; Bisiacchi, P.S. It's a matter of mind! Cognitive functioning predicts the athletic performance in ultra-marathon runners. *PLoS ONE* **2015**, *10*, e0132943. [CrossRef] [PubMed]
2. Joyner, M.J.; Coyle, E.F. Endurance exercise performance: The physiology of champions. *J. Physiol.* **2008**, *586*, 35–44. [CrossRef]
3. Hausswirth, C.; Lehénaff, D. Physiological demands of running during long distance runs and triathlons. *Sports Med.* **2001**, *31*, 679–689. [CrossRef] [PubMed]
4. Deldicque, L.; Francaux, M. Recommendations for healthy nutrition in female endurance runners: An update. *Front. Nutr.* **2015**, *2*, 17. [CrossRef] [PubMed]
5. Ormsbee, M.J.; Bach, C.W.; Baur, D.A. Pre-exercise nutrition: The role of macronutrients, modified starches and supplements on metabolism and endurance performance. *Nutrients* **2014**, *6*, 1782–1808. [CrossRef]
6. Fuhrman, J.; Ferreri, D.M. Fueling the vegetarian (vegan) athlete. *Curr. Sports Med. Rep.* **2010**, *9*, 233–241. [CrossRef] [PubMed]
7. Melina, V.; Craig, W.; Levin, S. Position of the academy of nutrition and dietetics: Vegetarian diets. *J. Acad. Nutr. Diet.* **2016**, *116*, 1970–1980. [CrossRef]
8. Wirnitzer, K.C. Vegan nutrition: Latest boom in health and exercise. In *Therapeutic, Probiotic, and Unconventional Foods. Section 3: Unconventional Foods and Food Ingredients*; Grumezescu, A.M., Holban, A.M., Eds.; Academic Press: Cambridge, MA, USA; Elsevier: Amsterdam, The Netherlands, 2018; Chapter 21; pp. 387–453.
9. Turner, D.R.; Sinclair, W.H.; Knez, W.L. Nutritional adequacy of vegetarian and omnivore dietary intakes. *J. Nutr. Health Sci.* **2014**, *1*, 201. [CrossRef]
10. Berkow, S.E.; Barnard, N. Vegetarian diets and weight status. *Nutr. Rev.* **2006**, *64*, 175–188. [CrossRef]
11. Kahleova, H.; Levin, S.; Barnard, N.D. Vegetarian dietary patterns and cardiovascular disease. *Prog. Cardiovasc. Dis.* **2018**, *61*, 54–61. [CrossRef]
12. Kahleova, H.; Pelikanova, T. Vegetarian diets in the prevention and treatment of type 2 diabetes. *J. Am. Coll. Nutr.* **2015**, *34*, 448–458. [CrossRef]
13. Kahleova, H.; Levin, S.; Barnard, N. Cardio-metabolic benefits of plant-based diets. *Nutrients* **2017**, *9*, 848. [CrossRef]
14. Tuso, P.; Stoll, S.R.; Li, W.W. A plant-based diet, atherogenesis, and coronary artery disease prevention. *Perm J.* **2015**, *19*, 62–67. [CrossRef] [PubMed]
15. Liu, X.; Yan, Y.; Li, F.; Zhang, D. Fruit and vegetable consumption and the risk of depression: A meta-analysis. *Nutrition* **2016**, *32*, 296–302. [CrossRef] [PubMed]
16. Fraser, G.E. Associations between diet and cancer, ischemic heart disease, and all-cause mortality in non-Hispanic white California Seventh-day Adventists. *Am. J. Clin. Nutr.* **1999**, *70*, 532S–538S. [CrossRef] [PubMed]
17. Turner-McGrievy, G.M.; Moore, W.J.; Barr-Anderson, D. The interconnectedness of diet choice and distance running: Results of the research understanding the nutrition of endurance runners (runner) study. *Int. J. Sport Nutr. Exerc. Metab.* **2016**, *26*, 205–211. [CrossRef]
18. Wilson, P.B. Nutrition behaviors, perceptions, and belief of recent marathon finishers. *Phys. Sportsmed.* **2016**, *44*, 242–251. [CrossRef] [PubMed]
19. Diehl, K.; Thiel, A.; Zipfel, S.; Mayer, J.; Litaker, D.G.; Schneider, S. How healthy is the behavior of young athletes? A systematic literature review and meta-analyses. *J. Sports Sci. Med.* **2012**, *11*, 201–220.

20. Wirnitzer, K.; Seyfart, T.; Leitzmann, C.; Keller, M.; Wirnitzer, G.; Lechleitner, C.; Rüst, C.A.; Rosemann, T.; Knechtle, B. Prevalence in running events and running performance of endurance runners following a vegetarian or vegan diet compared to non-vegetarian endurance runners: The NURMI Study. *SpringerPlus* **2016**, *5*, 458. [CrossRef]
21. Word Health Organization (WHO). WHO Regional Office for Europe. Body Mass Index—BMI. Table 1. Nutritional Status. 2018. Available online: http://www.euro.who.int/en/health-topics/disease-prevention/nutrition/a-healthy-lifestyle/body-mass-index-bmi (accessed on 12 November 2018).
22. Word Health Organization (WHO). Global Health Observatory (GHO) Data. Mean Body Mass Index (BMI). Situation and Trends. 2018. Available online: http://www.who.int/gho/ncd/risk_factors/bmi_text/en/ (accessed on 12 November 2018).
23. Marc, A.; Sedeaud, A.; Guillaume, M.; Rizk, M.; Schipman, J.; Antero Jacquemin, J.; Haida, A.; Berthelot, G.; Toussaint, J.F. Marathon progress: Demography, morphology and environment. *J. Sports Sci.* **2014**, *32*, 524–532. [CrossRef]
24. Sedeaud, A.; Marc, A.; Marck, A.; Dor, F.; Schipman, J.; Dorsey, M.; Haida, A.; Berthelot, G.; Toussaint, J.F. BMI, a performance parameter for speed improvement. *PLoS ONE* **2014**, *9*, e90183. [CrossRef] [PubMed]
25. Barnard, N.D.; Levin, S.M.; Yokoyama, Y. A systematic review and meta-analysis of changes in body weight in clinical trials of vegetarian diets. *J. Acad. Nutr. Diet.* **2015**, *115*, 954–969. [CrossRef] [PubMed]
26. Clarys, P.; Deliens, T.; Huybrechts, I.; Deriemaeker, P.; Vanaelst, B.; De Keyzer, W.; Hebbelinck, M.; Mullie, P. Comparison of nutritional quality of the vegan, vegetarian, semi-vegetarian, pesco-vegetarian and omnivorous diet. *Nutrients* **2014**, *6*, 1318–1332. [CrossRef] [PubMed]
27. Venderley, A.M.; Campbell, W.W. Vegetarian diets: Nutritional considerations for athletes. *Sports Med.* **2006**, *36*, 293–305. [CrossRef] [PubMed]
28. Spencer, E.A.; Appleby, P.N.; Davey, G.K.; Key, T.J. Diet and body mass index in 38000 EPIC-Oxford meat-eaters, fish-eaters, vegetarians and vegans. *Int. J. Obes. Relat. Metab. Disord.* **2003**, *27*, 728–734. [CrossRef] [PubMed]
29. Appleby, P.N.; Thorogood, M.; Mann, J.I.; Key, T.J. The Oxford Vegetarian Study: An overview. *Am. J. Clin. Nutr.* **1999**, *70*, 525S–531S. [CrossRef] [PubMed]
30. Barr, S.I.; Rideout, C.A. Nutritional considerations for vegetarian athletes. *Nutrition* **2004**, *20*, 696–703. [CrossRef] [PubMed]
31. Fox, N.; Ward, K.J. You are what you eat? Vegetarianism, health and identity. *Soc. Sci. Med.* **2008**, *66*, 2585–2595. [CrossRef] [PubMed]
32. Leitzmann, C.; Keller, M. *Vegetarische Ernährung*, Aktualisierte Auflage, ed.; UTB: Stuttgart, Germany, 2013.
33. Waldmann, A.; Koschizke, J.W.; Leitzmann, C.; Hahn, A. Dietary intakes and lifestyle factors of a vegan population in Germany: Results from the German Vegan Study. *Eur. J. Clin. Nutr.* **2003**, *57*, 947–955. [CrossRef] [PubMed]
34. Bedford, J.L.; Barr, S.I. Diets and selected lifestyle practices of self-defined adult vegetarians from a population-based sample suggest they are more "health conscious". *Int. J. Behav. Nutr. Phys. Act.* **2005**, *2*, 4. [CrossRef]
35. Pate, R.R.; Trost, S.G.; Levin, S.; Dowda, M. Sports participation and health-related behaviors among us youth. *Arch. Pediatr. Adolesc. Med.* **2000**, *154*, 904–911. [CrossRef] [PubMed]
36. Chang-Claude, J.; Hermann, S.; Eilber, U.; Steindorf, K. Lifestyle determinants and mortality in German vegetarians and health-conscious persons: Results of a 21-year follow-up. *Cancer Epidemiol. Biomark. Prev.* **2005**, *14*, 963–968. [CrossRef] [PubMed]
37. Nieman, D.C.; Underwood, B.C.; Sherman, K.M.; Arabatzis, K.; Barbosa, J.C.; Johnson, M.; Shultz, T.D. Dietary status of Seventh-Day Adventist vegetarian and non-vegetarian elderly women. *J. Am. Diet. Assoc.* **1989**, *89*, 1763–1769. [PubMed]
38. Cappelletti, S.; Piacentino, D.; Sani, G.; Aromatario, M. Caffeine: Cognitive and physical performance enhancer or psychoactive drug? *Curr. Neuropharmacol.* **2015**, *13*, 71–88. [CrossRef] [PubMed]
39. Fernandez, M.L. Rethinking dietary cholesterol. *Curr. Opin. Clin. Nutr. Metab. Care* **2012**, *15*, 117–121. [CrossRef]
40. Berger, S.; Raman, G.; Vishwanathan, R.; Jacques, P.F.; Johnson, E.J. Dietary cholesterol and cardiovascular disease: A systematic review and meta-analysis. *Am. J. Clin. Nutr.* **2015**, *102*, 276–294. [CrossRef]

41. Rogerson, D. Vegan diets: Practical advice for athletes and exercisers. *J. Int. Soc. Sports Nutr.* **2017**, *14*, 36. [CrossRef]
42. Sharma, A.; Amarnath, S.; Thulasimani, M.; Ramaswamy, S. Artificial sweeteners as a sugar substitute: Are they really safe? *Indian J. Pharmacol.* **2016**, *48*, 237–240. [CrossRef]
43. Soffritti, M.; Padovani, M.; Tibaldi, E.; Falcioni, L.; Manservisi, F.; Belpoggi, F. The carcinogenic effects of aspartame: The urgent need for regulatory re-evaluation. *Am. J. Ind. Med.* **2014**, *57*, 383–397. [CrossRef]
44. Burkert, N.T.; Muckenhuber, J.; Großschädl, F.; Rásky, É.; Freidl, W. Nutrition and health—The association between eating behavior and various health parameters: A matched sample study. *PLoS ONE* **2014**, *9*, e88278. [CrossRef]
45. Romieu, I.; Varraso, R.; Avenel, V.; Leynaert, B.; Kauffmann, F.; Clavel-Chapelon, F. Fruit and vegetable intakes and asthma in the E3N study. *Thorax* **2006**, *61*, 209–215. [CrossRef]
46. Du Toit, G.; Tsakok, T.; Lack, S.; Lack, G. Prevention of food allergy. *J. Allergy Clin. Immunol.* **2016**, *137*, 998–1010. [CrossRef] [PubMed]
47. Glick-Bauer, M.; Yeh, M.-C. The health advantage of a vegan diet: Exploring the gut microbiota connection. *Nutrients* **2014**, *6*, 4822–4838. [CrossRef] [PubMed]
48. Van der Wall, E.E. Long-distance running: Running for a long life? *Neth. Heart J.* **2014**, *22*, 89–90. [CrossRef] [PubMed]
49. Hoffman, M.D.; Krishnan, E. Health and exercise-related medical issues among 1,212 ultramarathon runners: Baseline findings from the Ultrarunners longitudinal tracking (ULTRA) study. *PLoS ONE* **2014**, *9*, e83867. [CrossRef] [PubMed]
50. Buscail, C.; Sabate, J.-M.; Bouchoucha, M.; Torres, M.J.; Allès, B.; Hercberg, S.; Benamouzig, R.; Julia, C. Association between self-reported vegetarian diet and the irritable bowel syndrome in the French NutriNet cohort. *PLoS ONE* **2017**, *12*, e0183039. [CrossRef]
51. Miall, A.; Khoo, A.; Rauch, C.; Snipe, R.M.J.; Camões-Costa, V.L.; Gibson, P.R.; Costa, R.J.S. Two weeks of repetitive gut-challenge reduce exercise-associated gastrointestinal symptoms and malabsorption. *Scand. J. Med. Sci. Sports* **2018**, *28*, 630–640. [CrossRef]
52. Shipway, R.; Holloway, I. Running free: Embracing a healthy lifestyle through distance running. *Perspect. Public Health* **2010**, *130*, 270–276. [CrossRef]
53. Knechtle, B.; Quarella, A. Running helps—Or how you escape depression without a psychiatrist and end up running a marathon! *Praxis (Bern 1994)* **2007**, *96*, 1351–1356. [CrossRef]
54. Beezhold, B.; Radnitz, C.; Rinne, A.; DiMatteo, J. Vegans report less stress and anxiety than omnivores. *Nutr. Neurosci.* **2015**, *18*, 289–296. [CrossRef]
55. Boldt, P.; Knechtle, B.; Nikolaidis, P.; Lechleitner, C.; Wirnitzer, G.; Leitzmann, C.; Rosemann, T.; Wirnitzer, K. Half-Marathoners Report Best Health Status—Results from the NURMI Study (Step 2). Unpublished data from our laboratory. *Eur. J. Sports Sci.* under review.
56. Zeiger, J.S.; Zeiger, R.S. Mental toughness latent profiles in endurance athletes. *PLoS ONE* **2018**, *13*, e0193071. [CrossRef] [PubMed]
57. Dunn, D.; Turner, C. Hypothyroidism in women. *Nurs. Womens Health* **2016**, *20*, 93–98. [CrossRef]
58. Tonstad, S.; Nathan, E.; Oda, K.; Fraser, G. Vegan diets and hypothyroidism. *Nutrients* **2013**, *5*, 4642–4652. [CrossRef] [PubMed]
59. Marti, B.; Abelin, T.; Minder, C.E.; Vader, J.P. Smoking, alcohol consumption, and endurance capacity: An analysis of 6,500 19-year-old conscripts and 4,100 joggers. *Prev. Med.* **1988**, *17*, 79–92. [CrossRef]
60. Appleby, P.N.; Crowe, F.L.; Bradbury, K.E.; Travis, R.C.; Key, T.J. Mortality in vegetarians and comparable nonvegetarians in the United Kingdom. *Am. J. Clin. Nutr.* **2016**, *103*, 218–230. [CrossRef] [PubMed]
61. Peters, S.A.; Huxley, R.R.; Woodward, M. Do smoking habits differ between women and men in contemporary Western populations? Evidence from half a million people in the UK Biobank study. *BMJ Open* **2014**, *4*, e005663. [CrossRef] [PubMed]
62. Schüpbach, R.; Wegmüller, R.; Berguerand, C.; Bui, M.; Herter-Aeberli, I. Micronutrient status and intake in omnivores, vegetarians and vegans in Switzerland. *Eur. J. Nutr.* **2017**, *56*, 283–293. [CrossRef]
63. McDougall, C.; McDougall, J. Plant-Based Diets Are Not Nutritionally Deficient. *Perm J.* **2013**, *17*, 93. [CrossRef]
64. Baig, J.A.; Sheikh, S.A.; Islam, I.; Kumar, M. Vitamin D status among vegetarians and non-vegetarians. *J. Ayub Med. Coll. Abbottabad* **2013**, *25*, 152–155.

65. Gani, L.U.; How, C.H. Vitamin D deficiency. *Singap. Med. J.* **2015**, *56*, 433–437. [CrossRef] [PubMed]

66. Larson-Meyer, E. Vitamin D supplementation in athletes. *Nestle Nutr. Inst. Workshop Ser.* **2013**, *75*, 109–121. [CrossRef] [PubMed]

67. Schülein, S.; Taylor, K.J.; Schriefer, D.; Blettner, M.; Klug, S.J. Participation in preventive health check-ups among 19,351 women in Germany. *Prev. Med. Rep.* **2017**, *6*, 23–26. [CrossRef] [PubMed]

68. Ehling, M.; Pötzsch, O. Demographic changes in Germany up to 2060—Consequences for blood donation. *Transfus. Med. Hemother.* **2010**, *37*, 131–139. [CrossRef] [PubMed]

69. Leitzmann, C. Vegetarian nutrition: Past, present, future. *Am. J. Clin. Nutr.* **2014**, *100*, 496S–502S. [CrossRef] [PubMed]

70. Haeusler, K.G.; Herm, J.; Kunze, C.; Krüll, M.; Brechtel, L.; Lock, J.; Hohenhaus, M.; Heuschmann, P.U.; Fiebach, J.B.; Haverkamp, W.; et al. Rate of cardiac arrhythmias and silent brain lesions in experienced marathon runners: Rationale, design and baseline data of the Berlin Beat of Running study. *BMC Cardiovasc. Disord.* **2012**, *12*, 69. [CrossRef] [PubMed]

![nutrients logo] *nutrients*

MDPI

Article

Cardiometabolic Health in Relation to Lifestyle and Body Weight Changes 3–8 Years Earlier

Tessa M. Van Elten [1,2,3,4,5,*], **Mireille. N. M. Van Poppel** [1,4,6], **Reinoud J. B. J. Gemke** [1,4,5,7], **Henk Groen** [8], **Annemieke Hoek** [9], **Ben W. Mol** [10,11] and **Tessa J. Roseboom** [2,3,4,5]

1   Department of Public and Occupational Health, Amsterdam UMC, Vrije Universiteit Amsterdam, De Boelelaan, 1117 Amsterdam, The Netherlands; mireille.van-poppel@uni-graz.at (M.N.M.V.P.); rjbj.gemke@vumc.nl (R.J.B.J.G.)
2   Department of Clinical Epidemiology, Biostatistics and Bioinformatics, University of Amsterdam, Amsterdam UMC, Meibergdreef 9, 1105 AZ Amsterdam, The Netherlands; t.j.roseboom@amc.uva.nl
3   Department of Obstetrics and Gynecology, University of Amsterdam, Amsterdam UMC, Meibergdreef 9, 1105 AZ Amsterdam, The Netherlands
4   Amsterdam Public Health Research Institute, 1105 AZ Amsterdam, The Netherlands
5   Amsterdam Reproduction and Development, 1105 AZ Amsterdam, The Netherlands
6   Institute of Sport Science, University of Graz, 8010 Graz, Austria
7   Department of Pediatrics, Emma Childrens Hospital, Vrije Universiteit Amsterdam, Amsterdam UMC, 1081 HV Amsterdam, The Netherlands
8   Department of Epidemiology, University of Groningen, University Medical Centre Groningen, 9700 RB Groningen, The Netherlands; h.groen01@umcg.nl
9   Department of Obstetrics and Gynecology, University of Groningen, University Medical Centre, Groningen, 9700 RB Groningen, The Netherlands; a.hoek@umcg.nl
10 School of Medicine, The Robinson Institute, University of Adelaide, 5006 Adelaide, Australia; b.w.mol@amc.uva.nl
11 Department of Obstetrics and Gynecology, Monash University, 3800 Melbourne, Australia
*   Correspondence: t.vanelten@vumc.nl; Tel.: +31-20-444-5612

Received: 17 October 2018; Accepted: 26 November 2018; Published: 10 December 2018

Abstract: The degree to which individuals change their lifestyle in response to interventions differs and this variation could affect cardiometabolic health. We examined if changes in dietary intake, physical activity and weight of obese infertile women during the first six months of the LIFEstyle trial were associated with cardiometabolic health 3–8 years later (N = 50–78). Lifestyle was assessed using questionnaires and weight was measured at baseline, 3 and 6 months after randomization. BMI, blood pressure, body composition, pulse wave velocity, glycemic parameters and lipid profile were assessed 3–8 years after randomization. Decreases in savory and sweet snack intake were associated with lower HOMA-IR 3–8 years later, but these associations disappeared after adjustment for current lifestyle. No other associations between changes in lifestyle or body weight during the first six months after randomization with cardiovascular health 3–8 years later were observed. In conclusion, reductions in snack intake were associated with reduced insulin resistance 3–8 years later, but adjustment for current lifestyle reduced these associations. This indicates that changing lifestyle is an important first step, but maintaining this change is needed for improving cardiometabolic health in the long-term.

Keywords: dietary intake; physical activity; body weight; lifestyle change; cardiometabolic health; long-term follow-up

---

1. Introduction

Unhealthy diet, physical inactivity and a BMI over 25 kg/m^2 are well known risk factors for cardiovascular diseases (CVDs). It is therefore that primary prevention of CVDs focuses, among others, on improving these lifestyle factors and reducing body weight [1]. Observational studies showed graded relationships between healthy changes in dietary intake, physical activity and body weight over time and future cardiometabolic health. For example, improvements in diet [2], or transition to a more active lifestyle [3] lowered risk of coronary heart disease among women in the Nurses' Health Study.

Randomized controlled trials (RCTs) showed that improving lifestyle and reducing body weight improved cardiometabolic health [4,5]. However, although lifestyle interventions can improve cardiometabolic health, the effectiveness of lifestyle interventions varies [6,7]. For example, a meta-analysis including 54 RCTs to determine the effect of aerobic exercise on blood pressure showed that net change in blood pressure after following a physical activity intervention varied from −16.7 to 3.9 mmHg for systolic blood pressure and from −11.0 to 11.3 mmHg for diastolic blood pressure [8]. Despite the large variation in lifestyle and body weight change, the effects of a lifestyle intervention are often described as a randomized group comparison. By studying the effect of absolute change in lifestyle and body weight, more knowledge can be gathered on dose-response relationships and potential thresholds between lifestyle and body weight change with later life cardiometabolic health. This information will aid in improving primary prevention.

The LIFEstyle RCT enrolled obese infertile women, and allocated them to a six-month diet and physical activity intervention or to prompt infertility treatment [9–11]. We previously showed that the intervention lowered the intake of high caloric snacks and beverages and increased physical activity in the short term [12], reduced body weight [11] and improved cardiometabolic health at the end of the six-month intervention period [13]. Furthermore, women allocated to the intervention group reported a lower energy intake at 5.5 years after randomization [14]. Individual responses to the lifestyle intervention varied largely among study participants. The mean weight change in the intervention group was −4.4 kg and the standard deviation of 5.8 kg underlines this large variation [11]. Also women allocated to the control group changed their lifestyle: 10.5% of the women in the control group lost 5% or more of their original body weight [11]. Hence we here investigate individual changes in lifestyle and body weight and relate these changes in dietary intake, physical activity and weight during the first six months after randomization to cardiometabolic health 3–8 years later.

We hypothesized that women who increased their intake of vegetables and fruit, decreased their intake of sugary drinks and snacks, became more physically active and lost more weight during the first six months after randomization had a better cardiometabolic health 3–8 years after randomization compared to women who did not show these improvements in lifestyle and weight. We therefore examined if the change in dietary intake, physical activity and body weight of obese infertile women, combining the intervention and control group, over the first six months of a preconception lifestyle intervention study was associated with their cardiometabolic health 3–8 years after the start of the study.

2. Materials and Methods

2.1. Study Population

The study population comprises all women who participated in the follow-up of the LIFEstyle study. The LIFEstyle study was a multicenter randomized controlled trial (RCT), conducted between 2009 and 2014 in the Netherlands [9–11]. In total, 577 women between 18 and 39 years old, with a BMI of ≥ 29 kg/m^2 were randomized into a six-month structured lifestyle intervention (intervention group) or infertility care as usual (control group). The lifestyle intervention focused on eating a healthy diet according to the Dutch Dietary Guidelines 2006 [15], including a caloric reduction of 600 kcal/day but not below 1200 kcal/day, and being physically active 2–3 times a week for at least 30 min at moderate intensity (60–85% of maximum heart rate frequency). Women were additionally advised to increase physical activity in daily life by taking at least 10,000 steps per day.

At 3–8 years after randomization, 574 women were eligible to participate in the follow-up study of the LIFEstyle RCT [16]. During the follow-up study, data were collected using questionnaires and physical examinations. In the current study, we included all women that participated in the physical examinations (*N* = 111; Figure 1).

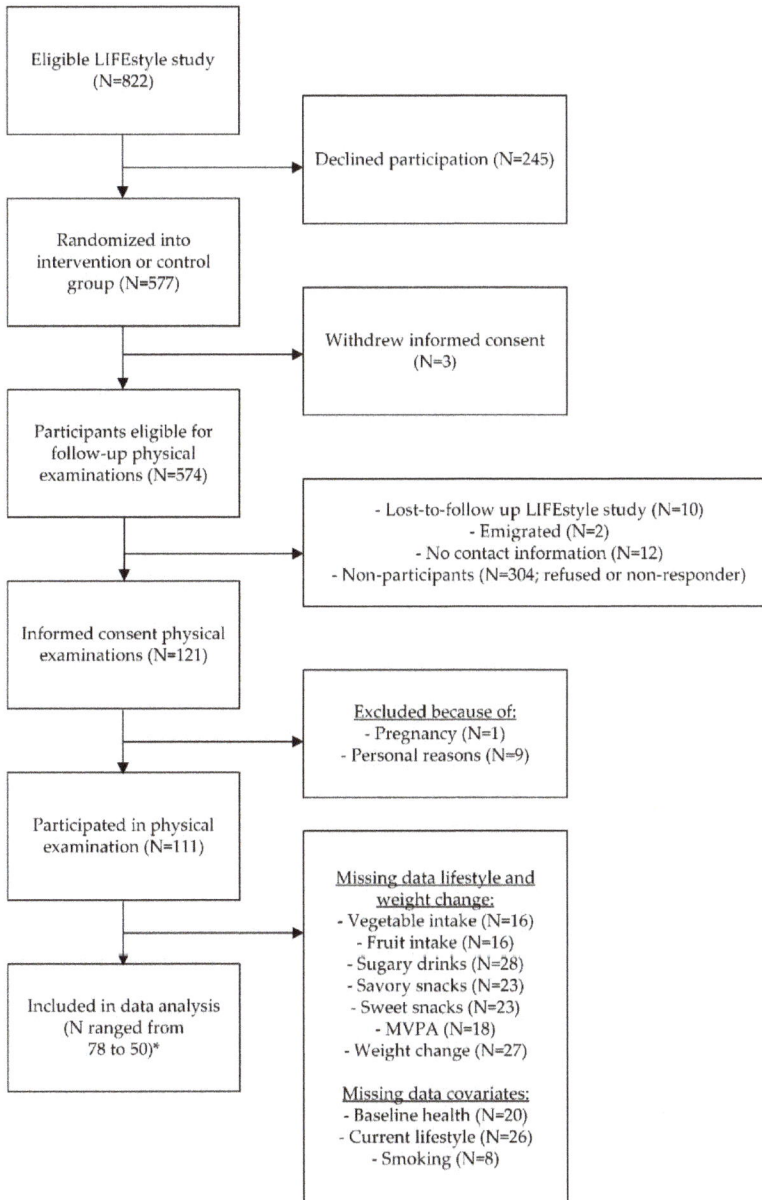

Figure 1. Flow-chart of the study participants. MVPA = Moderate to Vigorous Physical Activity * *N* depends on missing data in the combined models including independent variables and covariates.

The study was conducted in accordance with the Declaration of Helsinki and all procedures were approved by the Medical Ethics Committee of the University Medical Centre Groningen, the Netherlands (METc 2008/284). Written informed consent was obtained from all participants at the start of the LIFEstyle study as well as the start of the follow-up.

2.2. Assessment of Dietary Intake, Physical Activity and Body Weight to Calculate Change

In order to calculate change in lifestyle and body weight over the first six months of the study, we used dietary intake, physical activity and body weight measures collected at baseline, 3 and 6 months after randomization in the lifestyle intervention study. Dietary intake was examined using a 33-item food frequency questionnaire (FFQ) consisting of two parts. The first part was based on the standardized questionnaire on food consumption used for the Public Health Monitor in the Netherlands [17], asking about the type of cooking fats used, the consumed type of bread, frequency of breakfast use, frequency of consumption and portion size of vegetables, fruits and fruit juice. The questions on the intake of fruit, fruit juice and cooked vegetables were validated against two 24-h recalls. The estimated intake of fruit and fruit juice consumption based on the questionnaire showed fairly strong comparability with the intake based on the two 24-h recalls, however the comparability for cooked vegetables was weak [17]. The second part consisted of additional questions about savory and sweet snack intake and the intake of soda. For all foods, frequency of consumption per week or per months was asked and portion size was asked per standard household measure. The presented portion sizes and food groups in the current study were pre-specified in the questions of the FFQ. Dietary intake was studied as the intake of vegetables (raw as well as cooked; grams/day), fruits (grams/day), sugar containing beverages (fruit juice and soda; glasses/day), savory snacks (crisps, pretzels, nuts and peanuts; handful/week) and sweet snacks (biscuits, pieces of chocolate, candies or liquorices; portion/week). One portion of sweet snacks included 2 biscuits, or 2 pieces of chocolate, or 5 candies, or 5 pieces of liquorice.

Physical activity was examined using the validated Short QUestionnaire to ASsess Health-enhancing physical activity (SQUASH) questionnaire [18]. This questionnaire asked about the number of days per week, the average time per day or week (hours and/or minutes), and the intensity (low, moderate, high) of physical activity in four domains: commuting activities, leisure time activities, household activities, and activities at work and school. Physical activity was studied as total moderate to vigorous physical activity (MVPA; hours/week).

Body weight (kg) was measured during hospital visits at baseline, 3 months and 6 months after randomization by trained research nurses that were not involved in the lifestyle intervention coaching.

2.3. Cardiometabolic Health at Follow-Up

Cardiometabolic health 3–8 years after randomization was examined in a mobile research vehicle by two researchers, using a standardized research protocol. Women were asked not to eat or drink from 90 min onwards before the mobile research vehicle arrived at their home. They were additionally asked not to drink caffeine containing beverages or to smoke from 12 h onwards before the physical examination. Height and current weight were measured to calculate current BMI. Height was measured to the nearest 0.1 cm using a wall stadiometer (SECA 206; SECA, Germany) on bare feet, with heels flat on the ground at an angle of 90 degrees, head in Frankfort horizontal position and heels, back and shoulders straight against the wall. Current weight was measured in underwear to the nearest 0.1 kg using a digital weighting scale (SECA 877; SECA, Germany), while the participant was standing still and looking straight ahead. All measurements were done twice, and in case of >0.5 cm difference in height and >0.5 kg difference in weight, a third measurement was performed. After sitting quietly for 5 min, blood pressure was measured three times at heart level, at the non-dominant arm, using an automatic measurement device (Omron HBP-1300; OMRON Healthcare, The Netherlands) with appropriate cuff size. The three measurements were done using a 30 s time interval and women were not allowed to talk in between, move, cross their legs or tense their arm muscles. Body composition was measured twice by bio-electrical impedance (BIA; Bodystat 1500; Bodystat Ltd., Isle of Man, UK) after lying quietly for 5 min. On beforehand participants were asked to take of any jewelry, belts, piercings, etc. which could affect the BIA measurements. After cleaning the skin with alcohol, electrode strips were attached at the dorsal side of the left hand and foot with at least 3–5 cm in between the two electrodes. Women were instructed not to talk in between the measurements and attention was paid that arms and legs did not touch other parts of the body. A third measurement was performed in case the impedance or resistance differed >5Ω. Fat mass and fat free mass were calculated using equation of Kyle and colleagues [19]. Immediately after the BIA measurement, still in supine position, carotid-femoral pulse wave velocity (PWV) was measured twice using the Complior Analyse (Complior; Alam Medical, Saint-Quentin-Fallavier, France). Mechanotransducer censors were placed at the carotid artery on the right side and on the femoral artery on the left side. Blood pressure in lying position was measured once before the actual measurement started and entered in the Complior software. Directly after the measurement, distance between both censors was measured and also entered in the Complior software. In case of >10% difference in PWV between both measurements, a third measurement was done. The following equation was used to calculate PWV: PWV = 0.8 × (distance between the carotis and fermoralis measuring site/Δ time between upstrokes of pressure waves) [20].

Apart from the physical examinations, a trained nurse visited the participants at home to draw a venous blood sample after an overnight fast. All venous blood samples were analyzed at the biochemical laboratory of the Amsterdam UMC. We examined metabolic health by fasting serum concentrations of glucose (Roche cobas 8000, c702; Roche Diagnostics, Rotkreuz, Switzerland) and insulin (Centaur XP; Siemens, Munich, Germany), triglycerides (Roche cobas 8000, c702; Roche Diagnostics, Switzerland), total cholesterol (Roche cobas 8000, c502; Roche Diagnostics, Switzerland) and high density (HDL-C) lipoprotein cholesterol (Roche cobas 8000, c702; Roche Diagnostics, Switzerland). Low density (LDL-C) lipoprotein cholesterol was calculated using the following formula; (total cholesterol) − (high density lipoprotein cholesterol) − 0.45 × (triglycerides). Furthermore, insulin resistance (HOMA-IR) was calculated as fasting insulin concentration in μU/mL multiplied by fasting glucose concentration in mmol/L divided by 22.5. We additionally examined if metabolic syndrome was present or not. If present, participant met at least three of the following criteria determined by the American Heart Association: glucose ≥ 5.6 mmol/L, HDL-C < 1.3 mmol/L, triglycerides ≥ 1.7 mmol/L, waist circumference ≥ 88 cm or blood pressure ≥ 130/85 mmHg [21].

2.4. Statistical Analysis

We used linear regression models to study the association between the change in dietary intake, physical activity, and weight during the first six months after randomization and cardiometabolic health 3–8 years later. Results are displayed as betas (β) and 95% confidence intervals (C.I.). For metabolic syndrome, logistic regression was used and results are displayed as odds ratios (OR) and 95% C.I. We recalculated the intake of vegetables and fruits into 10 g/day dividing the change by ten. Our regression models therefore display the effect on cardiometabolic health per 10 g change of vegetable intake and fruit intake per day.

For the descriptive statistics the change in lifestyle and body weight over time was calculated by subtracting the baseline measurement from the last know measurement at preferably 6 months or otherwise 3 months after randomization (Table S1). In most women, the last known measurement was at six months after randomization, but when missing, the measurement at three months after randomization was used (*N* = 5). This means that a higher change score for vegetable intake, fruit intake, and MVPA is healthier, while a higher change score for sugary drink intake, savory and sweet snack intake, and weight change (higher means weight gain instead of weight loss) is unhealthier. In our regression models, the change in vegetable intake, fruit intake and MVPA was calculated by subtracting the baseline measurement from the last known measurement. The change in sugary drinks, savory and sweet snack intake and body weight was calculated by subtracting the last known measurement from the baseline measurement. An increase in change score in our regression models is favorable, reflecting an increase in vegetable intake, fruit intake and MVPA, and a decrease in sugary drink intake, savory and sweet snack intake and body weight.

All crude regression models were corrected for baseline cardiometabolic health, depending on the outcome variable (e.g., BMI at follow-up was corrected for BMI at baseline), with exception of fat mass, fat free mass, and PWV as these outcomes were not measured at baseline. We therefore corrected fat mass and fat free mass models for baseline BMI [22], and PWV models for baseline systolic blood pressure [23]. Based on literature [24] and the associations with both the change variables and the outcome variables, we corrected the adjusted regression models additionally for (1) smoking at follow-up (yes/no; self-reported by means of a questionnaire); (2) current dietary intake, depending on which food group is added into the model; e.g., if we examined the association between the change in vegetable intake during the first six months after randomization and current BMI, we corrected for current vegetable intake, and when examining the effect of MVPA we corrected for total current energy intake (kcal/day). Current dietary intake, i.e., dietary intake during follow-up, was assessed with the same 33-item FFQ as described previously [17]; (3) current MVPA (min/day), i.e., MVPA during follow-up, which was measured for seven consecutive days using the triaxial Actigraph wGT3X-BT or GT3X+ accelerometer [25]. Freedson cut-off points were used to determine the number of minutes per day in MVPA ($\geq$1952 counts/min) [26]. We additionally added randomization group (intervention/control group) and time between randomization and follow-up (years) into the adjusted model to see if this affected the effect estimates. As women got pregnant during the LIFEstyle study, some diet, physical activity and weight measurements were collected during early pregnancy. We therefore once excluded measurements collected during pregnancy to see if effect estimates changed.

Statistical analyses were performed using the software Statistical Package for the Social Sciences (SPSS) version 24 for Windows (SPSS, Chicago, IL, USA). *p*-values <0.05 were considered statistically significant.

3. Results

Of the 577 women randomly allocated to the intervention and control group during the trial, 574 women were eligible to participate in the physical examinations at 3–8 years after the intervention (Figure 1). Of these eligible women, 121 were willing to participate in the follow up study and signed informed consent (21.1%) and we collected data of 111 women. Because of missing data regarding the change in dietary intake, physical activity, weight and the covariates, we were able to include 50 up to 78 women in our regression analyses.

Mean age of the women during physical examination was 36.4 years (SD = 4.3), most of them were Caucasian (94.6%), had an intermediate vocational education (49.1%) and were obese at 3–8 years after the intervention (mean BMI = 35.5 kg/m2 (SD = 5.3); Table 1). Baseline characteristics, collected during the LIFEstyle RCT, did not differ between participants (*N* = 111) and non-participants (*N* = 463) of the follow-up study (*N* = 111), with exception of ethnicity and the change in sweet snack intake (Table S1). Participants were more often of Caucasian origin (94.6%) compared to non-participants (85.7%). Additionally, the change in sweet snacks during the first six months after randomization was lower in the participants (−0.1 portions/week (SD = 5.6)) compared to the non-participants (−3.3 portions/week (SD = 10.2)).

Table 1. Characteristics and cardiometabolic health of the study population (N = 111).

Age at follow-up (years; mean; SD)	36.4 (4.3)
Caucasian (N; %)	105 (94.6)
Education level (N; %)	
No education or primary school (4–12 years)	1 (0.9)
Secondary education	25 (23.6)
Intermediate Vocational Education	52 (49.1)
Higher Vocational Education or University	28 (26.4)
Body Mass Index at randomization (kg/m^2; mean; SD)	35.7 (3.0)
Current smoker at follow-up (yes; N; %)	16 (15.5)
PCOS (yes; N; %)	43 (38.7)
Nulliparous at follow-up (yes; N; %)	21 (20.8)
Familial predisposition cardiovascular diseases (yes; N; %)	92 (89.3)
Gestational diabetes (yes; N; %) *	18 (17.5)
(Pre-)eclampsia (yes; N; %) *	16 (15.5)
HELLP syndrome (yes; %; N) *	7 (6.8)
Randomization group (intervention group; N; %)	50 (45.0)
Cardiovascular outcomes at follow-up	
Body Mass Index 3–8 years after randomization (kg/m^2; mean; SD)	35.5 (5.3)
Systolic blood pressure (mmHg; mean; SD)	120.4 (14.4)
Diastolic blood pressure (mmHg; mean; SD)	81.7 (9.5)
Fat mass (% of total body weight; mean; SD)	43.1 (4.3)
Fat free mass (kg; mean; SD)	56.9 (6.5)
Pulse Wave Velocity (m/s; mean; SD)	7.2 (2.0)
Metabolic outcomes at follow-up	
Glucose (mmol/L; mean; SD)	5.3 (0.7)
Insulin (pmol/L; mean; SD)	81.7 (53.1)
HOMA-IR (mean; SD)	3.4 (2.8)
Triglycerides (mmol/L; mean; SD)	1.2 (0.8)
Total cholesterol (mmol/L; mean; SD)	4.6 (0.9)
LDL-C (mmol/L; mean; SD)	2.8 (0.8)
HDL-C (mmol/L; mean; SD)	1.3 (0.3)
Metabolic syndrome at follow-up (yes; N; %)	40 (40.0)

PCOS = Polycystic Ovary Syndrome; HELLP = syndrome characterized by hemolysis (H), elevated liver enzymes (EL) and low platelet count (LP); HOMA-IR = Homeostatic Model Assessment of Insulin Resistance; LDL-C = low-density lipoproteins cholesterol; HDL-C = high-density lipoproteins cholesterol. * Diagnosed with these pregnancy complications during any pregnancy in the past.

Table 2 shows the change in dietary intake, physical activity and body weight during the first six months of the LIFEstyle study in our study population.

We did not observe the hypothesized associations between increases in vegetable intake, fruit intake and MVPA during the first six months of the LIFEstyle study with a more favorable BMI, blood pressure, body composition, PWV and metabolic health 3–8 years later (Table 3). Furthermore, decreased sugary drink intake and body weight were not associated with a more favorable cardiovascular health 3–8 years later. A decrease in savory snacks intake during the first six months after randomization was associated with a lower HOMA-IR at follow-up (crude model: −0.16 (−0.32; −0.001); p = 0.049). This association disappeared after adjustment for smoking, current savory snack intake and current MVPA (adjusted model: −0.09 (−0.28; 0.09); p = 0.33). Changes in savory snack intake were not associated with women's BMI, blood pressure, body composition, PWV or other metabolic outcomes at follow-up. Furthermore, a decrease in sweet snacks intake during the first six months after randomization was associated with a lower HOMA-IR at follow-up (crude model: −0.16 (−0.06; −0.06); p = 0.003). Also this association disappeared after adjustment for smoking, current sweet snack intake and current MVPA (adjusted model: 0.01 (−0.09; 0.12); p = 0.84).

Table 2. Change in dietary intake, physical activity and weight during the first six months of the
LIFEstyle study in the study population included for follow-up *.

Change in	N	Mean (SD)	Median (IQR)
Vegetable intake (grams/day)	95	1.92 (58.70)	3.57 (−28.57; 35.71)
Fruit intake (grams/day)	95	25.56 (75.61)	14.29 (0.00; 85.71)
Sugary drinks (glasses/day)	83	−0.43 (1.81)	0.00 (−0.84; 0.21)
Savory snacks (handful/week)	88	−1.67 (5.57)	0.00 (−4.18; 0.57)
Sweet snacks (portion/week)	88	−0.15 (5.63)	0.00 (−2.61; 1.62)
Total MVPA (hour/week)	93	1.05 (12.1)	0.83 (−3.25; 6.13)
Body weight (kilograms)	84	−2.62 (5.10)	−2.10 (−4.68; 0.68)

MVPA – Moderate to Vigorous Physical Activity. * For all variables the change was calculated as preferably 6 months
or otherwise 3 months minus baseline. If a higher score on the change variable is favorable or not depends on the
independent variable of interest: A higher change score for vegetable intake, fruit intake, and MVPA is healthier,
while a higher change score for sugary drink intake, savory and sweet snack intake, and weight change (higher
means weight gain instead of weight loss) is unhealthier.

Table 3. Association between the change in dietary intake and physical activity during the
preconception lifestyle intervention and cardiometabolic health 3–8 years after randomization †.

	N	Crude Model * β (95% C.I.)	p	Adjusted Model ** β (95% C.I.)	p
Change in Vegetable Intake (10 g/day)					
Body Mass Index (kg/m²)	76	−0.02 (−0.23; 0.19)	0.85	−0.04 (−0.25; 0.17)	0.71
Systolic blood pressure (mmHg)	74	0.50 (−0.07; 1.07)	0.08	0.49 (−0.11; 1.09)	0.11
Diastolic blood pressure (mmHg)	74	0.31 (−0.05; 0.68)	0.09	0.32 (−0.06; 0.70)	0.10
Fat mass (% of total body weight)	76	−0.04 (−0.21; 0.13)	0.64	−0.04 (−0.21; 0.14)	0.68
Fat free mass (kg)	76	0.14 (−0.14; 0.41)	0.33	0.10 (−0.18; 0.38)	0.47
Pulse Wave Velocity (m/s)	60	0.02 (−0.09; 0.14)	0.66	−0.004 (−0.13; 0.12)	0.95
Glucose (mmol/L)	64	−0.01 (−0.05; 0.02)	0.46	−0.01 (−0.05; 0.02)	0.52
Insulin (pmol/L)	63	0.56 (−1.35; 2.47)	0.56	1.03 (−0.94; 3.01)	0.30
HOMA-IR	61	0.001 (−0.11; 0.11)	0.99	0.02 (−0.10; 0.14)	0.72
Triglycerides (mmol/L)	64	−0.02 (−0.04; 0.01)	0.16	−0.01 (−0.03; 0.01)	0.33
Total cholesterol (mmol/L)	64	0.01 (−0.02; 0.05)	0.42	0.02 (−0.01; 0.06)	0.24
LDL-C (mmol/L)	64	0.02 (−0.01; 0.05)	0.15	0.02 (−0.01; 0.05)	0.14
HDL-C (mmol/L)	64	−0.002 (−0.01; 0.01)	0.70	−0.001 (−0.01; 0.01)	0.92
Change in fruit intake (10 g/day)					
Body Mass Index (kg/m²)	78	0.08 (−0.10; 0.26)	0.37	0.13 (−0.07; 0.32)	0.20
Systolic blood pressure (mmHg)	76	−0.09 (−0.59; 0.41)	0.72	−0.31 (−0.87; 0.25)	0.28
Diastolic blood pressure (mmHg)	76	0.00 (−0.32; 0.33)	>0.99	−0.04 (−0.41; 0.32)	0.81
Fat mass (% of total body weight)	78	0.10 (−0.05; 0.24)	0.19	0.12 (−0.04; 0.28)	0.13
Fat free mass (kg)	78	0.20 (−0.04; 0.44)	0.09	0.25 (−0.01; 0.52)	0.06
Pulse Wave Velocity (m/s)	63	−0.05 (−0.16; 0.05)	0.30	−0.05 (−0.16; 0.07)	0.39
Glucose (mmol/L)	66	0.03 (−0.01; 0.06)	0.12	0.01 (−0.02; 0.05)	0.41
Insulin (pmol/L)	65	0.52 (−1.26; 2.30)	0.56	0.52 (−1.40; 2.44)	0.59
HOMA-IR	63	−0.01 (−0.11; 0.10)	0.89	−0.02 (−0.14; 0.09)	0.72
Triglycerides (mmol/L)	66	−0.004 (−0.03; 0.02)	0.67	0.00 (−0.02; 0.02)	0.97
Total cholesterol (mmol/L)	66	0.03 (−0.002; 0.06)	0.07	0.02 (−0.01; 0.05)	0.27
LDL-C (mmol/L)	66	0.02 (−0.003; 0.05)	0.08	0.02 (−0.01; 0.04)	0.26
HDL-C (mmol/L)	66	0.003 (−0.01; 0.01)	0.57	−0.002 (−0.01; 0.01)	0.73

Table 3. *Cont.*

	N	Crude Model *		Adjusted Model **	
		β (95% C.I.)	p	β (95% C.I.)	p
Change in sugary drink intake (glass/day)					
Body Mass Index (kg/m^2)	62	0.46 (−0.71; 1.62)	0.44	0.42 (−0.79; 1.62)	0.49
Systolic blood pressure (mmHg)	60	2.04 (−1.55; 5.63)	0.26	2.34 (−1.34; 6.02)	0.21
Diastolic blood pressure (mmHg)	60	1.39 (−0.96; 3.74)	0.24	1.55 (−0.91; 4.00)	0.21
Fat mass (% of total body weight)	62	0.56 (−0.38; 1.50)	0.24	0.54 (−0.46; 1.54)	0.28
Fat free mass (kg)	62	0.58 (−1.15; 2.31)	0.50	0.48 (−1.32; 2.28)	0.60
Pulse Wave Velocity (m/s)	50	0.19 (−0.37; 0.76)	0.50	0.16 (−0.44; 0.75)	0.60
Glucose (mmol/L)	54	−0.03 (−0.27; 0.22)	0.84	−0.04 (−0.29; 0.21)	0.76
Insulin (pmol/L)	53	2.01 (−8.17; 12.19)	0.69	3.70 (−7.04; 14.44)	0.49
HOMA-IR	51	−0.20 (−0.90; 0.51)	0.57	−0.02 (−0.77; 0.72)	0.95
Triglycerides (mmol/L)	54	0.02 (−0.13; 0.16)	0.84	0.03 (−0.13; 0.18)	0.74
Total cholesterol (mmol/L)	54	0.07 (−0.11; 0.26)	0.44	0.06 (−0.13; 0.25)	0.56
LDL-C (mmol/L)	54	0.12 (−0.04; 0.29)	0.13	0.10 (−0.06; 0.27)	0.21
HDL-C (mmol/L)	54	−0.06 (−0.13; 0.01)	0.07	−0.07 (−0.13; 0.003)	0.06
Change in savory snack intake (handful/week)					
Body Mass Index (kg/m^2)	72	0.05 (−0.20; 0.29)	0.72	0.21 (−0.06; 0.47)	0.12
Systolic blood pressure (mmHg)	70	0.25 (−0.47; 0.97)	0.49	0.45 (−0.35; 1.26)	0.26
Diastolic blood pressure (mmHg)	70	0.18 (−0.29; 0.65)	0.44	0.38 (−0.14; 0.90)	0.15
Fat mass (% of total body weight)	72	0.003 (−0.20; 0.20)	0.98	0.09 (−0.13; 0.31)	0.42
Fat free mass (kg)	72	−0.02 (−0.39; 0.36)	0.94	0.21 (−0.19; 0.60)	0.30
Pulse Wave Velocity (m/s)	58	0.04 (−0.06; 0.15)	0.44	0.04 (−0.08; 0.16)	0.49
Glucose (mmol/L)	62	−0.03 (−0.08; 0.02)	0.25	−0.02 (−0.08; 0.04)	0.57
Insulin (pmol/L)	61	−1.57 (−4.08; 0.95)	0.22	0.04 (−2.81; 2.88)	0.98
HOMA-IR	59	−0.16 (−0.32; −0.001)	0.049	−0.09 (−0.28; 0.09)	0.33
Triglycerides (mmol/L)	62	−0.02 (−0.06; 0.01)	0.17	−0.02 (−0.06; 0.02)	0.29
Total cholesterol (mmol/L)	62	−0.01 (−0.06; 0.04)	0.65	−0.02 (−0.08; 0.03)	0.42
LDL-C (mmol/L)	62	−0.001 (−0.04; 0.04)	0.95	−0.01 (−0.06; 0.04)	0.74
HDL-C (mmol/L)	62	−0.002 (−0.02; 0.02)	0.81	−0.01 (−0.03; 0.01)	0.42
Change in sweet snack intake (portion/week)					
Body Mass Index (kg/m^2)	72	−0.10 (−0.29; 0.09)	0.30	−0.04 (−0.26; 0.17)	0.69
Systolic blood pressure (mmHg)	70	0.34 (−0.23; 0.92)	0.23	0.30 (−0.38; 0.98)	0.38
Diastolic blood pressure (mmHg)	70	0.25 (−0.11; 0.62)	0.17	0.30 (−0.13; 0.74)	0.17
Fat mass (% of total body weight)	72	−0.09 (−0.24; 0.06)	0.24	−0.04 (−0.21; 0.13)	0.65
Fat free mass (kg)	72	−0.04 (−0.32; 0.24)	0.76	0.04 (−0.28; 0.36)	0.79
Pulse Wave Velocity (m/s)	58	0.05 (−0.05; 0.16)	0.29	0.04 (−0.07; 0.14)	0.47
Glucose (mmol/L)	62	−0.03 (−0.07; 0.001)	0.06	−0.002 (−0.04; 0.04)	0.90
Insulin (pmol/L)	61	−1.16 (−2.84; 0.53)	0.18	0.57 (−1.37; 2.52)	0.56
HOMA-IR	59	−0.16 (−0.26; −0.06)	0.003	0.01 (−0.09; 0.12)	0.84
Triglycerides (mmol/L)	62	−0.02 (−0.04; 0.01)	0.17	−0.02 (−0.05; 0.01)	0.12
Total cholesterol (mmol/L)	62	0.02 (−0.01; 0.06)	0.18	0.01 (−0.04; 0.05)	0.71
LDL-C (mmol/L)	62	0.02 (−0.01; 0.05)	0.12	0.02 (−0.02; 0.05)	0.41
HDL-C (mmol/L)	62	0.004 (−0.01; 0.02)	0.52	−0.002 (−0.02; 0.01)	0.82
Change in total MVPA (hour/week)					
Body Mass Index (kg/m^2)	76	−0.03 (−0.14; 0.09)	0.64	0.05 (−0.07; 0.16)	0.43
Systolic blood pressure (mmHg)	74	0.14 (−0.19; 0.46)	0.41	0.11 (−0.24; 0.47)	0.53
Diastolic blood pressure (mmHg)	74	−0.10 (−0.31; 0.10)	0.32	−0.09 (−0.31; 0.14)	0.44
Fat mass (% of total body weight)	76	−0.01 (−0.11; 0.08)	0.76	0.03 (−0.07; 0.13)	0.51
Fat free mass (kg)	76	0.09 (−0.06; 0.24)	0.25	0.14 (−0.02; 0.30)	0.09
Pulse Wave Velocity (m/s)	61	0.02 (−0.05; 0.09)	0.56	0.003 (−0.07; 0.07)	0.94
Glucose (mmol/L)	64	−0.01 (−0.03; 0.01)	0.26	−0.01 (−0.03; 0.02)	0.61
Insulin (pmol/L)	63	−0.59 (−1.74; 0.56)	0.31	−0.25 (−1.48; 0.99)	0.69
HOMA-IR	61	−0.03 (−0.10; 0.04)	0.39	−0.003 (−0.08; 0.07)	0.94
Triglycerides (mmol/L)	64	−0.001 (−0.02; 0.01)	0.86	0.002 (−0.01; 0.02)	0.76
Total cholesterol (mmol/L)	64	−0.01 (−0.03; 0.01)	0.38	−0.01 (−0.03; 0.02)	0.54
LDL-C (mmol/L)	64	−0.01 (−0.03; 0.01)	0.22	−0.01 (−0.03; 0.01)	0.42
HDL-C (mmol/L)	64	0.004 (−0.003; 0.01)	0.28	0.002 (−0.01; 0.01)	0.60

Table 3. Cont.

	N	Crude Model *		Adjusted Model **	
		β (95% C.I.)	p	β (95% C.I.)	p
Change in body weight during the intervention (kilograms)					
Body Mass Index (kg/m²)	66	−0.12 (−0.35; 0.11)	0.31	−0.15 (−0.36; 0.06)	0.17
Systolic blood pressure (mmHg)	65	−0.27 (−0.89; 0.35)	0.38	−0.32 (−0.95; 0.32)	0.32
Diastolic blood pressure (mmHg)	65	−0.19 (−0.61; 0.23)	0.38	−0.22 (−0.65; 0.21)	0.31
Fat mass (% of total body weight)	66	−0.02 (−0.20; 0.16)	0.79	−0.04 (−0.22; 0.14)	0.65
Fat free mass (kg)	66	−0.14 (−0.42; 0.14)	0.32	−0.18 (−0.44; 0.08)	0.17
Pulse Wave Velocity (m/s)	53	0.02 (−0.07; 0.11)	0.63	0.02 (−0.07; 0.10)	0.69
Glucose (mmol/L)	55	0.001 (−0.05; 0.05)	0.98	−0.004 (−0.05; 0.04)	0.88
Insulin (pmol/L)	54	1.97 (−0.67; 4.61)	0.14	1.53 (−0.94; 3.99)	0.22
HOMA-IR	53	0.12 (−0.04; 0.28)	0.13	0.09 (−0.05; 0.24)	0.21
Triglycerides (mmol/L)	55	−0.01 (−0.03; 0.02)	0.65	−0.01 (−0.03; 0.02)	0.58
Total cholesterol (mmol/L)	55	−0.01 (−0.05; 0.04)	0.74	−0.004 (−0.05; 0.04)	0.87
LDL-C (mmol/L)	55	−0.01 (−0.05; 0.03)	0.61	−0.01 (−0.05; 0.03)	0.61
HDL-C (mmol/L)	55	0.01 (−0.01; 0.03)	0.24	0.01 (−0.001; 0.03)	0.06

MVPA = Moderate to Vigorous Physical Activity; HOMA-IR = Homeostatic Model Assessment of Insulin Resistance; LDL-C = low-density lipoproteins cholesterol; HDL-C = high-density lipoproteins cholesterol. † Change was calculated as preferably 6 months or otherwise 3 months minus baseline for vegetable intake, fruit intake and MVPA, and for sugary drink intake, snack intake and body weight, change was calculated as baseline minus preferably 6 months or otherwise 3 months. * The crude model is adjusted for baseline health outcomes and baseline lifestyle behavior. ** Model further adjusted for smoking (yes/no), current diet behavior depending on the variable of interest (e.g., in case of change in vegetable intake the model is adjusted for current vegetable intake, in case of MVPA the current dietary behavior is defined as current kcal intake) and current MVPA (min/day).

Dietary intake, physical activity and weight change during the first six months after randomization were not associated with having metabolic syndrome at follow-up (Table 4).

Table 4. Association between the change in dietary intake and physical activity during the preconception lifestyle intervention and metabolic syndrome 3–8 years after randomization †.

Change in	N	Crude Model *		Adjusted Model **	
		OR (95% C.I.)	p-Value	OR (95% C.I.)	p-Value
Vegetable intake (10 g/day)	59	1.01 (0.90; 1.14)	0.87	1.01 (0.89; 1.14)	0.93
Fruit intake (10 g/day)	61	1.00 (0.90; 1.11)	0.99	1.01 (0.91; 1.12)	0.87
Sugary drink intake (glasses/day)	49	0.97 (0.38; 2.47)	0.94	0.75 (0.20; 2.87)	0.68
Savory snack intake (handful/week)	57	1.03 (0.89; 1.20)	0.67	1.17 (0.94; 1.44)	0.16
Sweet snack intake (portion/week)	57	1.01 (0.90; 1.12)	0.91	1.02 (0.88; 1.18)	0.76
Total MVPA (30 min/week)	59	1.00 (0.93; 1.07)	0.91	1.04 (0.95; 1.13)	0.42
Weight change (kg)	52	1.06 (0.90; 1.25)	0.49	0.98 (0.76; 1.27)	0.90

MVPA = Moderate to Vigorous Physical Activity. † Change was calculated as preferably 6 months or otherwise 3 months minus baseline for vegetable intake, fruit intake and MVPA, and for sugary drink intake, snack intake and body weight, change was calculated as baseline minus preferably 6 months or otherwise 3 months. * The crude model is adjusted for baseline metabolic syndrome (yes/no) and baseline lifestyle behavior. ** Model further adjusted for smoking (yes/no), current diet behavior depending on the variable of interest (e.g., in case of change in vegetable intake the model is adjusted for current vegetable intake, in case of MVPA the current dietary behavior is defined as current kcal intake) and current MVPA (min/day).

Adding randomization arm into the model (intervention or control group) and time between randomization and follow-up (years) did not change effect estimates. Additionally, effect estimates hardly changed when we excluded women in early pregnancy from our study sample (results not shown).

4. Discussion

A decrease in savory and sweet snacks during the first six months after randomization was associated with lower insulin resistance 3–8 years later. However, these associations became non-significant after adjustment for current lifestyle. No other associations between changes in lifestyle or body weight during the first six months after randomization with cardiovascular health 3–8 years later were observed.

A reason why we did not observe statistically significant associations between changes in lifestyle and body weight during the first six months after randomization with cardiovascular health 3–8 years later might be the low number of participants included in the follow-up study, and therefore we had low power to observe the hypothesized associations. Furthermore, it might be that there was not enough individual variation in the changes in lifestyle and body weight to find any associations with cardiovascular health at follow-up, which could be explained by the fact that women allocated to the control group participated to a larger extent in our follow-up study than women allocated to the intervention group.

There were multiple associations pointing towards our hypothesis that healthy changes in lifestyle are associated with more favorable cardiovascular health. Our findings that women with a higher intake of fruit ($p = 0.06$) and more MVPA ($p = 0.09$) had a higher fat free mass are in accordance with physiological mechanisms. Muscular activity stimulates the development and maintenance of lean muscle mass [27] and the high fiber content of the diet, associated with fruit intake, is related to lower body fat [28]. Furthermore, women who lost weight during the first six months after randomization tended to have higher HDL-cholesterol ($p = 0.06$). This is in line with findings in other studies and may relate to links between lower visceral fat mass and higher HDL-cholesterol [29].

However, we also observed that women who decreased their sugary drink intake tended to have a lower HDL-cholesterol ($p = 0.06$), which is not in line with our hypothesis and unexpected, assuming that a lower intake of sugary drinks causes a lower BMI, which is associated with improved HDL-cholesterol [29]. We do not know why we observed this: additional corrections for current BMI did not weaken this association and high sugary drink intake was not correlated with high MVPA levels, which is associated with higher HDL-cholesterol [30]. We were not able to analyze the effect of fresh fruit juice on HDL-cholesterol, because the 33-item FFQ do not ask specifically about fresh fruit juice. Evidence showed that fresh fruit juice might be associated with lower HDL-cholesterol [31]. It could therefore be that women specifically reduced their intake of sugar sweetened fruit juice, but not fresh fruit juice, which might have led to the unexpected association between decreases in sugary drinks and lower HDL-cholesterol. However, given the number of associations studied, it might also be that this is a chance finding.

Our results of reduced snack intake and improved HOMA-IR indicate that changing lifestyle is an important first step, but that maintaining a healthy lifestyle is needed for improving cardiometabolic health in the long-term. However, lifestyle change and maintaining those healthy changes on the long-term is notoriously difficult. To sustain long-term intervention adherence and thereby improve cardiometabolic health, it might be helpful to provide extended care by offering long-term individual or group contact to stimulate healthy behavior [32].

An important strength of the current study is the detailed information about the intake of specific foods and beverages, physical activity and body weight during the first six months after randomization into a preconception lifestyle program. This enabled us to gain more knowledge about lifestyle and body weight changes during the first six months after randomization in association with cardiometabolic health 3–8 years after the intervention instead of a randomized comparison between groups. We additionally had good quality data (measured by trained researchers) about cardiometabolic health at the start of the intervention and at 3–8 years after randomization, and were able to take into account women's baseline cardiometabolic health. There are also limitations that should be mentioned. Dietary intake as well as physical activity was measured using self-reported questionnaires. Obese women tend to under-report unhealthy behavior and over-report healthy behavior [33], and women allocated to the intervention group might do this to a larger extent because of social desirability bias [34,35]. However, adding randomization group into our regression models hardly changed the effect estimates, which indicates that the effect of social desirability bias induced by randomization group is minimal. Furthermore, there was a wide range (3–8 years) in the time between inclusion in the preconception lifestyle intervention study and our follow-up assessment of cardiometabolic health. This wide range might have affected the associations between lifestyle and body weight change with cardiometabolic health at follow-up. However, adding time in between randomization and follow-up into our models hardly changed the effect estimates. Finally, the 33-item FFQ pre-specified two food groups, fruit juice and savory snack intake, including foods known to have favorable as well as unfavorable effects on cardiometabolic health. The question on consumption of fruit juice does not distinguish between fresh fruit juice and sugar sweetened juice, while studies show that the consumption of fresh juice reduces cardiovascular risk factors due to, amongst others, the antioxidant effects and anti-inflammatory effects [31]. Furthermore, the question on savory snack consumption combines the intake of crisps, pretzels, nuts and peanuts into one question, while studies show that nuts and peanuts might be beneficial for cardiovascular health [36]. It might therefore be that an increase in these food groups represents a healthy change instead of an unhealthy change. We recommend future research to use a more extensive FFQ, not pre-specifying these foods into one food group. Future research should replicate our results in a larger study population, preferably with larger variations in lifestyle and weight changes.

5. Conclusions

To conclude, decreases in savory and sweet snack intake were associated with reduced insulin resistance 3–8 years later, but after adjustment for current lifestyle these associations disappeared. No other associations between changes in lifestyle or body weight during the first six months after randomization with cardiovascular health 3–8 years later were observed. Changing lifestyle is an important first step, but maintaining this change is needed to improve cardiometabolic health in the long-term.

Supplementary Materials: The following are available online at http://www.mdpi.com/2072-6643/10/12/1953/s1, Table S1. Characteristics and baseline cardiometabolic health of participants versus the non-participants.

Author Contributions: Conceptualization, T.M.V.E., M.N.M.V.P., R.J.B.J.G., H.G., A.H., B.W.M. and T.J.R.; Formal analysis, T.M.V.E.; Funding acquisition, A.H. and T.J.R.; Investigation, T.M.V.E.; Methodology, T.M.V.E., M.N.M.V.P. and T.J.R.; Project administration, T.M.V.E.; Supervision, M.N.M.V.P., R.J.B.J.G., H.G., A.H., B.W.M. and T.J.R.; Validation, M.N.M.V.P.; Visualization, T.M.V.E. and M.N.M.V.P.; Writing—original draft, T.M.V.E.; Writing—review & editing, M.N.M.V.P., R.J.B.J.G., H.G., A.H., B.W.M. and T.J.R..

Funding: The LIFEstyle study was funded by ZonMw, the Dutch Organization for Health Research and Development, grant number: 50-50110-96-518. The follow-up of the LIFEstyle trial was funded by grants from the Dutch Heart Foundation (2013T085) and the European Commission (Horizon2020 project 633595 DynaHealth). BWM is supported by a NHMRC Practitioner Fellowship (GNT1082548). None of these organizations had a role in data collection, analysis, interpretation of data or writing the report.

Acknowledgments: First of all, we would like to thank all women who participated in the LIFEstyle RCT and the WOMB project. And we thank all members of the WOMB consortium and of the Dutch NVOG Consortium (www.studies-obsgyn.nl) who were involved in the LIFEstyle RCT and the WOMB project.

Conflicts of Interest: A.H.: The department of Obstetrics and Gynecology of the UMCG received an unrestricted educational grant from Ferring pharmaceuticals BV, The Netherlands, outside the submitted work. BWM reports consultancy for ObsEva, Merck KGaA and Guerbet. All other authors declare that they have no competing interests. The funders had no role in the design of the study; in the collection, analyses, or interpretation of data; in the writing of the manuscript, or in the decision to publish the results.

References

1. Stewart, J.; Manmathan, G.; Wilkinson, P. Primary prevention of cardiovascular disease: A review of contemporary guidance and literature. *JRSM Cardiovasc. Dis.* **2017**, *6*, 2048004016687211. [CrossRef] [PubMed]
2. Hu, F.B.; Stampfer, M.J.; Manson, J.E.; Grodstein, F.; Colditz, G.A.; Speizer, F.E.; Willett, W.C. Trends in the Incidence of Coronary Heart Disease and Changes in Diet and Lifestyle in Women. *N. Engl. J. Med.* **2000**, *343*, 530–537. [CrossRef] [PubMed]

3.   Manson, J.E.; Hu, F.B.; Rich-Edwards, J.W.; Colditz, G.A.; Stampfer, M.J.; Willett, W.C.; Speizer, F.E.; Hennekens, C.H. A Prospective Study of Walking as Compared with Vigorous Exercise in the Prevention of Coronary Heart Disease in Women. *N. Engl. J. Med.* **1999**, *341*, 650–658. [CrossRef] [PubMed]

4.   Zhang, X.; Devlin, H.M.; Smith, B.; Imperatore, G.; Thomas, W.; Lobelo, F.; Ali, M.K.; Norris, K.; Gruss, S.; Bardenheier, B.; et al. Effect of lifestyle interventions on cardiovascular risk factors among adults without impaired glucose tolerance or diabetes: A systematic review and meta-analysis. *PLoS ONE* **2017**, *12*, e0176436. [CrossRef] [PubMed]

5.   Wu, T.; Gao, X.; Chen, M.; van Dam, R.M. Long-term effectiveness of diet-plus-exercise interventions vs. diet-only interventions for weight loss: A meta-analysis. *Obes. Rev.* **2009**, *10*, 313–323. [CrossRef]

6.   Wing, R.R.; Phelan, S. Long-term weight loss maintenance. *Am. J. Clin. Nutr.* **2005**, *82*, 222S–225S. [CrossRef]

7.   Wing, R.R.; Hill, J.O. Successful weight loss maintenance. *Annu. Rev. Nutr.* **2001**, *21*, 323–341. [CrossRef]

8.   Whelton, S.P.; Chin, A.; Xin, X.; He, J. Effect of aerobic exercise on blood pressure: A Meta-analysis of randomized, controlled trials. *Ann. Intern. Med.* **2002**, *136*, 493–503. [CrossRef]

9.   Dutch Trial Registration Trial Registration LIFEstyle RCT. Available online: http://www.trialregister.nl/trialreg/admin/rctview.asp?TC=1530 (accessed on 15 March 2018).

10.   Mutsaerts, M.A.; Groen, H.; ter Bogt, N.C.; Bolster, J.H.; Land, J.A.; Bemelmans, W.J.; Kuchenbecker, W.K.; Hompes, P.G.; Macklon, N.S.; Stolk, R.P.; et al. The LIFESTYLE study: Costs and effects of a structured lifestyle program in overweight and obese subfertile women to reduce the need for fertility treatment and improve reproductive outcome. A randomised controlled trial. *BMC Womens. Health* **2010**, *10*, 22. [CrossRef]

11.   Mutsaerts, M.A.Q.; van Oers, A.M.; Groen, H.; Burggraaff, J.M.; Kuchenbecker, W.K.H.; Perquin, D.A.M.; Koks, C.A.M.; van Golde, R.; Kaaijk, E.M.; Schierbeek, J.M.; et al. Randomized Trial of a Lifestyle Program in Obese Infertile Women. *N. Engl. J. Med.* **2016**, *374*, 1942–1953. [CrossRef]

12.   Van Elten, T.M.; Karsten, M.D.A.; Geelen, A.; van Oers, A.M.; van Poppel, M.N.M.; Groen, H.; Gemke, R.J.B.J.; Mol, B.W.; Mutsaerts, M.A.Q.; Roseboom, T.J.; et al. study Effects of a preconception lifestyle intervention in obese infertile women on diet and physical activity; A secondary analysis of a randomized controlled trial. *PLoS ONE* **2018**, *13*, e0206888. [CrossRef] [PubMed]

13.   Van Dammen, L.; Wekker, V.; van Oers, A.M.; Mutsaerts, M.A.Q.; Painter, R.C.; Zwinderman, A.H.; Groen, H.; van de Beek, C.; Muller Kobold, A.C.; Kuchenbecker, W.K.H.; et al. Effect of a lifestyle intervention in obese infertile women on cardiometabolic health and quality of life: A randomized controlled trial. *PLoS ONE* **2018**, *13*, e0190662. [CrossRef] [PubMed]

14.   Van Elten, T.M.; Karsten, M.D.A.; Geelen, A.; Gemke, R.J.B.J.; Groen, H.; Hoek, A.; van Poppel, M.N.M.; Roseboom, T.J. Preconception lifestyle intervention reduces long term energy intake in women with obesity and infertility: A randomised controlled trial. *Int. J. Behav. Nutr. Phys. Act.* Under Rev. **2018**, in press.

15.   Gezondheidsraad. Richtlijnen Goede Voeding 2006. Available online: https://www.gezondheidsraad.nl/documenten/adviezen/2006/12/18/richtlijnen-goede-voeding-2006 (accessed on 22 May 2018).

16.   Van de Beek, C.; Hoek, A.; Painter, R.C.; Gemke, R.J.B.J.; van Poppel, M.N.M.; Geelen, A.; Groen, H.; Willem Mol, B.; Roseboom, T.J. Women, their Offspring and iMproving lifestyle for Better cardiovascular health of both (WOMB project): A protocol of the follow-up of a multicentre randomised controlled trial. *BMJ Open* **2018**, *8*, e016579. [CrossRef] [PubMed]

17.   Van den Brink, C.; Ocké, M.; Houben, A.; van Nierop, P.; Droomers, M. *Validation of a Community Health Services Food Consumption Questionnaire in the Netherlands*; RIVM rapport 260854008/2005; RIVM: Bilthoven, The Netherlands, 2005.

18.   Wendel-Vos, G. Reproducibility and relative validity of the short questionnaire to assess health-enhancing physical activity. *J. Clin. Epidemiol.* **2003**, *56*, 1163–1169. [CrossRef]

19.   Kyle, U.G.; Genton, L.; Karsegard, L.; Slosman, D.O.; Pichard, C. Single prediction equation for bioelectrical impedance analysis in adults aged 20–94 years. *Nutrition* **2001**, *17*, 248–253. [CrossRef]

20.   Reference Values for Arterial Stiffness' Collaboration. Determinants of pulse wave velocity in healthy people and in the presence of cardiovascular risk factors: 'Establishing normal and reference values'. *Eur. Heart J.* **2010**, *31*, 2338–2350. [CrossRef]

21.   Grundy, S.M.; Cleeman, J.I.; Daniels, S.R.; Donato, K.A.; Eckel, R.H.; Franklin, B.A.; Gordon, D.J.; Krauss, R.M.; Savage, P.J.; Smith, S.C.; et al. Diagnosis and management of the metabolic syndrome: An American Heart Association/National Heart, Lung, and Blood Institute scientific statement. *Circulation* **2005**, *112*, 2735–2752. [CrossRef]

22. Deurenberg, P.; Andreoli, A.; Borg, P.; Kukkonen-Harjula, K.; de Lorenzo, A.; van Marken Lichtenbelt, W.; Testolin, G.; Vigano, R.; Vollaard, N. The validity of predicted body fat percentage from body mass index and from impedance in samples of five European populations. *Eur. J. Clin. Nutr.* **2001**, *55*, 973–979. [CrossRef]

23. Kim, E.J.; Park, C.G.; Park, J.S.; Suh, S.Y.; Choi, C.U.; Kim, J.W.; Kim, S.H.; Lim, H.E.; Rha, S.W.; Seo, H.S.; et al. Relationship between blood pressure parameters and pulse wave velocity in normotensive and hypertensive subjects: Invasive study. *J. Hum. Hypertens.* **2007**, *21*, 141–148. [CrossRef]

24. De Backer, G.; Ambrosioni, E.; Borch-Johnsen, K.; Brotons, C.; Cifkova, R.; Dallongeville, J.; Ebrahim, S.; Faergeman, O.; Graham, I.; Mancia, G.; et al. European guidelines on cardiovascular disease prevention in clinical practice Third Joint Task Force of European and other Societies on Cardiovascular Disease Prevention in Clinical Practice (constituted by representatives of eight societies and by invited experts). *Eur. Heart J.* **2003**, *24*, 1601–1610. [CrossRef] [PubMed]

25. Santos-Lozano, A.; Santín-Medeiros, F.; Cardon, G.; Torres-Luque, G.; Bailón, R.; Bergmeir, C.; Ruiz, J.; Lucia, A.; Garatachea, N. Actigraph GT3X: Validation and Determination of Physical Activity Intensity Cut Points. *Int. J. Sports Med.* **2013**, *34*, 975–982. [CrossRef] [PubMed]

26. Freedson, P.S.; Melanson, E.; Sirard, J. Calibration of the Computer Science and Applications, Inc. accelerometer. *Med. Sci. Sports Exerc.* **1998**, *30*, 777–781. [CrossRef] [PubMed]

27. Haskell, W.L.; Lee, I.M.; Pate, R.R.; Powell, K.E.; Blair, S.N.; Franklin, B.A.; Macera, C.A.; Heath, G.W.; Thompson, P.D.; Bauman, A. Physical Activity and Public Health: Updated recommendation for adults from the American College of Sports Medicine and the American Heart Association. *Med. Sci. Sports Exerc.* **2007**, *39*, 1423–1434. [CrossRef]

28. Slavin, J.L. Dietary fiber and body weight. *Nutrition* **2005**, *21*, 411–418. [CrossRef]

29. Rashid, S.; Genest, J. Effect of Obesity on High-density Lipoprotein Metabolism. *Obesity* **2007**, *15*, 2875–2888. [CrossRef]

30. Loprinzi, P.D.; Addoh, O. The association of physical activity and cholesterol concentrations across different combinations of central adiposity and body mass index. *Health Promot. Perspect.* **2016**, *6*, 128–136. [CrossRef]

31. Zheng, J.; Zhou, Y.; Li, S.; Zhang, P.; Zhou, T.; Xu, D.-P.; Li, H.-B. Effects and Mechanisms of Fruit and Vegetable Juices on Cardiovascular Diseases. *Int. J. Mol. Sci.* **2017**, *18*, 555. [CrossRef]

32. Middleton, K.R.; Anton, S.D.; Perri, M.G. Long-Term Adherence to Health Behavior Change. *Am. J. Lifestyle Med.* **2013**, *7*, 395–404. [CrossRef]

33. Scagliusi, F.B.; Ferriolli, E.; Pfrimer, K.; Laureano, C.; Cunha, C.S.F.; Gualano, B.; Lourenço, B.H.; Lancha, A.H. Characteristics of women who frequently under report their energy intake: A doubly labelled water study. *Eur. J. Clin. Nutr.* **2009**, *63*, 1192–1199. [CrossRef]

34. Hebert, J.R.; Hurley, T.G.; Peterson, K.E.; Resnicow, K.; Thompson, F.E.; Yaroch, A.L.; Ehlers, M.; Midthune, D.; Williams, G.C.; Greene, G.W.; et al. Social Desirability Trait Influences on Self-Reported Dietary Measures among Diverse Participants in a Multicenter Multiple Risk Factor Trial. *J. Nutr.* **2008**, *138*, 226S–234S. [CrossRef] [PubMed]

35. Adams, S.A.; Matthews, C.E.; Ebbeling, C.B.; Moore, C.G.; Cunningham, J.E.; Fulton, J.; Hebert, J.R. The Effect of Social Desirability and Social Approval on Self-Reports of Physical Activity. *Am. J. Epidemiol.* **2005**, *161*, 389–398. [CrossRef] [PubMed]

36. De Souza, R.G.M.; Schincaglia, R.M.; Pimentel, G.D.; Mota, J.F. Nuts and Human Health Outcomes: A Systematic Review. *Nutrients* **2017**, *9*, 1311. [CrossRef] [PubMed]

nutrients

MDPI

Article

Exercise and the Timing of Snack Choice: Healthy Snack Choice is Reduced in the Post-Exercise State

Christopher R. Gustafson [1], Nigina Rakhmatullaeva [1], Safiya E. Beckford [2], Ajai Ammachathram [2], Alexander Cristobal [2] and Karsten Koehler [2,*]

[1]   Department of Agricultural Economics, University of Nebraska-Lincoln, Lincoln, NE 68583, USA; cgustafson6@unl.edu (C.R.G.); n.rakhmatullaeva@gmail.com (N.R.)
[2]   Department of Nutrition and Health Sciences, University of Nebraska-Lincoln, Lincoln, NE 68583, USA; safiya.beckford@huskers.unl.edu (S.E.B.); ajai@unl.edu (A.A.); acristobal0474@huskers.unl.edu (A.C.)
*   Correspondence: kkoehler3@unl.edu; Tel.: +1-402-472-7521

Received: 15 November 2018; Accepted: 5 December 2018; Published: 7 December 2018

Abstract: Acute exercise can induce either a compensatory increase in food intake or a reduction in food intake, which results from appetite suppression in the post-exercise state. The timing of food choice—choosing for immediate or later consumption—has been found to influence the healthfulness of foods consumed. To examine both of these effects, we tested in our study whether the timing of food choice interacts with exposure to exercise to impact food choices such that choices would differ when made prior to or following an exercise bout. Visitors to a university recreational center were equipped with an accelerometer prior to their habitual workout regime, masking the true study purpose. As a reward, participants were presented with a snack for consumption after workout completion. Participants made their snack choice from either an apple or chocolate brownie after being pseudo-randomly assigned to choose prior to ("before") or following workout completion ("after"). Complete data were available for 256 participants (54.7% male, 22.1 $\pm$ 3.1 years, 24.7 $\pm$ 3.7 kg/m^2) who exercised 65.3 $\pm$ 22.5 min/session. When compared with "before," the choice of an apple decreased (73.7% vs. 54.6%) and the choices of brownie (13.9% vs. 20.2%) or no snack (12.4% vs. 25.2%) increased in the "after" condition ($\chi^2 = 26.578$, $p < 0.001$). Our results provide support for both compensatory eating and exercise-induced anorexia. More importantly, our findings suggest that the choice of food for post-exercise consumption can be altered through a simple behavioral intervention.

Keywords: compensatory eating; exercise-induced anorexia; food choice; acute exercise; behavioral economics; nudges

---

1. Introduction

Regular exercise and a healthy diet are important staples of a healthy lifestyle. The beneficial effects of exercise for the treatment and prevention of many physiological and psychological conditions, including diabetes, cardiovascular disease, certain cancers, recovery from stroke, emotional well-being, depression, anxiety, and suicidal behaviors [1–8], are well documented. However, the impact of exercise on overweight and obesity, and particularly its ability to produce meaningful weight loss remains under debate [9]. It is undisputed that exercise increases energy expenditure and thereby has the potential to induce weight loss. However, individual weight loss responses are mixed [10], suggesting that the success of exercise as a weight loss strategy is largely dependent on its effects on the other component of energy balance, i.e., dietary energy intake. A primary barrier to exercise-induced weight loss is compensatory eating, which is defined as an increase in food intake following exercise or physical activity [11].

It is a common belief that exercise stimulates appetite and food intake [12], and as much as 77% of college students report engaging in compensatory eating [11,13]. The mechanisms for this compensatory increase in food intake following exercise are manifold, but most likely include endocrine pathways that favor food intake to ensure the maintenance of body weight, and more specifically lean body mass [14,15]. On the other hand, there is also evidence in the literature contrary to compensation. Evidence shows that only 19% of intervention studies reported an increase in energy intake after exercise, whereas 65% show no change [16]. Furthermore, it has repeatedly been demonstrated that appetite and hunger are suppressed following exercise, particularly in the immediate post-exercise state [12]. This reduction in perceived hunger has been termed "exercise-induced anorexia" and has been linked to the suppression of orexigenic hormones, such as ghrelin, and concomitant increases in satiety hormones, including peptide YY and glucagon-like peptide 1 [17–19].

While the impact of exercise on energy intake is important from an energy balance perspective, it is critical to understand that energy intake is ultimately the product of food choices, as individuals select foods rather than nutrients under free-living conditions [12]. It has been proposed that exercise participation can impact food selection, modify the sensitivity to sensory cues [12], and alter the reward value of foods with particular sensory and/or macronutrient profiles [20,21]. Hedonic mechanisms controlling food intake are stimulated by the sensory pleasure of eating palatable food and may result in increased food intake [22]. This increase in food intake has been linked to the neuroendocrine factor dopamine, which has been shown to reinforce the pleasure derived from eating highly palatable foods [23,24]. Exercise has been found to reduce addictive behaviors such as drug and alcohol consumption [23], and as such it is possible that exercise may also elicit a reduction in food intake as a result of the rewarding value of food. Previous experiments have shown that the responsiveness of brain regions related to food reward is altered in the post-exercise state [25], such that the brain's neural reward system's response to low energy density foods is increased and the reaction to high energy density is reduced [26–28].

In support of the contradictory findings in the literature, it has been proposed that exercise could either increase or decrease the reinforcing value of energy-dense, palatable foods. For example, the deliberate choice of highly palatable, energy-dense foods (e.g., fatty and/or sweet "treats") in the post-exercise state has been linked to compensatory eating and reduced weight loss success [21]. Others have reported that carbohydrate-rich foods are rated more palatable in the post-exercise state [25,26], possibly reflecting increased carbohydrate utilization during aerobic exercise [29]. Alternatively, exercise could also reduce the consumption of these foods as a result of improved appetite control coupled with a higher motivation to engage in healthy behaviors [28]. These diametrically opposed effects of exercise on food choices could ultimately explain the inter-individual variation in the degree of compensatory eating following exercise interventions [21].

Another important influence on dietary quality is the timing of food choice relative to consumption. Prior research from behavioral economics and psychology suggests that changing the timing of food choice relative to consumption—whether food choices will be consumed immediately or at some later point—influences the healthfulness of the foods chosen [30,31]. Behavioral economic models of choices over time (the most prominent of which is hyperbolic discounting [32]) have been formulated to represent individuals who have inconsistent preferences. These models predict that choices will differ depending on whether the chosen item is to be received immediately or after a delay. When the decision-maker will receive their choice immediately, the models predict that the individual will make less healthy, more impatient decisions than if the receipt is delayed. Individuals whose choices fit this pattern are said to have "present-biased preferences." An upshot of present-biased preferences is that people will often be willing to pre-commit to a healthier behavior if given the opportunity to do so [33,34].

To integrate the impact of biophysical and behavioral effects on post-exercise food intake in a real-life scenario, the overall goal of the present study was to test whether manipulating the timing of the choice of a snack to be received after exposure to exercise would alter food choices such that in the post-exercise state the choice of snacks with varying energy density and health attributes would differ from choices made prior to exercise. Our intervention addresses a simple but important question: *Can a simple nudge—changing the time when the choice of a post-exercise snack is made—help individuals select healthier food options?* Given the importance of understanding why people make unhealthy dietary choices and how these choices can be discouraged [35], examining how factors such as the timing of food choice that may promote healthy eating in the context of exercising would be beneficial [28]. This is particularly true since behavioral weight loss strategies typically employ a combination of diet and/or exercise. However, as mentioned previously, evidence suggests that increases in food consumption in response to exercise can derail weight loss efforts [16]. Knowledge about the degree to which food choices are altered from pre- to post-exercise could ultimately allow individuals to pre-commit to healthier food choices by selecting food in a state which favors their choice of healthier options. Our primary hypothesis was that individuals are more likely to choose an "unhealthy," energy-dense snack in the post-exercise state when compared to choosing a snack for post-exercise consumption prior to exercising, supporting previous evidence of compensatory eating [11] as well as findings from behavioral studies of food choice [30]. To also account for previously reported reductions in appetite and hunger in the immediate post-exercise state, we further hypothesized that the number of participants who would decline a snack would also increase in the post-exercise state.

2. Materials and Methods

2.1. Study Design

The experiment was conducted at the University of Nebraska-Lincoln Recreational and Wellness Center and was focused on the effect of the timing of food choice—which was assigned by the researchers to take place before or after the participant exercised—on selection of a food item. The food item was received after the exercise was completed in both conditions. The experiment took place on randomly chosen weekdays between February 16 and April 30, 2018, and implementation of conditions was balanced across the study period and days of the week. Individuals were recruited to participate as they entered the recreational center. All participants had to be at least 19 years of age and to have come to the Recreational and Wellness Center to exercise. When individuals were invited to participate, the purpose of the study was presented as being focused on calibrating activity sensors to various exercise types. As an incentive for participation in this "calibration study," participants were offered a choice of food items upon completion of the study. In actuality, participants' choice of a food item was of central interest to the study, but this interest in participants' food choices was not highlighted to minimize experimenter demand effects [36] and social desirability biases [37,38]. Participants signed a written informed consent form prior to the experiment. The study was approved by the University of Nebraska-Lincoln Institutional Review Board.

2.2. Procedures

After participants signed the informed consent form, study staff measured their height and weight on a digital column scale (Seca, Hamburg, Germany). Next, participants were fitted with an accelerometer (GT3X, ActiGraph, Pensacola, FL, USA) on their non-dominant wrist, which they wore for the duration of their workout. Study staff instructed all participants to proceed with the workout that the participant had planned to complete at the Recreation and Wellness Center prior to study recruitment.

When subjects returned the accelerometers upon completion of their workout, additional information was collected, including each participant's date of birth, the type of activity or activities conducted during the workout, and whether participants ate or drank anything during the workout. Participants were told that this information was needed to personalize the accelerometer data.

2.3. Food Choice Paradigm

In order to examine changes in food choices over the course of a self-selected workout, participants were given a snack choice to be received upon completion of their workout as a reward for study participation. In one condition, prior to beginning their workout ("before"), participants were asked to choose which snack item they wanted to receive. Upon completion of their workout and them returning the accelerometer, participants were given the food item they had selected. In the other condition ("after"), participants were asked to choose their snack right as they came back to return their accelerometer after their workout. In the "after" condition, participants received their snack immediately after making their choice. For logistical purposes, one condition ("before" or "after") was implemented per study day.

In order to present two snack options with distinguishable perceived health attributes but similar taste attributes (sweet), participants were able to choose between an apple (deemed "healthy") and a brownie (deemed "unhealthy"). Participants also had the option to decline both snacks ("neither"). Both snack options were visible to participants as they checked in ("before" condition) or as they checked out ("after" condition). Prior to giving the chosen snack to the participant, study staff inquired about food allergies and intolerances.

All food used in the experiment was purchased and prepared in a large batch by a food and beverage management specialist in a department certified kitchen and stored at appropriate conditions (temperature checks were performed twice per day) prior to presenting it to study participants. The apple variety offered in the study was Fuji with an average energy content of 121 kcal per medium-sized apple [39]. Brownies were prepared from a prepackaged commercial brownie mix (Ghiradelli, San Leandro, CA, USA) and had an energy content of 140 kcal per piece.

2.4. Data Analyses

Data were analyzed using R Statistical Software (R Core Team, R Foundation, Vienna, Austria). Prior to data analysis, data from individuals who participated in the study multiple times were eliminated from the dataset such that only the first study visit was included in the analyses. In addition, data from participants who reported food allergies or intolerances that could have affected the food choice were eliminated from the dataset. Participants' food item choices were analyzed to evaluate whether the proportion of choices of the different food items (including "neither") differed by condition using a chi-squared test. We then used multinomial, multivariate logistic regression models to examine the relationship between food choice and condition while controlling for potentially confounding variables comprising body mass index (BMI) category ($\leq$25 kg/m^2; >25 kg/m^2), age (in years), gender, workout duration (in minutes), whether the participant consumed any food during their workout, and mode of exercise (aerobic, resistance, and other). BMI categories were aggregated from four initial categories (underweight, normal weight, overweight, and obese) into two (underweight/normal weight, and overweight/obese) because few participants fell into the underweight (3 participants) and obese (18 participants) categories. Results of the multinomial logistic regression models are presented as odds ratios (OR) and 95-percent confidence intervals (95% CI) for independent variables. Statistical significance was considered for *p*-values < 0.05.

3. Results

A total of 299 data points were initially collected. Data from 31 participants (42 observations) were eliminated because individuals had participated in the study more than once. In addition, data from one participant who reported having celiac disease were omitted. The final dataset used for analysis contained observations from 256 unique participants. Of these, 137 participants (53.5%) completed the "before" condition and 119 completed the "after" condition (46.5%). On average, 54.7% of participants were male, 22.1 ± 3.1 years old, had an average BMI of 24.7 ± 3.7 kg/m^2, and exercised for 65.3 ± 22.5 min. There were no significant differences between conditions in gender distribution, age, BMI or BMI categories, workout duration, food consumption, or mode of exercise (Table 1).

Table 1. Characteristics of the study sample.

	All Participants (*n* = 256)	"Before" Condition (*n* = 137)	"After" Condition (*n* = 119)	*p*-Value *
Age	22.1 ± 3.1	22.0 ± 2.9	22.1 ± 3.4	0.72
Gender (Male)	135 (52.7%)	75 (54.7%)	60 (50.4%)	0.49
BMI (kg/m^2)	24.7 ± 3.7	24.8 ± 3.6	24.6 ± 3.8	0.55
Underweight/Normal weight	148 (57.8%)	75 (54.8%)	73 (62.3%)	0.29
Overweight/Obese	108 (42.2%)	62 (45.2%)	46 (38.7%)	0.29
Workout Duration (min)	65.3 ± 22.5	67.3 ± 25.5	63.0 ± 18.3	0.12
Food Consumption (Yes)	8 (3.1%)	5 (3.6%)	3 (2.5%)	0.60
Aerobic Exercise (Yes)	172 (67.2%)	87 (63.5%)	85 (73.4%)	0.18
Resistance Exercise (Yes)	192 (75%)	102 (74.5%)	90 (75.6%)	0.83
Other Exercise (Yes)	9 (3.5%)	7 (5.1%)	2 (1.7%)	0.13

* "before" vs. "after".

In the "before" condition, 101 participants (73.7%) selected an apple, 19 (13.9%) selected a brownie, and 17 (12.4%) declined a snack upon completion of their workout. In the "after" condition, 65 participants (54.6%) selected an apple, which is an ~20% decrease from the "before" condition. Twenty-four participants (20.2%) selected a brownie in the "after" condition, and 30 participants (25.2%) declined a snack option upon completion of their workout. The patterns of choices (Figure 1) differed significantly between the "before" and "after" condition (χ-squared = 26.578, df = 2; $p < 0.001$).

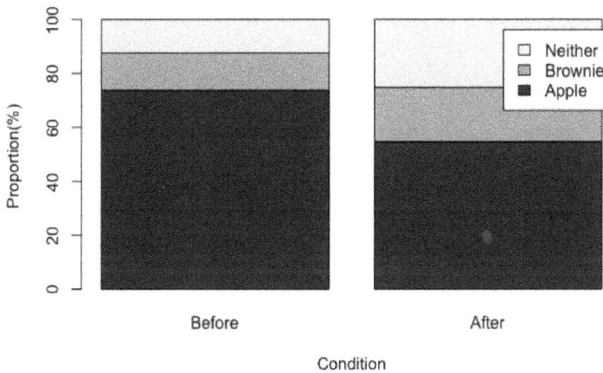

Figure 1. Proportion of snack choices (apple, brownie, or neither) for consumption after completion of a workout chosen either before the beginning of the workout ("before" condition) or after completion of the workout ("after" condition).

Results from the multinomial, multivariate logistic regression model are reported in Table 2. The omitted snack choice was "neither" in the regression model. Results confirm that the odds of a participant choosing an apple decreased significantly in the "after" condition relative to the choice of neither (OR: 0.33; 95% CI: 0.16–0.66), even after controlling for other variables. The estimated odds ratio for the effect of the "after" condition on apple choice remained unchanged regardless of other independent variables included in the regression model, including BMI status, age, gender, and workout duration. There were no significant differences in brownie choice (relative to choosing neither) based on condition (OR: 0.64; 95% CI: 0.26–1.55). In addition to the condition, only the BMI category significantly contributed to the regression model of snack choice. Individuals who were classified as overweight or obese (BMI > 25 kg/m^2) were significantly less likely to select a brownie than individuals with a BMI < 25 kg/m^2 (OR: 0.37; 95% CI: 0.15–0.92).

Table 2. Odds ratios (OR) and 95% confidence intervals (CI) from multinomial logistic regression results of snack choice, relative to neither.

	Apple		Brownie	
	OR	95% CI	OR	95% CI
Condition: "After"	0.33	(0.16–0.66)	0.64	(0.26–1.55)
BMI Status (>25 kg/m^2)	0.61	(0.31–1.23)	0.37	(0.15–0.92)
Age	0.98	(0.87–1.10)	1.09	(0.96–1.24)
Gender (male)	1.17	(0.56–2.43)	1.81	(0.71–4.64)
Workout Duration (min)	1.00	(0.98–1.02)	1.00	(0.98–1.02)
Food Consumption (Yes)	1.51	(0.17–13.62)	1.16	(0.07–20.00)
Aerobic Exercise (Yes)	1.11	(0.48–2.59)	0.85	(0.30–2.40)
Resistance Exercise (Yes)	1.03	(0.43–2.43)	0.91	(0.30–2.77)
Other Exercise (Yes)	0.47	(0.04–1.56)	0.47	(0.05–4.45)
Intercept	10.65	(0.53–214.60)	0.31	(0.01–9.75)

Number of observations = 256, Akaike Information Criterion = 470.9.

4. Discussion

The goal of the present study was to assess whether the timing of snack choice would interact with exposure to a single exercise bout to alter food choices. Using a behavioral intervention approach, we show that a very simple modification—making a choice about a post-exercise snack either prior to or following the completion of the exercise bout—significantly alters the snack choice. Our findings indicate that the likelihood of choosing an apple, a food typically considered as "healthy", is about one third (33.5%) greater when the choice is presented prior to engaging in exercise; however, when the choice is presented following the exercise bout, individuals are approximately 39% more likely to choose a brownie, a food typically considered as "unhealthy", and 112% more likely to decline either snack option. These findings exhibit elements of two previously identified effects of exercise on food choice: compensatory eating, which refers to the increase in food intake following exercise of physical activity [11], and exercise-induced anorexia, which refers to a temporary reduction in appetite immediately following exercise [40]. They also correspond to patterns seen in previous behavioral research, in which individuals are more likely to make healthier choices if the food will be delivered in the future rather than immediately [30].

4.1. Interindividual Variation in Post-Exercise Food Choices: Compensatory Eating and Exercise-Induced Anorexia

The increase in the preference for a brownie, an energy-dense and palatable snack, can be seen as evidence for compensatory eating. Finlayson et al. speculated that hedonic processes are modulated by increased energy expended during exercise, thereby promoting overconsumption in individuals prone to compensatory eating [22]. Considering that a greater tendency for compensatory eating has been linked to attenuated weight loss during exercise interventions [21], our findings highlight the importance of timing of food choices for individuals who exercise to lose weight. In contrast, others have reported a reduced preference for energy-dense food following exercise. For example, an acute bout of resistance exercise decreased the preference for high-fat food [27], and 2 weeks of aerobic exercise were found to reduce the reinforcing value of high energy density foods [28]. However, it is noteworthy that there was a dose-dependent relationship between habitual exercise and the reduction in the reinforcing value of high energy density foods, which was most pronounced in individuals who exercised 5 days per week when compared to individuals who exercised less regularly [28].

In addition to an increased preference for an "unhealthy" option, we further observed an increase in the number of individuals who declined either food option among participants who were given a snack choice in the post-exercise state. While we acknowledge that there may be other reasons for this finding, this finding is in accordance with previous reports of exercise-induced anorexia. This phenomenon, which describes a transient reduction in appetite and hunger in the immediate post-exercise state, has been linked to reductions in ghrelin, an appetite-stimulating hormone, and increased concentrations of satiety hormones, including peptide YY and glucagon-like peptide 1 [17–19,41]. However, hunger is not necessarily indicative of actual food intake, as shown in a study involving exposure to exercise message commercials, which resulted in higher ratings of hunger but lower caloric intake [42].

Taken together, increases in compensatory eating and exercise-induced anorexia result in a greater inter-individual variability in post-exercise food intake. These competing effects, i.e., an increase in consumption of an "unhealthy" option along with a greater number of individuals declining either food option, negate an overall group effect. In fact, when determining the caloric intake based on food choices made (brownie, apple or neither), average caloric intake did not change dramatically from "before" (109 kcal/participant) to "after" (94 kcal/participant), which is in support of a previous meta-analysis by Schubert et al. who failed to identify a definite effect of an exercise task on post-exercise caloric intake [10]. This finding is also in agreement with observations from our laboratory, according to which inter-individual differences in food choices are greater in the post-exercise state when compared to the rested state (Koehler, unpublished observation).

4.2. Impact of Gender and Body Mass Index

In contrast to previous studies, we failed to observe a significant gender effect in our food choice paradigm. Although most laboratory experiments addressing the impact of a single exercise bout on food intake were not specifically designed to detect gender differences, it is well established in the literature that food intake patterns differ between men and women and that women consistently make healthier choices than men [43]. Furthermore, the motivation to exercise tends to differ between genders, as women are more likely to exercise in attempts to lose weight [44]. As such, it is not surprising that when compared to their male counterparts, female athletes exhibit more self-control for high-calorie and sweet food options [45]. However, despite these gender differences, a recent meta-regression showed no impact of gender on post-exercise calorie intake [10], which is in agreement with the lack of a gender effect in our study. This lack of gender difference may be further explained by the fact that a greater proportion of our female participants conducted aerobic exercise (84% vs. 52% in males, $p < 0.001$), which is reflective of the general exercise motives of young adults [46]. It is possible that the lower energy expenditure of resistance exercise, when compared to aerobic exercise [47], reduced the likelihood of compensatory eating in the male participants. Furthermore, a recent study reported that resistance exercise, but not aerobic exercise, reduced explicit liking for high-fat food [27].

We did, however, find an effect of BMI on overall food choice, whereby individuals with a BMI indicative of overweight or obesity (>25 kg/m^2) were less likely to choose our "unhealthy" option. This finding is contrary to previous literature suggesting that obesity is associated with an increased drive to eat palatable foods [48], specifically in the absence of hunger [49], and greater impulsivity toward food reward [50]. However, it should be noted that the current study was selectively conducted in individuals who voluntarily exercised, a group not necessarily representative of individuals who are overweight and obese [51]. As such, our overweight or obese participants may have exhibited a greater motivation to lose weight compared to non-exercising individuals [52]. Furthermore, it is also possible that our measure of BMI overestimated levels of overweight and obesity in our group of exercisers, as it is well known that BMI fails to differentiate between individuals with excess adiposity vs. individuals with increased muscle mass [53].

4.3. Limitations

As the study presents a simple behavioral manipulation in a real-life setting, there are various limitations to our investigation. First, we failed to assess changes in ratings of appetite or hunger over the course of the exercise bout as well as appetite-regulating peptides, and more specifically whether inter-individual differences in these outcomes would explain differences in food choices from pre- to post-exercise. However, as previously mentioned, it is well known that hunger levels alone fail to explain differences in the food intake response following exercise [22].

In order to test the impact of exercise on food choice in a real-life scenario, we chose to allow participants to follow their regular exercise regimen. As such, exercise intensity was self-selected and consequently varied among our participants. High-intensity exercise has been shown to favor a negative energy balance to a greater extent than low-intensity exercise [54]. Although each participant wore an accelerometer during their exercise bout, we chose not to include this data in the present analysis due to previous studies demonstrating that the Actigraph tends to perform poorly during vigorous activities [55], which we presumed to be the primary intensity range in the present study. Regardless, future studies should attempt to assess exercise intensity using objective (e.g., heart rate) or subjective (e.g., ratings of perceived exertion) measures. Another limitation was that we did not assess the prandial state, such as the time of the last meal or current hunger levels. While intriguing, these measures were not taken in order to avoid revealing the true purpose of this investigation. We did, however, assess food intake (in excess of water) during the workouts, but this variable did not have a significant effect on food choice as shown by our regression analysis (Table 2). While future studies should more carefully monitor food intake and hunger levels prior to the experiment, we are confident that using a relatively large sample and conducting all experimental trials at the same time of day minimized the impact of the prandial state on our outcomes.

Our findings clearly highlight that food choices change depending on whether a post-exercise snack is chosen prior to or following the exercise bout. Behavioral economic models of intertemporal choice provide another perspective on this pattern of choice [32]. Many people have inconsistent preferences for outcomes that occur at different time points relative to when the choice is made. While most of the evidence is for choices made over monetary outcomes, there is some evidence about food choice as well. When people make choices for foods that they will receive immediately, time-inconsistent preferences tend to lead them to be more indulgent—choosing less healthy foods—than if receipt of the selected food is in the future. This model predicts that people are more likely to make a healthy choice for their future self than they are for their current self [56]. While this phenomenon may have explained the increased preference for the "unhealthy" option (brownie) following the workout (immediately before the snack) compared to before the workout (on average 65 min before their snack), it fails to explain the increase in the number of participants who declined either snack. A second behavioral economic influence may shed light on increases in both unhealthy and neither food option: projection bias [57]. Projection bias refers to the failure of a decision-maker to correctly predict their preferences for outcomes that occur in the future. This bias is thought to be particularly likely to occur when the individual is making a choice in one state that will be experienced in another state [58]. It is well established that exercise induces changes in an individual's state by, for instance, suppressing appetite-inducing hormones [17–19,41]. Future studies can feature more sophisticated designs that separately identify these various influences on decision-making, such as including non-exercising control activities to disentangle the effects of time-inconsistent preferences from compensatory eating and exercise-induced anorexia and varying the state in which participants make food choices for immediate and future receipt.

5. Conclusions

Overall, our study demonstrates that the choice of a post-exercise snack can be shifted through a simple behavioral intervention, i.e., choosing the snack prior to or after the exercise bout. As such, our results provide support for both an increased preference for compensatory eating as well as an increased degree of exercise-induced anorexia. Our findings have important practical implications for individuals who are attempting to lose or control their weight through exercise, as well as for health professionals providing guidance and support to these individuals. Corroborating previous research [34,58], participants in this study who chose their snack before they exercised—the "before" condition—were more likely to select a healthier snack than participants who chose immediately prior to receipt of the snack (the "after" condition). A simple strategy such as encouraging individuals to make choices about foods that they will eat post-exercise prior to their workout may help those who are attempting to lose weight through diet and exercise pre-commit to healthier foods. Pre-commitment

Nutrients **2018**, *10*, 1941

can prevent individuals from offsetting the gains in exercise-related caloric expenditure through compensatory eating.

Author Contributions: Conceptualization, K.K. and C.R.G.; Methodology, K.K., C.R.G. and A.A.; Formal Analysis, C.R.G. and N.R.; Investigation, N.R., S.E.B., A.C.; Resources, K.K., C.R.G., A.A.; Data Curation, N.R. and C.R.G.; Writing-Original Draft Preparation, K.K., C.R.G., S.E.B., A.C., A.A.; Writing-Review & Editing, K.K. and C.R.G.; Visualization, C.R.G., N.R.; Supervision, K.K. and C.R.G.; Project Administration, K.K. and C.R.G.; Funding Acquisition, K.K. and C.R.G.

Funding: This work was funded by a Food for Health Collaboration Initiative grant by the University of Nebraska awarded to K.K. and C.R.G.

Acknowledgments: The authors would like to thank Brian Smith, Chaise Murphy and Alexandra Martin for their support during data collection.

Conflicts of Interest: The authors declare no conflict of interest.

References

1. Cornelissen, V.A.; Smart, N.A. Exercise training for blood pressure: A systematic review and meta-analysis. *J. Am. Heart Assoc.* **2013**, *2*, e004473. [CrossRef] [PubMed]
2. Goodyear, L.J.; Kahn, B.B. Exercise, glucose transport, and insulin sensitivity. *Annu. Rev. Med.* **1998**, *49*, 235–261. [CrossRef] [PubMed]
3. Moore, S.C.; Lee, I.-M.; Weiderpass, E.; Campbell, P.T.; Sampson, J.N.; Kitahara, C.M.; Keadle, S.K.; Arem, H.; Berrington de Gonzalez, A.; Hartge, P.; et al. Association of leisure-time physical activity with risk of 26 types of cancer in 1.44 million adults. *JAMA Internal Medicine* **2016**, *176*, 816–825. [CrossRef] [PubMed]
4. Cox, E.P.; O'Dwyer, N.; Cook, R.; Vetter, M.; Cheng, H.L.; Rooney, K.; O'Connor, H. Relationship between physical activity and cognitive function in apparently healthy young to middle-aged adults: A systematic review. *J. Sci. Med. Sport.* **2016**, *19*, 616–628. [CrossRef] [PubMed]
5. Strohle, A. Physical activity, exercise, depression and anxiety disorders. *J. Neural. Transm. (Vienna)* **2009**, *116*, 777–784. [CrossRef] [PubMed]
6. Simpson, D.; Callisaya, M.L.; English, C.; Thrift, A.G.; Gall, S.L. Self-Reported Exercise Prevalence and Determinants in the Long Term After Stroke: The North East Melbourne Stroke Incidence Study. *J. Stroke Cerebrovasc. Dis.* **2017**, *26*, 2855–2863. [CrossRef]
7. Gutierrez, P.M.; Davidson, C.L.; Friese, A.H.; Forster, J.E. Physical Activity, Suicide Risk Factors, and Suicidal Ideation in a Veteran Sample. *Suicide Life Threat Behav.* **2016**, *46*, 284–292. [CrossRef] [PubMed]
8. Pompili, M.; Venturini, P.; Campi, S.; Seretti, M.E.; Montebovi, F.; Lamis, D.A.; Serafini, G.; Amore, M.; Girardi, P. Do Stroke Patients have an Increased Risk of Developing Suicidal Ideation or Dying by Suicide? An Overview of the Current Literature. *CNS Neurosci. Ther.* **2012**, *18*, 711–721. [CrossRef] [PubMed]
9. Dombrowski, S.U.; Knittle, K.; Avenell, A.; Araujo-Soares, V.; Sniehotta, F.F. Long term maintenance of weight loss with non-surgical interventions in obese adults: Systematic review and meta-analyses of randomised controlled trials. *BMJ* **2014**, *348*, g2646. [CrossRef]
10. Schubert, M.M.; Desbrow, B.; Sabapathy, S.; Leveritt, M. Acute exercise and subsequent energy intake. A meta-analysis *Appetite* **2013**, *63*, 92–104. [CrossRef] [PubMed]
11. Stein, A.T.; Greathouse, L.J.; Otto, M.W. Eating in response to exercise cues: Role of self-control fatigue, exercise habits, and eating restraint. *Appetite* **2016**, *96*, 56–61. [CrossRef] [PubMed]
12. Bellisle, F. Food choice, appetite and physical activity. *Public Health Nutr.* **1999**, *2*, 357–361. [CrossRef] [PubMed]
13. Moshier, S.J.; Landau, A.J.; Hearon, B.A.; Stein, A.T.; Greathouse, L.; Smits, J.A.J.; Otto, M.W. The Development of a Novel Measure to Assess Motives for Compensatory Eating in Response to Exercise: The CEMQ. *Behav. Med.* **2016**, *42*, 93–104. [CrossRef] [PubMed]
14. Blundell, J.E.; Gibbons, C.; Caudwell, P.; Finlayson, G.; Hopkins, M. Appetite control and energy balance: Impact of exercise. *Obes. Rev.* **2015**, *16*, 67–76. [CrossRef] [PubMed]
15. Thomas, D.M.; Bouchard, C.; Church, T.; Slentz, C.; Kraus, W.E.; Redman, L.M.; Martin, C.K.; Silva, A.M.; Vossen, M.; Westerterp, K. Why do individuals not lose more weight from an exercise intervention at a defined dose? An energy balance analysis. *Obes. Rev.* **2012**, *13*, 835–847. [PubMed]

16. Blundell, J.E.; King, N.A. Physical activity and regulation of food intake: Current evidence. *Med. Sci. Sports Exerc.* **1999**, *31*, S573–S583. [CrossRef] [PubMed]
17. Vatansever-Ozen, S.; Tiryaki-Sonmez, G.; Bugdayci, G.; Ozen, G. The effects of exercise on food intake and hunger: Relationship with acylated ghrelin and leptin. *J. Sports Sci. Med.* **2011**, *10*, 283–291. [PubMed]
18. Ueda, S.Y.; Yoshikawa, T.; Katsura, Y.; Usui, T.; Nakao, H.; Fujimoto, S. Changes in gut hormone levels and negative energy balance during aerobic exercise in obese young males. *The Journal of Endocrinology* **2009**, *201*, 151–159. [CrossRef]
19. Ueda, S.Y.; Yoshikawa, T.; Katsura, Y.; Usui, T.; Fujimoto, S. Comparable effects of moderate intensity exercise on changes in anorectic gut hormone levels and energy intake to high intensity exercise. *J. Endocrinol.* **2009**, *203*, 357–364. [CrossRef]
20. Blundell, J.E.; Stubbs, R.J.; Hughes, D.A.; Whybrow, S.; King, N.A. Cross talk between physical activity and appetite control: Does physical activity stimulate appetite? *Proceedings Nutr. Soc.* **2003**, *62*, 651–661. [CrossRef]
21. Finlayson, G.; Arlotti, A.; Dalton, M.; King, N.; Blundell, J.E. Implicit wanting and explicit liking are markers for trait binge eating. A susceptible phenotype for overeating. *Appetite* **2011**, *57*, 722–728. [CrossRef] [PubMed]
22. Finlayson, G.; Bryant, E.; Blundell, J.E.; King, N.A. Acute compensatory eating following exercise is associated with implicit hedonic wanting for food. *Physiol. Behav.* **2009**, *97*, 62–67. [CrossRef] [PubMed]
23. Alonso-Alonso, M.; Woods, S.C.; Pelchat, M.; Grigson, P.S.; Stice, E.; Farooqi, S. Food reward system: Current perspectives and future research needs. *Nutr. Rev.* **2015**, *73*, 296–307. [CrossRef] [PubMed]
24. Singh, M. Mood, food, and obesity. *Front Psychol.* **2014**, *5*, 925. [CrossRef] [PubMed]
25. Evero, N.; Hackett, L.C.; Clark, R.D.; Phelan, S.; Hagobian, T.A. Aerobic exercise reduces neuronal responses in food reward brain regions. *J. Appl. Physiol. (1985)* **2012**, *112*, 1612–1619. [CrossRef] [PubMed]
26. Crabtree, D.R.; Chambers, E.S.; Hardwick, R.M.; Blannin, A.K. The effects of high-intensity exercise on neural responses to images of food. *Am. J. Clin. Nutr.* **2014**, *99*, 258–267. [CrossRef] [PubMed]
27. McNeil, J.; Cadieux, S.; Finlayson, G.; Blundell, J.E.; Doucet, É. The effects of a single bout of aerobic or resistance exercise on food reward. *Appetite* **2015**, *84*, 264–270. [CrossRef]
28. Panek, L.M.; Jones, K.R.; Temple, J.L. Short term aerobic exercise alters the reinforcing value of food in inactive adults. *Appetite* **2014**, *81*, 320–329. [CrossRef]
29. King, J.A.; Wasse, L.K.; Ewens, J.; Crystallis, K.; Emmanuel, J.; Batterham, R.L.; Stensel, D.J. Differential acylated ghrelin, peptide yy3–36, appetite, and food intake responses to equivalent energy deficits created by exercise and food restriction. *J. Clin. Endocrinol. Metab.* **2011**, *96*, 1114–1121. [CrossRef]
30. Read, D.; van Leeuwen, B. Predicting Hunger: The Effects of Appetite and Delay on Choice. *Organ Behav. Hum. Decis. Process.* **1998**, *76*, 189–205. [CrossRef]
31. Ikeda, S.; Kang, M.I.; Ohtake, F. Hyperbolic discounting, the sign effect, and the body mass index. *J. Health Econ.* **2010**, *29*, 268–284. [CrossRef] [PubMed]
32. Laibson, D. Golden Eggs and Hyperbolic Discounting*. *Q. J. Econ.* **1997**, *112*, 443–478. [CrossRef]
33. Loewenstein, G.; Brennan, T.; Volpp, K.G. Asymmetric paternalism to improve health behaviors. *JAMA* **2007**, *298*, 2415–2417. [CrossRef] [PubMed]
34. Schwartz, J.; Mochon, D.; Wyper, L.; Maroba, J.; Patel, D.; Ariely, D. Healthier by Precommitment. *Psychol. Sci.* **2014**, *25*, 538–546. [CrossRef] [PubMed]
35. Barlow, P.; Reeves, A.; McKee, M.; Galea, G.; Stuckler, D. Unhealthy diets, obesity and time discounting: A systematic literature review and network analysis. *Obes. Rev.* **2016**, *17*, 810–819. [CrossRef] [PubMed]
36. Nichols, A.L.; Maner, J.K. The good-subject effect: Investigating participant demand characteristics. *J. Gen Psychol.* **2008**, *135*, 151–165. [CrossRef] [PubMed]
37. Adams, S.A.; Matthews, C.E.; Ebbeling, C.B.; Moore, C.G.; Cunningham, J.E.; Fulton, J.; Hebert, J.R. The effect of social desirability and social approval on self-reports of physical activity. *Am. J. Epidemiol.* **2005**, *161*, 389–398. [CrossRef]
38. Hebert, J.R.; Clemow, L.; Pbert, L.; Ockene, I.S.; Ockene, J.K. Social desirability bias in dietary self-report may compromise the validity of dietary intake measures. *Int. J. Epidemiol.* **1995**, *24*, 389–398. [CrossRef]
39. United States Department of Agriculture Agricultural Research Service. USDA Food Composition Databases 2018. Available online: https://ndb.nal.usda.gov/ndb/ (accessed on 29 October 2018).

40. King, N.A.; Burley, V.J.; Blundell, J.E. Exercise-induced suppression of appetite: Effects on food intake and implications for energy balance. *Eur. J. Clin. Nutr.* **1994**, *48*, 715–724.

41. Broom, D.R.; Batterham, R.L.; King, J.A.; Stensel, D.J. Influence of resistance and aerobic exercise on hunger, circulating levels of acylated ghrelin, and peptide yy in healthy males. *AJP: Regul. Integr. Comp. Physiol.* **2009**, *296*, R29–R35. [CrossRef]

42. Van Kleef, E.; Shimizu, M.; Wansink, B. Food compensation: Do exercise ads change food intake? *Int. J. Behav. Nutr. Phys. Act.* **2011**, *8*, 6. [CrossRef] [PubMed]

43. Cobb-Clark, D.A.; Kassenboehmer, S.C.; Schurer, S. Healthy habits: The connection between diet, exercise, and locus of control. *J. Econ. Behav. Organ.* **2014**, *98*, 1–28. [CrossRef]

44. Middleman, A.B.; Vazquez, I.; Durant, R.H. Eating patterns, physical activity, and attempts to change weight among adolescents. *J. Adolesc. Health* **1998**, *22*, 37–42. [CrossRef]

45. Privitera, G.J.; Dickinson, E.K. Control your cravings: Self-controlled food choice varies by eating attitudes, sex, and food type among Division I collegiate athletes. *Psychol. Sport Exerc.* **2015**, *19*, 18–22. [CrossRef]

46. Bryan, A.D.; Rocheleau, C.A. Predicting aerobic versus resistance exercise using the theory of planned behavior. *Am. J. Health Behav.* **2002**, *26*, 83–94. [CrossRef] [PubMed]

47. Ainsworth, B.E.; Haskell, W.L.; Herrmann, S.D.; Meckes, N.; Bassett, D.R., Jr.; Tudor-Locke, C. 2011 Compendium of Physical Activities: A second update of codes and MET values. *Med. Sci. Sports Exerc.* **2011**, *43*, 1575–1581. [CrossRef] [PubMed]

48. Schultes, B.; Ernst, B.; Wilms, B.; Thurnheer, M.; Hallschmid, M. Hedonic hunger is increased in severely obese patients and is reduced after gastric bypass surgery. *Am. J. Clin. Nutr.* **2010**, *92*, 277–283. [CrossRef]

49. Barkeling, B.; King, N.A.; Näslund, E.; Blundell, J.E. Characterization of obese individuals who claim to detect no relationship between their eating pattern and sensations of hunger or fullness. *Int. J. Obes. (Lond.)* **2007**, *31*, 435–439. [CrossRef] [PubMed]

50. Schiff, S.; Amodio, P.; Testa, G.; Nardi, M.; Montagnese, S.; Caregaro, L.; di Pellegrino, G.; Sellitto, M. Impulsivity toward food reward is related to BMI: Evidence from intertemporal choice in obese and normal-weight individuals. *Brain Cogn.* **2016**, *110*, 112–119. [CrossRef] [PubMed]

51. Garland, T.; Schutz, H.; Chappell, M.A.; Keeney, B.K.; Meek, T.H.; Copes, L.E.; Acosta, W.; Drenowatz, C.; Maciel, R.C.; van Dijk, G.; et al. The biological control of voluntary exercise, spontaneous physical activity and daily energy expenditure in relation to obesity: Human and rodent perspectives. *J. Exp. Biol.* **2011**, *214*, 206–229. [CrossRef] [PubMed]

52. Sharifi, N.; Mahdavi, R.; Ebrahimi-Mameghani, M. Perceived Barriers to Weight loss Programs for Overweight or Obese Women. *Health Promot. Perspect.* **2013**, *3*, 11–22. [PubMed]

53. Goonasegaran, A.R.; Nabila, F.N.; Shuhada, N.S. Comparison of the effectiveness of body mass index and body fat percentage in defining body composition. *Singapore Med. J.* **2012**, *53*, 403–408. [PubMed]

54. Pomerleau, M.; Imbeault, P.; Parker, T.; Doucet, E. Effects of exercise intensity on food intake and appetite in women. *Am. J. Clin. Nutr.* **2004**, *80*, 1230–1236. [CrossRef] [PubMed]

55. Gastin, P.B.; Cayzer, C.; Robertson, S.; Dwyer, D. Validity of the ActiGraph GT3X+ and BodyMedia SenseWear Armband to estimate energy expenditure during physical activity and sport. *J. Sci. Med. Sport* **2018**, *21*, 291–295. [CrossRef] [PubMed]

56. Scharff, R.L. Obesity and Hyperbolic Discounting: Evidence and Implications. *J. Consum. Policy* **2009**, *32*, 3–21. [CrossRef]

57. Loewenstein, G.; O'Donoghue, T.; Rabin, M. Projection Bias in Predicting Future Utility*. *Q. J. Econ.* **2003**, *118*, 1209–1248. [CrossRef]

58. De-Magistris, T.; Gracia, A. Assessing Projection Bias in Consumers' Food Preferences. *PLoS ONE* **2016**, *11*, e0146308. [CrossRef] [PubMed]

![nutrients logo] *nutrients*

MDPI

Article

The Effectiveness of Nutrition Education for Overweight/Obese Mother with Stunted Children (NEO-MOM) in Reducing the Double Burden of Malnutrition

Trias Mahmudiono [1,2,*], Abdullah Al Mamun [3], Triska Susila Nindya [1], Dini Ririn Andrias [1], Hario Megatsari [4] and Richard R. Rosenkranz [5]

1 Department of Nutrition, Faculty of Public Health, Universitas Airlangga, Surabaya 60115, Indonesia; triskasnindya@yahoo.com (T.S.N.); dien_ra@yahoo.com (D.R.A.)
2 Southeast Asian Ministers of Education Organization Regional Center for Food and Nutrition (SEAMEO RECFON), Pusat Kajian Gizi Regional (PKGR), Universitas Indonesia, Jakarta 10430, Indonesia
3 Institute for Social Science Research, The University of Queensland, Indooroopilly, Queensland 4068, Australia; mamun@sph.uq.edu.au
4 Department of Health Promotion and Behavior Sciences, Faculty of Public Health, Universitas Airlangga, Surabaya 60115, Indonesia; hario.megatsari@gmail.com
5 Department of Food, Nutrition, Dietetics and Health, Kansas State University, Manhattan, KS 66506, USA; ricardo@ksu.edu
* Correspondence: trias-m@fkm.unair.ac.id; Tel.: +62-31-5964808

Received: 15 November 2018; Accepted: 1 December 2018; Published: 4 December 2018

Abstract: (1) Background: In households experiencing the double burden of malnutrition, stunted children are in a better position for growth improvement when parents are able to direct their resources to support nutrition requirements. This study assesses the effectiveness of maternal nutrition education to reduce child stunting. (2) Methods: This was a Randomized Controlled Trial involving pairs of overweight/obese mothers with stunted children aged 2 to 5 years old in urban Indonesia. Methods: Seventy-one mother-child pairs were randomly assigned to receive either a 12-week nutrition education or printed educational materials. Mixed factorial ANOVA was used to test for between-group differences over time in relation to child's height, weight, maternal self-efficacy, outcome expectation, and caloric intake. (3) Results: Across groups, there was a significant effect of time on child height and weight but no significant differences were observed between-groups. Maternal self-efficacy, outcome expectations in providing animal protein for the children (p-value = 0.025) and mother's total caloric intake (p-value = 0.017) favored the intervention group over the comparison group. (4) Conclusions: The behavioral intervention produced strong improvement in maternal self-efficacy to engage in physical activity, eat fruits and vegetables and to provide children with growth-promoting animal protein, but did not significantly influence child height gain.

Keywords: nutrition education; health promotion; behavioral intervention; self-efficacy; stunting; overweight; obesity; physical activity; dual burden of malnutrition

---

1. Introduction

In developing countries, one fourth of children under the age of five fail to grow normally because of a condition known as stunting [1]. Stunting is a condition where the child is shorter than their normal peers as measured using the height-for-age z-score (HAZ) of less than minus two according to the child growth standard from the WHO-Anthro 2005. Child stunting is a public health nutrition problem that hinders the development of future generations. Compared to their non-stunted peer, stunted children have shown to be more susceptible to gain more fat mass than lean mass in a cohort in Brazil [2]. After 7 to 9 years follow up, previously stunted children at 2 years of age were significantly shorter and lighter but their body mass index (BMI) or centralization of body fat was not significantly different from non-stunted South African children [3]. Beyond physiological effects, stunting may limit a child's cognitive abilities and productivity [4]. In light of these damaging consequences, the WHO and its member countries are working to achieve a 40% reduction in child stunting by 2025 through the Scaling-Up Nutrition (SUN) program [4].

Effective community-based interventions must be developed to ameliorate child stunting and support WHO and UNICEF programs to combat child growth problems worldwide. A systematic review of the literature to explore the impact of education and complementary feeding on growth of children under 2 years of age in developing countries showed positive results [5]. In a subgroup analysis of the food secure population, child-feeding education alone yielded a significant improvement in height gain in children under the age of 2 years [5]. A previous study in Bangladesh that assessed a 3-month nutrition education intervention along with complementary feeding showed promising results for height gain [6,7]. The effect of providing complementary food and intensive nutrition education on height gain (cm) in Bangladesh was 0.80 (95% Confidence Interval (CI) = 0.007–1.53) [7]. A systematic review of community-based nutrition education programs revealed significant results when community leaders met with caregivers twice a week in their home, to deliver nutrition education programs and cooking demonstrations [8].

A demographic shift in conjunction with an epidemiological and nutrition transition has created an unusual situation in which both over- and under-nutrition occur within the same population. Child stunting is a persistent feature of this problem, known as the double burden of malnutrition. A study in a Guatemalan population informed our hypothesis that in households suffering from the double burden of malnutrition, stunted children are less likely to experience food insecurity. Results of that study revealed that the prevalence of coexistence of under-nutrition (child stunting) and over-nutrition (maternal overweight/obesity) was highest (22.7%) among those in the middle (third) quintile of socioeconomic status (SES) [9]. The study showed that maternal overweight was positively related to higher economic status while child stunting was negatively associated with higher household economic status. Lack of access to food geared to the fulfilment of dietary energy was influential for the high prevalence of child stunting but not playing the major role to double burden of malnutrition as mothers exceeded their energy consumption. Households that was suffering from double burden of malnutrition did not necessarily lacking in food access in term of energy intake. It is believed that the difference was coming from unequal food distribution in terms among household's member. Larger number of family member or having extended family would increase the change of unmet nutrient requirement among member of the household as it varies across age groups. Top with low level of maternal nutritional literacy the problem of children having less nutrient intake resulted in their failure to grow (stunted) but the adults having excess energy intake ending up with overweight and obesity. This evidence suggests that in this socioeconomic group, relative to the others, in the absence of food insecurity and economic deprivation, modifiable factors such as food distribution and dietary diversity within the household were associated with the double burden of malnutrition. Furthermore, these households appeared to lack the capacity to direct resources properly to prevent child stunting. More specifically, we hypothesize that mothers were unable to make healthy food choices and manage food supply and distribution within the household.

A behavioral intervention was developed to target modifiable behaviors related to the double burden of malnutrition and to equip mothers with skills necessary to overcome these problems. The study was conducted in an urban setting in Indonesia through the Nutrition Education for Overweight/Obese Mother with Stunted Children (NEO-MOM) intervention. Drawing on concepts from Social Cognitive Theory (SCT), participants were prompted to set goals for themselves to improve their dietary habits and child feeding behaviors, with a focus on self-efficacy, nutrition literacy and dietary diversity.

This study was designed to test the hypothesis that for households facing the double burden of malnutrition in urban Indonesia, a behavioral intervention, coupled with a government food supplementation program, would be more effective than standard care combined with print educational materials for improving child outcomes for height and height-for-age z-score, maternal outcomes for weight, waist circumference, BMI, dietary diversity, dietary intake, self-efficacy, outcome expectations and nutrition literacy.

2. Materials and Methods

This randomized controlled trial (RCT) assessed the effectiveness of a behavioral intervention aimed at empowering mothers to address the double burden of malnutrition within the household.

2.1. Theory

The intervention was based on concepts of Bandura's Social Cognitive Theory (SCT) [10], which are briefly mentioned here. Following the constructs of Social Cognitive Theory, we measured the mother's self-efficacy, outcome expectations and knowledge measured as nutrition literacy. We developed eight measures of maternal self-efficacy, according to Bandura's guidelines for constructing self-efficacy scales [11].

2.2. Sample Size and Allocation

Based on the previous study in Bangladesh [7], which found 0.8 effect size of a three months length nutrition education intervention accompanied with complementary feeding and using 90% power, a minimum of 66 total samples were required to detect changes in height gain with two tailed alphas of 0.05. At baseline this study involved 71 eligible samples that was randomly allocated to 35 in the intervention group and 36 in the comparison group/usual care. Details about the methodology and protocols of the study can be found elsewhere [12]. This study did not compare the effect of the intervention group (NEO-MOM group) with a true control group, but with a comparison group that received printed educational materials (PRINT group) plus government supplementation on child stunting and maternal overweight/obesity (see Figure 1).

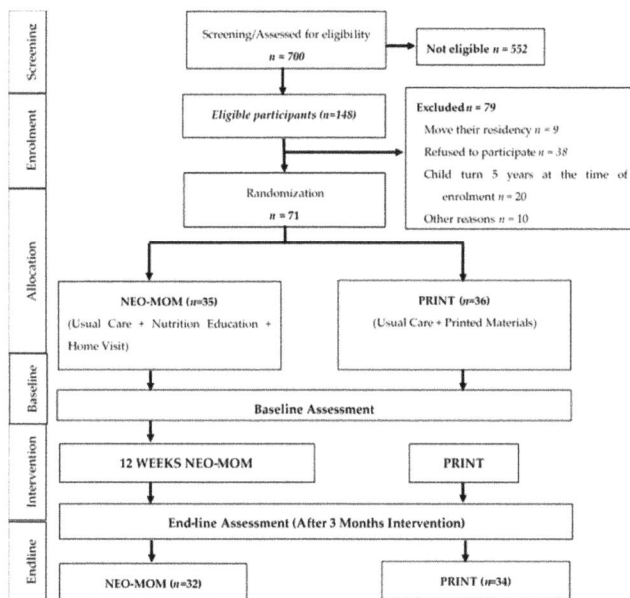

Figure 1. Adapted CONSORT diagram of the Nutrition Education for Overweight/Obese Mother with Stunted Children (NEO-MOM) study.

2.3. Dietary Data

We collected two 24-h dietary food recalls per mother at set times, baseline, and at the end of intervention (three months after baseline). A portion-size guide and food models were provided to parents to assist them in estimating portion sizes. Dietary data was analyzed using NutriSurvey, a software that draws on a database containing nutrition information on typical Indonesian Food. The database is updated yearly by the Department of Nutrition, Universitas Airlangga (UA)–Indonesia. Dietary diversity was calculated following the guidance developed by the Food and Agriculture Organization of the United Nations (FAO) where mothers were asked to recall dietary consumption in the past 24 h. The answer was then aggregated into 12 food groups to create the household's dietary diversity score (HDDS). The 12 food groups were cereals, tubers/roots, vegetables, fruits, fish, meat and poultry, eggs, nuts and seeds, dairy products, spices, oils and fats, and sweets. The dietary diversity score was ranging from 0 to 12.

2.4. Behavioral Measures

We measure maternal self-efficacy based on the Likert-scale ranged from 0 to 100 and covered barriers and tasks for mothers related to being physically active, to eating fruits and vegetables and providing children with animal protein in their meals. The animal protein in question was referring to any source of protein coming from animal-based products that include fish, meat and poultry, eggs, and dairy products. Outcome expectations were measured with a series of questions for the same tasks rated on a scale from 1 to 5, with 1 representing a strong disagreement and 5 representing a strong agreement. Nutrition literacy was measured in three domains: knowledge of macronutrients, skill in household food measures, and skills in grouping food in categories. There were 6-item of close-ended questions in the macronutrient's domain, and also there were 6-item questionnaire in the household food measure domain that was reflecting the common household measurements used in Indonesia. The food groups domain was adapted from the original American "*MyPlate*" to the Indonesian version of MyPlate known in Indonesian language as "*Piring Makanku*" or recently promoted as "*Isi Piringku*".

2.5. Statistical Analysis

For all variables that were normally distributed or transformed to normality, we analyzed the difference in the outcome from control and intervention group using a mixed factorial ANOVA. The within-subjects variables were the outcome variables in this research, and the between-subject variable was the group of intervention (NEO-MOM and PRINT). We used the household food insecurity access scale (HFIAS) score as covariates in the analysis. Furthermore, we conducted the ANCOVA test to see the difference in changes of primary and secondary outcome adjusted for its baseline value and the HFIAS score. For nonparametric statistics, we employed the two related-samples Mann Whitney U test and analyzed the data separately for the NEO-MOM group and the PRINT group with Bonferroni correction. All data analyses were performed in IBM SPSS Statistics 22 (Armonk, NY, USA). The statistical significance for all the tests was set at an alpha level of 0.05.

2.6. Ethical Clearance

This study was approved by The Institutional Review Board (IRB) at Kansas State University (reference number: 7894) as well as approved by the Surabaya City Review Board (Bakesbangpol No: 1366/LIT/2015) in Indonesia. All participants were explained about the study and signed the informed consent following the World Health Organization procedure, this study obtained the Universal Trial Number (UTN) U1111-1175-5834 and also registered in the Australian New Zealand Clinical Trials Registry (ANZCTR) and allocated the registration number: ACTRN12615001243505.

3. Results

3.1. Characteristics of Participants and Groups

Table 1 summarizes characteristics of the participants (children, mother, and household) at baseline in the NEO-MOM and PRINT groups. Almost all of the variables were similar and did not have significant different with the exception of child's weight (p-value = 0.045) and monthly food expenditure (p-value = 0.010). The average of child's weight in the NEO-MOM group was significantly lower (mean = 11.32 kg) compare to the PRINT group (mean = 12.25 kg). Monthly food expenditure in the NEO-MOM group was also around IDR 500,000 lower than those in PRINT group.

Table 1. Participants characteristics at baseline (n = 71).

Variable	NEO-MOM (n = 35)		PRINT (n = 36)		p-Value
	Mean	(SD)	Mean	(SD)	
Child characteristics					
Age (months)	39.57	7.82	40.24	8.11	0.679
Weight (kg)	11.32	1.92	12.25	2.44	0.045 *
Height (cm)	86.84	6.34	89.43	5.34	0.060
Height-for-age Z-score	−2.998	0.85	−2.674	0.57	0.071
Maternal characteristics					
Age (years)	34.09	6.86	31.47	6.76	0.123
Education (years)	7.29	7.75	7.89	3.62	0.417
Weight (kg)	64.59	8.83	67.99	10.15	0.233
Height (cm)	147.43	5.11	148.23	4.67	0.441
Waist circumference (cm)	92.61	8.30	93.91	9.72	0.364
Household characteristics					
Dietary diversity score	7.29	1.86	7.22	1.80	0.737
HFIAS score	8.94	5.75	5.92	5.49	0.061
Monthly Income (IDR)	1,532,857	512,769	2,116,666	1,448,562	0.054
Monthly Food expenditure (IDR)	974,074	343,726	1,217,307	316,418	0.010 *
Maternal physical activity					
Average daily step	3156	2134	2899	2356	0.196

Note. Significance based on α = 0.05; * p-value < 0.05. The analysis is based on Independent t-test.

3.2. Intervention Effect on Outcomes and Mediators

We used the HFIAS score as covariates when testing for an intervention effect in a mixed repeated measure ANOVA.

3.2.1. Child's Outcomes

There were no significant effects observed in the group-by-linear-time trend interaction for any of the child health outcomes, but we observed a significant time effect for child weight (p-value = 0.023) and child height (p-value = 0.001). There were significant increases in weight (p-value < 0.001) and child height (p-value < 0.001) for all groups in a pairwise comparison from baseline to 3-month after baseline evaluation (Figure 2). The mean group difference for child height was 2.47 cm (95% CI = 1.55 to 3.39) and for child weight it was 0.58 kg (95% CI = 0.32 to 0.85). The ANCOVA test showed that the change in child's height and weight was not significantly different between NEO-MOM and PRINT group (p-value = 0.526 and p-value = 0.431 respectively). In terms of child height-for-age z-score (HAZ), the observed improved value was not statistically significant using the related-samples Mann Whitney U test for both the NEO-MOM and PRINT groups (p-value = 0.183 and p-value = 0.051, respectively).

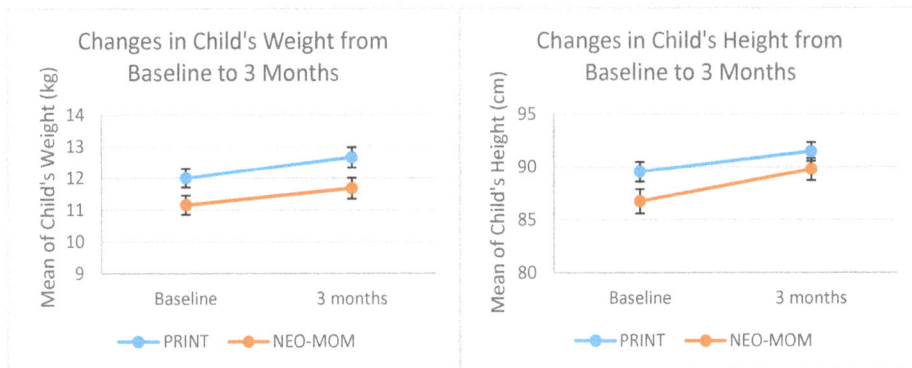

Figure 2. Profile plot of child's weight and height change from baseline to three months evaluation.

3.2.2. Maternal Outcomes

There were no significant effects in the group-by-linear-time trend interaction for maternal anthropometric outcomes such as weight, waist circumference, and the BMI. Similarly, there were no significant mean group differences from baseline to three months after the baseline evaluation for maternal weight (p-value = 0.223), waist circumference (p-value = 0.929), and the BMI (p-value = 0.066). In this analysis, to achieve normal distribution of the data, BMI was transformed using logistic transformation. There was no significant difference in the effect of study condition for any of the maternal outcome measures. As seen in Table 2, after three months intervention, the ANCOVA test revealed that the change in mother's weight and waist circumference was not significantly different between NEO-MOM and PRINT group (p-value = 0.871 and p-value = 0.397 respectively).

3.2.3. Household Dietary Diversity

The household dietary diversity score decreased for both NEO-MOM and PRINT group after the 3-month period. In the non-parametric related samples Mann Whitney U test, the results showed statistical significance at Z = −2,847 (p-value = 0.004) and Z = −3.380 (p-value < 0.001). The decline in the dietary diversity score was steeper in the PRINT group (from 7.29 at baseline to 5.68 after three months) than in the NEO-MOM group (from 7.44 at baseline to 6.50 after the 3-month intervention).

Table 2. The ANCOVA test results on primary outcomes & maternal self-efficacy ($n = 66$).

Variable	F	p-Value	Partial Eta Squared	Adjusted R Squared
Child's outcomes				
Weight (kg)	0.629	0.431	0.010	0.151
Height (cm)	0.407	0.526	0.007	−0.022
Maternal outcomes				
Weight (kg)	0.027	0.871	0.000	−0.028
Waist circumference (mm)	0.726	0.397	0.012	−0.025
BMI [#]	0.115	0.736	0.002	−0.023
Maternal self-efficacy (Barrier)				
Being physically active	2.035	0.159	0.032	0.323
Eating fruit	10.011	0.002 *	0.139	0.404
Eating vegetables	10.238	0.002 *	0.142	0.236
Providing animal protein for kids	5.224	0.026 *	0.078	0.474
Maternal self-efficacy (Task)				
Being physically active	3.922	0.052	0.059	0.276
Eating fruit	4.624	0.036 *	0.070	0.096
Eating vegetables	3.137	0.081	0.048	0.179
Providing animal protein for kids	4.468	0.039 *	0.067	0.081

Note. Significance based on $\alpha = 0.05$; * p-value < 0.05. [#] The analysis is based on log-transformed variable.

3.2.4. Maternal Self-Efficacy

We measured maternal self-efficacy in terms of barriers and task performance of four behaviors: being physically active, eating fruit, eating vegetables, and providing children with animal protein. All measures of maternal self-efficacy were having good internal consistency indicated by having Cronbach alpha > 0.65. The group by time interaction on maternal self-efficacy in dealing with barriers was significant for all measures, with rates of increase being more strongly positive in the NEO-MOM group than in the PRINT group (Table 3). The group by time interaction on maternal self-efficacy barriers for being physically active, eating fruits, eating vegetables, and providing the child with animal protein were all statistically significant (p-value = 0.030, 0.006, 0.002, and 0.042, respectively). As seen in Figure 3 the improvement in maternal self-efficacy barriers to provide their child with animal protein was in the right direction for the NEO-MOM group (from 60.36 at baseline to 67.24 after the 3-month evaluation) in contrast to the PRINT group, which showed a decrease (from 63.47 at baseline to 57.78 after a 3-month evaluation). There was a significant time effect for the maternal self-efficacy barrier of eating vegetables (p-value < 0.001). However, a similar time effect was not observed within subjects for the other three measures. There was no significant result in the between-subjects test. The group by linear time trend interaction effects on maternal self-efficacy to perform certain tasks was only significant for the task of eating fruit (p-value = 0.043) and for the task of providing animal protein for the child (p-value = 0.032) (See Figures 4 and 5). The rate of increase in the maternal self-efficacy in the task of eating fruit was strongly positive in the intervention condition (from 49.16 at baseline to 58.19 after the 3-month evaluation) than the comparison condition, which showed a negative trend (from 50.08 at baseline to 47.66 after a 3-month evaluation).

Table 3. Group means and test of within- and between- subject effect (*n* = 66).

| Outcome/Mediator | Mean (SE) | | | | Test of Within-Subject Effect | | Test of Between-Subject Effect |
| | Intervention (*n* = 32) | | Comparison (*n* = 34) | | | Partial Eta Square (*p*) | |
	Baseline	3 Month	Baseline	3 Month	Time	Time by Group Interaction	Group Difference
Child's outcomes							
Weight (kg)	11.19 (0.31)	11.69 (0.34)	11.98 (0.29)	12.64 (0.33)	0.080 (0.023) *	0.005 (0.578)	0.059 (0.050) *
Height (cm)	86.80 (1.07)	89.90 (1.01)	89.49 (1.04)	91.34 (0.98)	0.173 (0.001) **	0.027 (0.189)	0.034 (0.143)
Maternal outcomes							
Weight (kg)	65.17 (1.49)	64.87 (1.54)	68.35 (1.44)	67.91 (1.49)	0.034 (0.139)	0.001 (0.819)	0.033 (0.146)
Waist circumference (mm)	92.18 (1.60)	91.73 (1.80)	95.08 (1.56)	95.67 (1.74)	0.000 (0.992)	0.008 (0.505)	0.034 (0.141)
BMI [a]	30.13 (0.62)	29.95 (0.59)	31.01 (0.66)	30.63 (0.75)	0.054 (0.062)	0.006 (0.543)	0.022 (0.235)
Maternal self-efficacy (Barrier)							
Being physically active	43.95 (2.61)	49.11 (2.88)	50.55 (2.53)	45.34 (2.79)	0.008 (0.472)	0.072 (0.030) *	0.003 (0.650)
Eating fruit	53.04 (3.80)	62.49 (3.41)	55.66 (3.69)	49.82 (3.30)	0.025 (0.207)	0.115 (0.006) *	0.021 (0.251)
Eating vegetables	35.29 (2.62)	61.13 (3.39)	38.03 (2.54)	48.54 (3.28)	0.376 (<0.001) ***	0.146 (0.002) **	0.029 (0.174)
Providing animal protein for kids	60.36 (3.95)	67.24 (3.31)	63.47 (3.83)	57.78 (3.21)	0.008 (0.491)	0.064 (0.042) *	0.009 (0.448)
Maternal self-efficacy (Task)							
Being physically active	34.7 (3.99)	56.19 (4.17)	30.53 (3.86)	43.62 (4.04)	0.203 (<0.001) ***	0.026 (0.199)	0.047 (0.083)
Eating fruit	49.16 (2.12)	58.19 (3.73)	50.08 (2.09)	47.66 (3.67)	0.028 (0.185)	0.071 (0.034) *	0.031 (0.167)
Eating vegetables	55.17 (2.19)	58.57 (3.48)	55.30 (2.12)	49.98 (3.38)	0.014 (0.355)	0.041 (0.105)	0.028 (0.182)
Providing animal protein for kids	63.5 (1.83)	63.18 (3.42)	64.78 (1.78)	54.20 (3.32)	0.002 (0.717)	0.071 (0.032) *	0.024 (0.217)

Note. Significance based on α = 0.05; *** *p*-value < 0.001, ** *p*-value < 0.01, * *p*-value < 0.05. [a] The analysis is based on log-transformed variable.

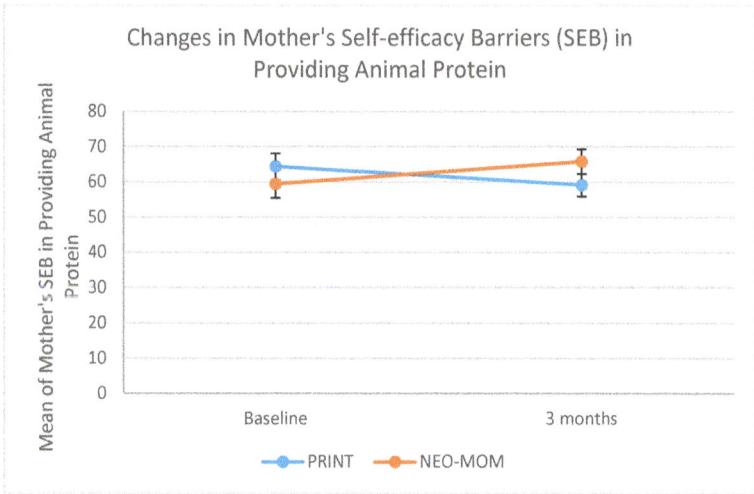

Figure 3. Profile plot of maternal self-efficacy (barriers) for providing animal protein.

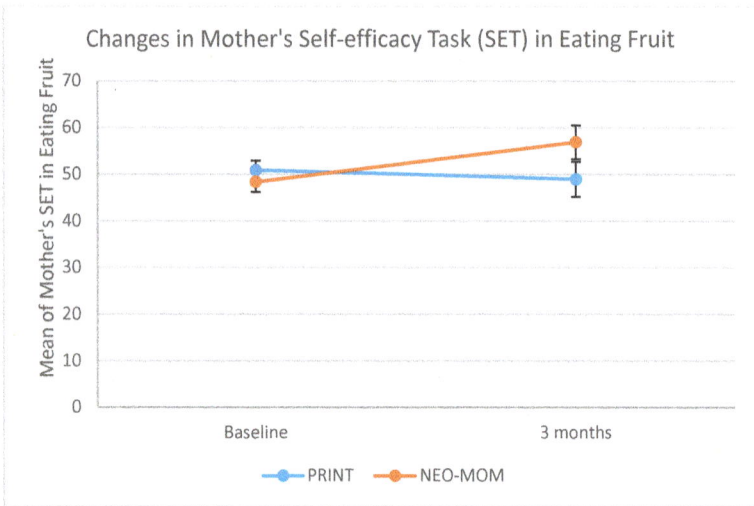

Figure 4. Profile plot of maternal self-efficacy (task) for eating fruit.

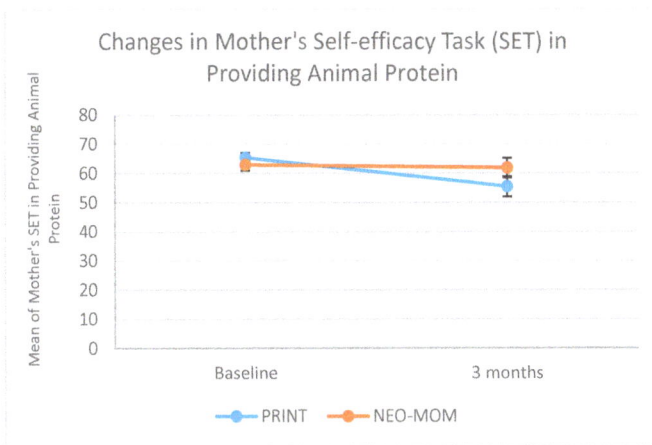

Figure 5. Profile plot of maternal self-efficacy (task) for providing animal protein.

The mother's self-efficacy (task) in providing animal protein for their children, showed a negative trend for both the NEO-MOM and PRINT group with a steeper rate of decline in the latter group (Figure 5). In the NEO-MOM group, maternal self-efficacy (task) for providing animal protein was 63.51 at baseline and was 63.18 after the 3-month evaluation, while in the PRINT group it declined from 64.78 at baseline to 54.20 after 3-months. There was also significant time effect for maternal self-efficacy (task) for being physically active (p-value < 0.001). No significant between-subjects' effects were revealed for any of the maternal self-efficacy tasks. As seen in Table 2, after three months intervention, the ANCOVA test revealed that the change in mother's self-efficacy was shown to be significantly different between NEO-MOM and PRINT group especially in the barriers self-efficacy in eating fruit (p-value = 0.002), barriers self-efficacy in eating vegetables (p-value = 0.002), barriers self-efficacy in serving the children with animal protein (p-value = 0.026), task self-efficacy in eating fruit (p-value = 0.036), task self-efficacy in providing their children with animal protein (p-value = 0.039).

3.2.5. Maternal Outcome Expectation

All measures of maternal outcome expectation were having good internal consistency such as: being physically active (Cronbach alpha = 0.86), eating fruit and vegetables (Cronbach alpha = 0.84), providing animal protein for their kids (Cronbach alpha = 0.89). Measures of maternal outcome expectation were not normally distributed. Even though all measures showed positive increase in the NEO-MOM group relative to the PRINT group (Table 4), there was only one measure, providing animal protein for the child that increased significantly in the NEO-MOM group ($Z = -2.242$; p-value = 0.025). Maternal outcome expectation for providing the children with animal protein was increasing for both NEO-MOM group (from 5.12 at baseline to 5.33 at 3-month evaluation) and PRINT group (from 5.13 at baseline to 5.17 at three months) even though it was not statistically significant.

Table 4. Group means and differences between group means for all outcomes in non-parametric statistics (*n* = 66).

Outcome/Mediator	Intervention				Comparison			
	Mean (SD)		Z	p-Value	Mean (SD)		Z	p-Value
	Baseline (n = 32)	3 Month (n = 32)			Baseline (n = 34)	3 Month (n = 34)		
Child's HAZ	−2.99 (0.85)	−2.85 (0.79)	−1.333	0.183	−2.67 (0.57)	−2.55 (0.61)	−1.951	0.051
Mother's BMI	30.13 (3.52)	29.95 (3.34)	−0.895	0.371	31.01 (3.84)	30.63 (4.36)	−1.646	0.100
Household dietary diversity	7.44 (1.70)	6.50 (2.11)	−2.847	0.004 *	7.29 (1.75)	5.68 (1.61)	−3.380	<0.001 **
Maternal outcome expectation								
Being physically active	5.04 (0.65)	5.21 (0.48)	−1.381	0.167	5.01 (0.72)	4.86 (0.72)	−1.082	0.279
Eating fruit & vegetables	5.06 (0.56)	5.18 (0.49)	−1.312	0.190	4.98 (0.54)	4.97 (0.57)	−0.152	0.879
Providing animal protein for kids	5.12 (0.54)	5.33 (0.48)	−2.242	0.025 *	5.13 (0.59)	5.17 (0.56)	−0.320	0.749
Maternal nutrition literacy								
Macronutrient	2.63 (1.62)	2.47 (1.48)	−0.630	0.529	3.15 (1.50)	3.03 (1.62)	−0.367	0.714
Household food measures	1.78 (1.01)	1.75 (0.76)	−0.339	0.735	1.53 (1.11)	1.32 (0.98)	−0.920	0.358
MyPlate categorization	12.72 (1.78)	13.19 (2.39)	−1.442	0.149	12.79 (1.90)	12.82 (1.96)	−0.039	0.969
Maternal dietary intake								
Energy	1075 (538)	845 (559)	−2.393	0.017 *	1029 (774)	840 (502)	−1.135	0.257
Protein	57.29 (42.11)	44.69 (39.59)	−1.627	0.104	46.50 (38.99)	39.28 (32.47)	−1.027	0.304
Fat	49.95 (51.99)	37.82 (48.80)	−1.646	0.100	56.21 (62.00)	43.29 (48.23)	−0.652	0.514
Carbohydrate	103.18 (74.50)	90.75 (69.79)	−0.926	0.355	91.71 (59.06)	77.71 (39.49)	−1.349	0.177
Iron	6.62 (8.04)	6.58 (11.54)	−1.197	0.231	6.32 (7.69)	5.22 (3.40)	−0.527	0.598
Zinc	3.93 (2.13)	3.18 (2.27)	−1.833	0.067	3.62 (2.51)	3.35 (1.89)	−0.225	0.822
Calcium	140.17 (96.21)	144.31 (200.88)	−1.141	0.254	157.05 (190.80)	159.81 (186.58)	−0.527	0.698
Vitamin A	842 (559)	680 (1466)	−1.496	0.135	982 (2441)	503 (738)	−1.930	0.054
Fiber	6.77 (7.29)	6.75 (10.60)	−1.029	0.303	8.47 (14.88)	5.45 (5.22)	−1.524	0.127

Note. The test was based on repeated measures non-parametric statistic of Mann Whitney U. Note. Significance based on α = 0.05; ** *p*-value < 0.001, * *p*-value < 0.05.

3.2.6. Maternal Nutrition Literacy

All measures of maternal nutrition literacy were not showing good internal consistency with Cronbach alpha <0.65. At baseline, Cronbach alpha obtained for maternal nutrition literacy for macronutrient domain was 0.56, for household food measures was 0.14, and for grouping foods according to Indonesian version of MyPlate was 0.41. After three months intervention, the internal consistency was also not good for all domains of nutrition literacy: macronutrient (Cronbach alpha = 0.53), household food measures (Cronbach alpha = 0.33), grouping foods (Cronbach alpha = 0.64).

Because the data for maternal literacy was not normally distributed, we employed repeated Mann Whitney test for statistical analysis. Results showed no significant effect of the intervention on the mother's nutrition literacy measures (Table 4). The greatest change in the nutrition literacy was observed in the NEO-MOM group for the literacy test for food group categorization test using the Indonesian version of MyPlate called *"Piring Makanku"* ($Z = -1.442$; *p*-value = 0.149).

3.2.7. Maternal Dietary Intake

Based on the results of the Mann Whitney test, almost all measures of maternal dietary intake showed no significant effect with only total energy (caloric) intake that was statistically significant in the NEO-MOM group (Table 4). Mother's total energy intake in the NEO-MOM group decreased from 1075 kcal at baseline to 845 kcal at 3 months ($Z = -2.393$; *p*-value = 0.017) and declined in the PRINT group from 1029 kcal at baseline to 840 at 3 months ($Z = -1.135$; *p*-value = 0.257).

3.2.8. Moderation of Intervention Effects

For all variables that passed the normality assumption we used the household food insecurity access scale (HFIAS) score as the covariate. Our results showed that the HFIAS score was a significant moderator of some of the significant outcome variables measures. Treated as a covariate in the mixed method ANOVA analysis, the HFIAS score revealed significant between-subjects effects for maternal self-efficacy (barriers) in providing children with animal protein ($F(1, 64) = 5.534$, *p*-value = 0.022), self-efficacy (task) in eating fruit ($F(1, 64) = 4.943$, *p*-value = 0.030), self-efficacy (task) in eating vegetables ($F(1, 64) = 4.781$, *p*-value = 0.033), and self-efficacy (task) in providing their child with animal protein ($F(1, 64) = 6.802$, *p*-value = 0.011).

4. Discussion

The goal of this study was to empower and equip overweight or obese mothers to overcome the double burden of malnutrition. Mothers received training on strategies in overcoming the double burden of malnutrition. They were trained through behavioral strategies such as mastery experience, vicarious experience, goal setting, and verbal motivation to achieve better health outcomes for themselves, as well as for their children. The hypothesis of this randomized controlled trial was that applying behavioral intervention strategies based on Bandura's Social Cognitive Theory for three months would be effective in improving child growth to address the issue of child stunting in a household facing the double burden of malnutrition. Results revealed that after a 3-month intervention there was a positive increase in child height in both groups but no catch-up in terms of the HAZ score. The lack of significance in the time and group interactions as well as the test of the between-subjects effect indicated that the significant time effect observed could well be attributable to the natural growth of the child and not the intervention.

Another possible explanation for the observed results might be related to the compliance of the mothers in implementing the Indonesian government food supplementation program for the underweight children. Compared to the previous study in Bangladesh that included food supplementation as a part of the intervention [6,7], in our study we relied on the food supplementation part from the government program and use it as an inclusion criterion for eligible participants. Hence, the amount of effort in ensuring participant's compliance in the use of food supplementation in the Bangladeshi study was likely to be more rigorous than this one. But, because we did not measure the compliance rate related to the consumption of food supplementation from the government, a comparison with previous study in Bangladesh was not possible.

Furthermore, children 2 to 5 years old have passed the optimum time for rapid linear growth that occurs during the first 1000 days of life. However, stunted children in this age range needed an intervention to catch up with their normal peers. In terms of the height-for-age z-score (HAZ) measure, we did not find a significant effect in both the intervention and comparison group. The fact that our intervention targeted children from the age of 24 months to less than 5 years old might have hindered the effect in comparison to one targeting children during the first 1000 days of life or from the womb up to 24 months old when they have the best opportunity for growth improvement. Results of a meta-analysis on interventions aimed at improving child nutritional status revealed that interventions were generally more effective for children under the age of 2 years, and for those who were nutritionally deprived [13].

We saw a significant increase in child's weight overtime for both group (*p*-value = 0.023), and there was borderline significant difference in the weight change from baseline between intervention and comparison group (*p*-value = 0.050). While for child's height, no significant difference was observed between the effect of intervention and comparison condition (*p*-value = 0.143). This supports a previous study in a developing country that reported greater effect size in increasing a child's weight than in improving a child's linear growth [6].

All of the maternal primary outcome measures showed no significant improvement after a 3-month intervention for both groups. For anthropometric outcomes such as weight, waist circumference and BMI it might take longer for the intervention to have significant effects. The length of time employed in the current study was calculated based on the time needed to improve child's height from a previous study and was not based on the maternal anthropometric measures. Therefore, the lack of significant effects in our study for maternal primary outcomes might be due to the insufficient length of the intervention.

However, we saw significant improvement on almost all measures of maternal self-efficacy for both tasks and barriers. There was significant increase for all four maternal self-efficacy (barriers) and two of the maternal self-efficacy (tasks) overtime. This result aligned with a study in Australia that showed maternal self-efficacy was a good predictor for the quality of diet among children aged 3 to 5 years [14]. Mother's child feeding behavior was indirectly related to child vegetable intake through maternal feeding self-efficacy in an Australian population [15]. Results from qualitative studies also support the importance of self-efficacy in influencing healthy eating decision-making [16] and food preparation behavior [17]. Even though we did not directly measure mother's behavior towards child feeding practices, we found significant effect of the intervention for both NEO-MOM and PRINT group in household's dietary diversity as an indirect measure of maternal healthy food choice behavior.

We measured behavior in terms of dietary diversity as an indication of healthy food choice in the household. The results revealed a significant effect but in a negative direction. After the 3-month intervention, the average household dietary diversity score significantly decreased for both the NEO-MOM group and comparison group. This may have been affected by the time the interview was conducted between baseline and the 3-month evaluation. Baseline data were collected at the beginning of the month, while the evaluation data was collected at the end of the month. Results may have been influenced by food budget availability and the fact that most Indonesian people receive their salary at the beginning of the month and may have had more money to spend on food at that time, as compared to the end of the month [18].

The effect of our intervention on maternal dietary intake was significant only for total energy (caloric) intake in the intervention group. These results align with previous RCTs on weight loss that suggest that as the first step in trying to lose weight, women tend to reduce their caloric intake. Other nutrient intake did not show significant results, perhaps because our two times in 24 h dietary recall may not have been sufficient to capture the variability of micronutrient intake as compared to total energy [19].

There was no significant effect revealed in the three domains of maternal nutrition literacy used in this study. These results might be related to the fact that our intervention was designed based on the behavioral strategies that followed the tenets of Social Cognitive Theory. Even though we provided six weeks (over 600 h) of nutrition education classes to improve mothers' knowledge, the content might not necessarily fit with the questions included in the validated nutrition literacy questionnaire [20]. The tools for nutrition literacy developed in the more highly educated settings in the U.S. might be too difficult to for mothers with less education in developing countries such as Indonesia.

Strengths and Limitation

To the best of our knowledge, this study was the first to conduct a randomized controlled trial (RCT) on households experiencing the problem of double burden of malnutrition in the form of the coexistence of stunted children and overweight/obese mother pairs (SCOWT). The strength of the study was a solid methodological approach and RCT design, a small attrition rate of around 5.5% in each group, and the high participation from local community health workers to deliver the intervention that promote adoption of our strategies. However, with limited resources, we could not incorporate supplementary feeding as part of our intervention, but we made use of the Indonesian government's 3-month supplementary feeding program to overcome severely underweight children as inclusion criteria. Therefore, a limitation of the study was the absence of a true control group. Without it, it is impossible to know whether the observed time effect was significantly different from natural growth following the age increase. For this reason, we may have underestimated the effect of print educational materials and ongoing food supplementation by the Indonesian government. Our observed effect could have been higher if we did not use the HFIAS score (measure of food insecurity) as covariates in the analysis. In this study we also assume that the mother was the focal member of the household responsible for purchasing food and its distribution in the family, which might not always be true. Other concerns might arise from the use of relatively high effect size from previous study [7] in calculating the sample size that might be attributable to under power the study. However, we minimize the effect by using fairly substantial bigger power relatively compared to the traditional 80%. The application of our results might be limited to an urban population in a developing country setting.

5. Conclusions

The behavioral-based nutrition education intervention produced strong improvement in maternal self-efficacy to engage in physical activity, eat fruits and vegetables, and to provide children with growth-promoting animal protein, but did not significantly influence child height gain. Although both of our interventions (NEO-MOM and PRINT) allowed significant increases in child growth overtime, no catch-up growth was observed in either group. Relative to the PRINT comparison group, our intervention improved almost all maternal self-efficacy measures, which are viewed as necessary steps for engaging in healthy behaviors. The behavioral intervention in this study was deemed feasible and it had a good retention rate. This study provides a basis for potential strategies to reduce the rate of child stunting in households undergoing double burden of malnutrition.

Author Contributions: T.M. was responsible for overall and/or sectional scientific management, formulating research question, making concept and design of the study, preparation of draft manuscript, doing revisions. A.A.M. was responsible for data cleaning, statistical analysis, providing critiques and revision of the manuscript. T.S.N. lead the data collection, coordinate the participants, setting up ground work for nutrition education session, H.M. develop hands on activities for nutrition education sessions and supervise data collection. D.R.A. was responsible in managing data input and preliminary cleaning of the data and writing the first draft of manuscript in Bahasa Indonesia. R.R.R. responsible for substantial contributions in design and conception of the study, and was involved in data analysis, manuscript preparation, providing critique, revision of the manuscript, and supervises the data collection. T.M. gave final approval of the version to be published; and agreed to be accountable for all aspects of the work in ensuring that questions related to the accuracy or integrity of any part of the work are appropriately investigated and resolved. All authors have given approval of the final manuscript.

Funding: The manuscript writing and the APC was funded by the Ministry of Research, Technology and Higher Education of the Republic of Indonesia through the Program of Academic Recharging for World Class University (PAR-WCU) and partially supported by the Southeast Asian Ministers of Education Organization Regional Center for Food and Nutrition (SEAMEO RECFON), Pusat Kajian Gizi Regional (PKGR), Universitas Indonesia.

Acknowledgments: This study was self-funded study with no support from external agencies. The first author thanks the financial support for finishing the manuscript from the Program of Academic Recharging for World Class University (PAR-WCU) by the Indonesian Ministry of Research, Technology, and Higher Degree (Kemenristekdikti) in conjunction with Universitas Airlangga and the Institute for Social Science Research (ISSR), The University of Queensland, and most importantly Department of Food, Nutrition, Dietetics and Health, Kansas State University.

Conflicts of Interest: We have no conflict of interest to report for this study entitled "The Effectiveness of Nutrition Education for Overweight/Obese Mother with Stunted Children (NEO-MOM) in Reducing Double Burden of Malnutrition".

References

1. Child Growth Standards: Methods and Development. WHO2014. Available online: http://www.who.int/childgrowth/standards/Technical_report.pdf (accessed on 23 March 2018).
2. Martins, P.A.; Hoffman, D.J.; Fernandes, M.T.B.; Nascimento, C.R.; Roberts, S.B.; Sesso, R.; Sawaya, A.L. Stunted children gain less lean body mass and more fat mass than their non-stunted counterparts: A prospective study. *Br. J. Nutr.* **2004**, *92*, 819–825. [CrossRef] [PubMed]
3. Cameron, N.; Wright, M.M.; Griffiths, P.L.; Norris, S.A.; Petiffor, J.M. Stunting at 2 years in relation to body composition at 9 years in African urban children. *Obesity* **2005**, *13*, 131–136. [CrossRef] [PubMed]
4. Improving Child Nutrition. The Achievable Imperative for Global Progress. Available online: https://www.unicef.org/gambia/Improving_Child_Nutrition_-_the_achievable_imperative_for_global_progress.pdf (accessed on 23 March 2018).
5. Lassi, Z.S.; Das, J.K.; Zahid, G.; Imdad, A.; Bhutta, Z.A. Impact of education and provision of complementary feeding on growth and morbidity in children less than 2 years of age in developing countries: A systematic review. *BMC Public Health* **2013**, *13*, S13. [CrossRef] [PubMed]
6. Imdad, A.; Yakoob, M.Y.; Bhutta, Z.A. Impact of maternal education about complementary feeding and provision of complementary foods on child growth in developing countries. *BMC Public Health* **2011**, *11*, S4. [CrossRef] [PubMed]
7. Roy, S.K.; Fuchs, G.J.; Mahmud, Z.; Ara, G.; Islam, S.; Shafique, S.; Akter, S.S.; Chakraborty, B. Intensive nutrition education with or without supplementary feeding improves the nutritional status of moderately-malnourished children in Bangladesh. *J. Health. Popul. Nutr.* **2005**, *23*, 320–330. [PubMed]
8. Majamanda, J.; Maureen, D.; Munkhondia, T.M.; Carrier, J. The Effectiveness of Community-Based Nutrition Education on the Nutrition Status of Under-five Children in Developing Countries. A Systematic Review. *Malawi Med. J.* **2014**, *26*, 115–118. [PubMed]
9. Lee, J.; Houser, R.F.; Must, A.; de Fulladolsa, P.P.; Bermudez, O.I. Socioeconomic disparities and the familial coexistence of child stunting and maternal overweight in Guatemala. *Econ. Hum. Biol.* **2012**, *10*, 232–241. [CrossRef] [PubMed]
10. Bandura, A. Human agency in social cognitive theory. *Am. Psycho.* **1989**, *44*, 1175–1184. [CrossRef]
11. Bandura, A. Guide for constructing self-efficacy scales. In *Self-Efficacy Beliefs of Adolescents*; Urdan, T., Pajares, F., Eds.; Information Age Publishing Inc.: Charlotte, CA, USA, 2006; pp. 307–337. ISBN 978-1593113667.
12. Mahmudiono, T.; Nindya, T.S.; Andrias, D.R.; Megatsari, H.; Rosenkranz, R.R. The effectiveness of nutrition education for overweight/obese mothers with stunted children (NEO-MOM) in reducing the double burden of malnutrition in Indonesia: Study protocol for a randomized controlled trial. *BMC Public Health* **2016**, *8*. [CrossRef] [PubMed]
13. Kristjansson, E.; Francis, D.K.; Liberato, S.; Benkhalti Jandu, M.; Welch, V.; Batal, M.; Greenhalgh, T.; Rader, T.; Noonan, E.; Shea, B.; et al. Food supplementation for improving the physical and psychosocial health of socio-economically disadvantaged children aged three months to five years. *Cochrane Dat. Sys. Rev.* **2015**, *3*, CD009924. [CrossRef] [PubMed]
14. Collins, L.J.; Lacy, K.E.; Campbell, K.J.; McNaughton, S.A. The Predictors of Diet Quality among Australian Children Aged 3.5 Years. *J. Ac. Nutr. Diet.* **2016**, *116*, 1114–1126.e2. [CrossRef] [PubMed]

15. Koh, G.A.; Scott, J.A.; Woodman, R.J.; Kim, S.W.; Daniels, L.A.; Magarey, A.M. Maternal feeding self-efficacy and fruit and vegetable intakes in infants. Results from the SAIDI study. *Appetite* **2014**, *81*, 44–51. [CrossRef] [PubMed]

16. Kilanowski, J.F. Influences on Healthy-Eating Decision Making in Latino Adolescent Children of Migrant and Seasonal Agricultural Workers. *J. Ped. Health Care* **2015**, *30*, 224–230. [CrossRef] [PubMed]

17. Smith, T.M.; Dunton, G.F.; Pinard, C.A.; Yaroch, A.L. Factors influencing food preparation behaviors: Findings from focus groups with Mexican-American mothers in southern California. *Public Health Nutr.* **2016**, *19*, 841–850. [CrossRef] [PubMed]

18. Wolfe, W.S.; Frongillo, E.A.; Valois, P. Understanding the experience of food insecurity by elders suggests ways to improve its measurement. *J. Nutr.* **2003**, *133*, 2762–2769. [CrossRef] [PubMed]

19. Gibson, R. *Principles of Nutritional Assessment*, 2nd ed.; Oxford University Press: New York, NY, USA, 2005; pp. 123–125. ISBN 978-0195171693.

20. Gibbs, H.; Chapman-Novakofski, K. Establishing content validity for the nutrition literacy assessment instrument. *Prev. Chronic Dis.* **2013**, *10*, E109. [CrossRef] [PubMed]

nutrients

MDPI

Article

Association of Sports Participation and Diet with Motor Competence in Austrian Middle School Students

Clemens Drenowatz [1,*] and Klaus Greier [2,3]

[1] Division of Physical Education, University of Education Upper Austria, 4020 Linz, Austria
[2] Division of Physical Education, Private University of Education (KPH-ES), 6422 Stams, Austria; nikolaus.greier@kph-es.at
[3] Department of Sport Science, University of Innsbruck, 6020 Innsbruck, Austria
* Correspondence: clemens.drenowatz@ph-ooe.at; Tel.: +43-732-7470-7426

Received: 9 November 2018; Accepted: 26 November 2018; Published: 29 November 2018

Abstract: Physical activity and diet are important contributors to overall health and development in adolescents. There remains, however, limited research on the combined association of sports participation and dietary pattern on motor competence, which is crucial for an active lifestyle during and beyond adolescence. The present study, therefore, examined the association between sports participation, dietary pattern, and motor competence in 165 middle school students (55% male) between 11 and 14 years of age. Body weight and height were measured, and motor competence was determined via the German motor test during regular Physical Education (PE). Further, participants completed a food frequency questionnaire and reported their engagement in club sports. Of the total sample 20% were overweight/obese and 49% reported participation in club sports, with no differences between boys and girls. Interaction effects of sports participation and dietary pattern on motor competence were limited, but sports participation and healthy diet were independently associated with higher motor performance. Healthy dietary choices, along with participation in club sports, therefore, should be promoted in adolescents in order to facilitate motor development. As adolescence is a crucial time for the establishment of lifelong behaviors, such efforts could facilitate a healthy lifestyle throughout adulthood.

Keywords: physical activity; food intake; adolescents; movement skills; fitness

---

1. Introduction

Low levels of physical activity (PA) and poor dietary choices are considered key health risk factors in youth [1]. In fact, 70% of premature deaths are attributed to behavioral choices begun during adolescent years [2]. Accordingly, adolescence provides a crucial window of opportunity for sustainable health promotion, including sufficient PA and healthy dietary choices [3]. Current guidelines emphasize a minimum of 60 min of moderate-to-vigorous PA [4], along with a diverse dietary intake that includes high consumption of leafy greens, fruits and vegetables, poultry, fish, and dairy while the consumption of fat and sugars should be limited [5]. Nevertheless, sedentary choices during leisure time, along with prolonged sitting times during school time are common in youth [6–8], and many children and adolescents do not meet current dietary recommendations [9,10]. Low PA levels most likely also contributed to a decline in physical fitness and motor competence in children and adolescents [11,12], which is a crucial contributor to a sustainable active lifestyle, as well as overall health and well-being [13,14].

In light of these trends, organized PA, including club sports, provides important opportunities for PA and the establishment of healthy behaviors [15,16]. In fact, a study in Australian adolescents between 10 and 16 years of age showed that they accrued the majority of their total PA during organized PA such as club sports [17]. Participation in sports during childhood and adolescence has also been associated with higher PA levels during adulthood [18,19]. With more than two-thirds of children and adolescents in various European countries participating in club sports, this may also be a viable setting for interventions targeting an active and healthy lifestyle [20]. In addition to beneficial associations with overall PA [21–23], participation in club sports has been associated with increased physical fitness and motor performance [24,25], as well as beneficial socio-emotional outcomes [26] and higher academic achievement [27,28]. A recent study further indicates beneficial associations between club sports participation and food intake [29], but the overall evidence on this relationship remains equivocal [30]. This may, at least partially, be attributed to higher energy needs in more active youth [31]. Accordingly, there appears to be a positive association between physical fitness and energy intake in children and adolescents [31,32]. Nevertheless, higher physical fitness has been associated with healthier dietary choices in adolescents, while this association is less clear in children [33–35].

Limited research, however, is available on the combined association of dietary pattern and sports participation with motor competence. Given the complex interaction between behavioral choices (i.e., diet and sports) and motor development, and the importance of motor competence in the promotion of an active lifestyle [36], such information may help with the refinement of current interventions, as well as the development of new strategies for the promotion of motor development in youth. The present study, therefore, examines the combined and independent association of club sports participation and dietary habits with motor competence in Austrian middle-school students. It was hypothesized that club sports participation, as well as healthy dietary patterns, are positively associated with motor competence. Further, it was hypothesized that the association between dietary pattern and motor competence is more pronounced in adolescents not participating in club sports.

2. Materials and Methods

A convenience sample of nine classes between grades 6 and 8 from middle schools in the Federal State of Tyrol, Austria, were selected for participation, resulting in 172 eligible participants between 11 and 14 years of age. Parents were informed about the nature of the study via mail, and provided written informed consent. Oral assent was obtained from the participants at the time of data collection, which occurred during May and June of 2018. The study was performed according to the ethical standards of the 2008 Declaration of Helsinki, with the study protocol being approved by the Institutional Review Board of the University of Innsbruck, the school board of the Federal State of Tyrol and the principals of the participating schools (approval number: 16/2017).

Anthropometric measurements. Body weight (kg) and height (cm) were measured according to standard procedures during a physical education class, by trained technicians, with participants wearing gym clothes and being barefoot. Body weight was measured to the nearest 0.1 kg with a gauged body scale (SECA® 803, Seca, Hamburg, Germany), and height was measured to the nearest 0.1 cm with a mobile stadiometer (SECA® 217, Seca, Hamburg, Germany). Body mass index (BMI, kg/m²) was calculated and converted to BMI percentiles based on the German reference system [37]. Subsequently, participants were classified as non-overweight/obese or overweight/obese using the 90th percentile.

Motor Competence. Following anthropometric measurements, motor competence was assessed via the German motor test (Deutscher Motorik Test, DMT6-18), which consists of eight test items to evaluate endurance, power, speed, coordination, and agility [38]. Specifically, the DMT6-18 consists of a 20 m sprint, sideways jumping, standing long jump, sit ups, push ups, backwards balancing, a stand and reach test, and a 6 min run. After a standardized 5 min warm-up, participants started the test with the 20 m sprint. Other tests were completed in random order, except for the 6 min run, which was completed at the end of the session. All tests, including practice trials, were administered in accordance with the specifications provided by the test manual [38]. In addition to raw performance scores, age- and sex-specific standardized values were calculated based on a German reference sample, which were used in the statistical analyses. The mean of these standardized scores was further used to calculate an overall motor competence score.

Dietary assessment. Dietary information was obtained via a standardized food frequency questionnaire that has been used previously with Austrian adolescents [39]. The questionnaire was administered by a trained technician during regular class time. Participants reported the frequency (days/week) of the consumption of 42 foods that were subsequently summarized into food groups. Principal component analysis was used to identify dietary patterns. The analysis revealed three factors with an Eigenvalue >1, which explained 55.9% of the total variance of dietary intake. Specifically, factor 1 was characterized by high loadings of meat, fish, bread (white and/or wholemeal), pasta and sweets consumption (meat/carbohydrates (CHO)); factor 2 was characterized by high loadings of milk, cereal, nuts and fruits (milk/cereal); and factor 3 was characterized by high consumption of water and vegetables, as well as low consumption of fast food (FF) and soft drinks (water/low FF).

Club sports participation. Participants also reported whether they participated in club sports, and how much time was spent in organized PA. Due to the limited variability in time spent in club sports, participants were stratified into club sports or no club sports for subsequent analyses.

Statistical Analysis. Descriptive statistics were calculated, and data was checked for normal distribution. Sex differences for club sports participation and weight status were determined via Chi-square tests, while differences in motor competence and dietary pattern were examined via multivariate analysis of variance (MANOVA). Tertiles of diet factor scores were used to examine the interaction and the main effects of club sports participation and dietary pattern on motor competence and body weight via 2×3 MANOVA (sports participation $\times$ diet factor tertiles). In a secondary multivariate analysis of covariance (MANCOVA), sex and BMI percentiles were included as covariates (2×3 MANCOVA).

3. Results

A total of 165 middle school students (55% male) provided complete data on food intake and motor competence. Descriptive data of the sample are shown in Table 1. Boys were significantly older than girls, and accordingly, were taller and heavier. There was, however, no sex difference in BMI percentile and the prevalence of overweight/obesity (girls: 17.1%, boys: 23.2%, $p = 0.358$). Almost half of the sample (48.5%) reported participation in club sports, with no difference in the sports participation rate between boys and girls (boys: 46.2%, girls: 51.4%, $p = 0.506$). Based on the absolute performance scores, boys performed significantly better than girls in the standing long jump and sit ups ($p < 0.001$) while girls performed better than boys in the stand and reach test ($p < 0.001$). Using age- and sex-normalized values, girls displayed higher scores compared to boys in the stand and reach test, sit ups, and 6-min run (Table 1). Boys, on the other hand, performed significantly better than girls at the 20 m sprint when using age- and sex-normalized values. Sex differences in total motor competence were borderline significant ($p = 0.056$), with girls having higher values than boys.

Dietary pattern also differed between boys and girls. Specifically, girls reported less frequent consumption of meat and soft drinks compared to boys, while their consumption of fruits and vegetables was more frequent (Table 2). This resulted in significantly lower scores on the meat/CHO factor ($p = 0.020$) in girls, while their score was higher for the water/low FF factor, compared to boys ($p = 0.006$). No significant sex difference occurred for the milk/cereal factor ($p = 0.225$).

Table 1. Anthropometric characteristics and motor competence for the total sample and separately for girls and boys. Values are mean ± SD.

	Total Sample (N = 165)	Girls (N = 74)	Boys (N = 91)	p *
Age (years)	12.9 ± 1.2	12.6 ± 1.1	13.1 ± 1.2	0.009
Height (cm)	161.3 ± 8.9	158.8 ± 6.8	163.4 ± 9.9	0.001
Weight (kg)	53.8 ± 14.3	51.3 ± 9.9	55.9 ± 16.9	0.043
BMI percentile	59.4 ± 29.4	60.9 ± 27.5	58.2 ± 31.0	0.563
20 m sprint (sec)	3.8 ± 0.4	3.9 ± 0.4	3.7 ± 0.4	0.023
Balance (steps)	38.2 ± 9.1	39.5 ± 7.5	37.2 ± 10.2	0.755
Sideways jump (# in 15 s)	42.1 ± 7.5	41.4 ± 7.0	42.6 ± 7.9	0.554
Stand and reach (cm)	−0.3 ± 9.7	3.8 ± 8.5	−3.6 ± 9.3	0.011
Push-ups (# in 40 s)	15.6 ± 3.9	15.6 ± 3.4	15.6 ± 4.2	0.067
Sit ups (# in 40 s)	23.9 ± 4.6	22.4 ± 3.8	25.1 ± 4.9	0.011
Standing long jump (cm)	171.0 ± 29.7	159.5 ± 25.2	180.3 ± 29.8	0.648
6 min run (m)	997 ± 157	988 ± 119	1005 ± 183	<0.001
Total motor competence	103.3 ± 7.7	104.6 ± 6.9	102.3 ± 8.1	0.056

* *p*-value based on age- and sex-normalized values. # number of repetitions performed. BMI: Body mass index.

Table 2. Frequency of consumption of food groups (days/week). Values are Means ± SD.

	Total Sample	Girls	Boys	p
Meat	2.5 ± 1.2	2.1 ± 1.1	2.7 ± 1.2	0.002
Fish & Eggs	1.8 ± 1.2	1.6 ± 1.0	2.0 ± 1.3	0.051
Milk	2.2 ± 1.0	2.2 ± 1.1	2.3 ± 1.0	0.458
Carbs	2.2 ± 1.1	2.2 ± 1.0	2.2 ± 1.1	0.702
Bread	3.2 ± 0.9	3.3 ± 0.8	3.1 ± 1.0	0.371
Nuts & Seeds	1.2 ± 0.9	1.2 ± 1.0	1.2 ± 0.9	0.662
Fast Food	1.6 ± 0.9	1.5 ± 0.8	1.7 ± 0.9	0.105
Sweets	2.6 ± 1.1	2.5 ± 1.1	2.7 ± 1.2	0.420
Fruits	2.7 ± 1.2	2.9 ± 1.3	2.6 ± 1.1	0.044
Veggie	2.5 ± 1.5	2.9 ± 1.6	2.2 ± 1.4	0.003
Soft Drink	2.3 ± 1.3	2.0 ± 1.1	2.6 ± 1.3	0.001
Water	4.4 ± 1.6	4.6 ± 1.7	4.2 ± 1.5	0.126

Dietary patterns did not differ by sports participation. The association between sports participation and meat/CHO intake, however, reached borderline significance ($p = 0.057$), with lower values occurring in adolescents participating in club sports. There was no difference in the prevalence of overweight/obesity by club sports participation, and across tertiles of diet factors. Also, no interaction effects of diet pattern and club sports participation on BMI percentile were observed.

Motor competence, however, was significantly associated with club sports participation and dietary pattern. Combined associations of diet and sports participation with motor competence, however, were limited. The only significant combined association was observed between club sports participation and milk/cereal consumption on backwards balancing ($p = 0.008$), with a stronger association between dietary pattern and motor competence in adolescents not participating in club sports (Table 3).

Several independent associations of club sports participation and dietary pattern on motor performance were observed. Specifically, club sports participation was associated with better performance on all individual motor competence test items ($p < 0.050$), except for the stand and reach test, resulting in better overall motor competence in participants reporting club sports, compared to those not reporting club sports (Figure 1). Regarding dietary patterns, higher water/low FF consumption was associated with better performance on sideways jumping (p for trend = 0.022), push-ups (p for trend = 0.020), sit ups (p for trend = 0.030), and 6-min run (p for trend = 0.032) (Figure 2). Further, lower scores on the milk/cereal factor were associated with better standing long jump performance and total motor competence (p for trend < 0.050). There was no significant association between meat/CHO consumption and motor competence. All previously reported results remained essentially unchanged after adjusting for sex and BMI percentile.

Table 3. Motor competence by milk/veggie consumption and sports participation. Values are sex-and age-normalized Means [†] ± SD.

	Low Milk/Cereal		Moderate Milk/Cereal		High Milk/Cereal	
	Club Sports	No Club Sports	Club Sports	No Club Sports	Club Sports	No Club Sports
20 m sprint [1]	108.9 ± 13.0	107.5 ± 9.5	106.9 ± 10.7	103.8 ± 14.1	108.1 ± 12.5	99.5 ± 13.0
Balance [1,3,*]	104.6 ± 9.6	107.6 ± 8.2	107.8 ± 8.8	99.4 ± 10.2	108.4 ± 9.8	103.6 ± 11.1
Sideways jump [1,*]	118.0 ± 6.9	111.6 ± 9.7	116.7 ± 7.9	111.7 ± 12.6	117.6 ± 7.9	110.1 ± 12.3
Stand and reach	101.5 ± 12.1	102.0 ± 9.3	101.9 ± 12.0	97.8 ± 12.3	100.3 ± 11.8	98.7 ± 11.1
Push ups [1,*]	111.2 ± 9.6	109.1 ± 7.9	110.1 ± 11.5	104.3 ± 13.3	111.5 ± 10.8	102.2 ± 10.9
Sit ups [1]	97.3 ± 8.7	96.9 ± 7.7	96.0 ± 9.2	92.1 ± 10.4	97.0 ± 7.4	91.8 ± 7.5
Standing long jump [1,*,2]	107.2 ± 9.2	104.2 ± 9.2	105.5 ± 12.6	100.6 ± 13.7	105.1 ± 11.0	97.4 ± 11.7
6 min run [1,*]	97.0 ± 10.1	94.1 ± 11.4	98.3 ± 9.1	89.3 ± 12.2	97.0 ± 9.6	89.8 ± 11.6
Total motor competence [†,1,*,2]	105.8 ± 6.4	104.2 ± 5.8	105.5 ± 6.3	100.0 ± 9.4	105.8 ± 7.1	98.4 ± 7.5

[†] Values >100 indicate above-average performance, while values <100 indicate below-average performance. [1] Significant main effect for club sport ($p < 0.050$; * $p < 0.010$). [2] Significant main effect for milk/veggie consumption (p for trend < 0.050; * $p < 0.010$). [3] Significant interaction effect of milk/veggie consumption and club sport participation ($p < 0.050$; * $p < 0.010$).

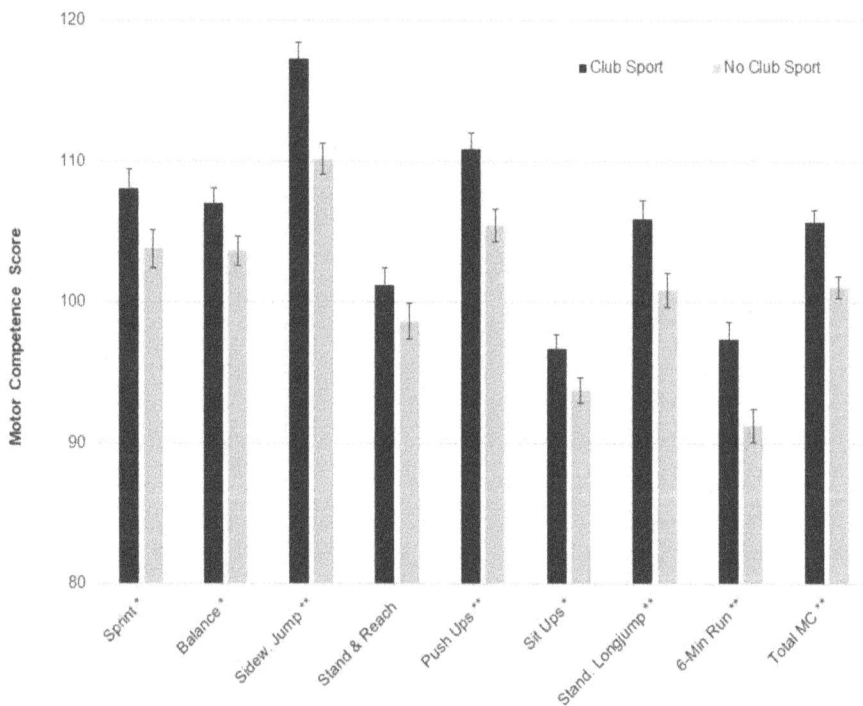

Figure 1. Main effects of club sports participation on motor competence based on 2 × 3 multivariate analysis of variance (MANOVA) (club sports by Water/low fast food (FF)). Values are sex-and age-normalized means with S.E.; * $p < 0.050$ ** $p < 0.010$. MC: motor competence.

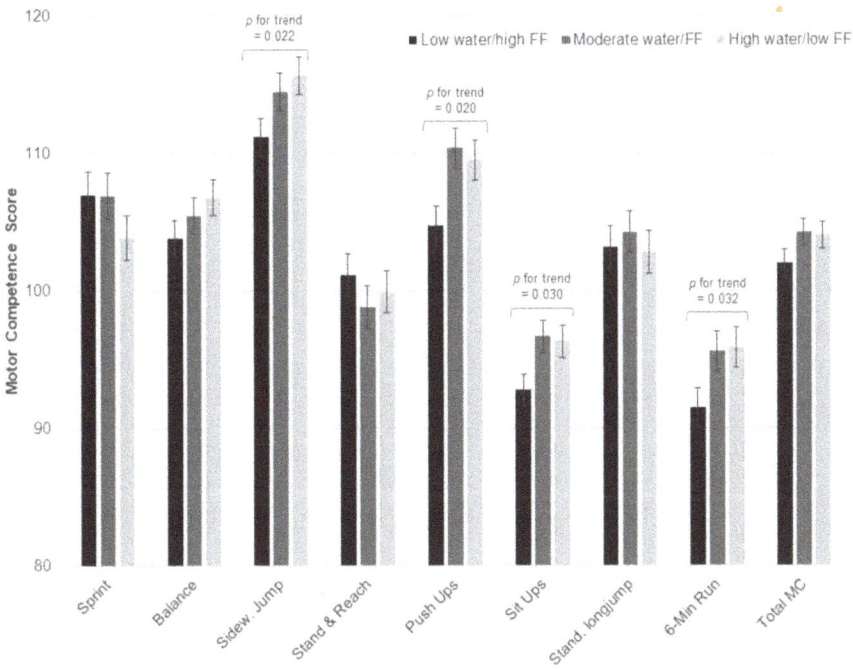

Figure 2. Main effects of club sports participation on motor competence based on 2 × 3 MANOVA (club sports by Water/low FF). Values are sex-and age-normalized means with S.E.

4. Discussion

Even though several studies have examined the association between sports/PA and motor competence [40–43], there exists limited research on the association between motor competence and dietary pattern. To the authors' knowledge, this was also the first study that examined the combined association of dietary pattern and club sports participation with motor competence in Austrian adolescents. While there were no significant associations between dietary patterns and club sports participation, the present study showed independent associations of dietary pattern, as well as club sports participation with motor competence in middle school students. Specifically, club sports participation and healthier dietary choices (i.e., high water and low fast food/soft drink consumption) were associated with higher motor competence. High milk/cereal consumption, on the other hand, was associated with lower motor competence, particularly in participants not reporting club sports. These associations were independent of body weight, and neither club sports participation nor dietary pattern was associated with body weight in the present study.

The positive association between club sports participation and motor competence is consistent with previous research [43–45]. Longitudinal studies further indicate that the strength and directionality of the association between motor competence and club sports participation, as well as PA change over time [46,47]. Particularly during adolescence, motor competence appears to be an important facilitator and precursor for participation in sports clubs, while high PA and sports participation may be a prerequisite for motor development during childhood [45–47]. Accordingly, children and adolescents enter either a positive spiral of high PA, including sports participation and increased motor competence, or a vicious cycle of low motor competence and disengagement from sports. The importance of motor competence for sustainable participation in sports during adolescence may be attributed to an easier acquisition of sport-specific skills in children with higher motor competence. Higher motor competence also enhances self-efficacy, which facilitates participation in PA and sports [46]. In addition to actual motor competence, perceived competence appears to play an important role in the motivation for participation in sports [48], which is an important component to continuous engagement in various forms of PA.

Participation in organized sports has also been associated with other healthy lifestyle choices, including diet [29,49–51]. The present study, however, did not show healthier dietary patterns in club sports participants. Other studies also reported inconsistent results for the association between club sports participation and dietary pattern [29,52,53]. Even though sports participation has been associated with higher intake of fruits and vegetables, club sports participants also have been shown to consume high amounts of fast food and sugar-sweetened beverages [52]. In fact, it has been argued that sports participation during middle school is a strong risk factor for high fast food consumption during high-school years [54]. The higher fast food consumption in sports participants may be attributed to a more irregular eating pattern and a lower amount of meals consumed at home. The higher energy needs of more active adolescents may also contribute to the consumption of more energy dense foods, including fast foods. The results of the present study, nevertheless, indicate beneficial associations of a healthy dietary pattern with motor competence, independent of club sports participation. Specifically, healthier dietary choices were associated with better performance on agility, strength, and endurance tests. Previous studies also reported increased cardiorespiratory fitness with healthier dietary choices, particularly during adolescence [33,35,55]. While this may, at least partially, be attributed to an indirect association between body weight and motor competence [56], there was no difference in motor competence between overweight/obese and normal weight adolescents in the present study. Another possible explanation, therefore, could be that healthier dietary patterns indicate a greater parental support for a healthy lifestyle in general, including the facilitation of PA. Accordingly, parents may facilitate exposure to diverse movement experiences, which would facilitate motor development, even in the absence of club sports participation. Further, dietary pattern has been associated with sedentary choices, which also affect motor development [47]. Specifically, high media time has been associated with poorer diet quality [57,58] as well as low motor competence [47,59].

Sedentary behavior and total PA, rather than participation in club sports, are also crucial correlates of body weight [60]. Neither sports participation nor dietary pattern, however, was associated with body weight in the present study. Results on the association between club sports and body weight have generally been inconsistent [52], which may emphasize the importance of total PA rather than sports in weight management. In fact, it has been argued that a large amount of time in youth sport is spent sedentary or in only light PA [61,62]. Similarly, controversy remains on the relationship between diet and adiposity in youth [30]. Even though several studies showed an inverse association between dietary intake and body weight [63,64], there are also studies that did not show any association [65], or even direct associations between diet an body weight [66,67]. At least partially, this may be attributed to problems in obtaining accurate dietary data, particularly in youth [32]. It should, however, also be considered that more active youth have higher energy needs [31]. Accordingly, children and adolescents with high caloric intake may be able to maintain a healthy body weight as long as they are sufficiently active.

Several limitations of this study, however, need to be considered when interpreting the results. There was no objective measurement of total PA in the present study. Due to the reliance on self-reporting, club sports participation was used as an indicator for PA, as this may be reported more accurately than total PA. Previous research also indicated that participation in club sports is directly associated with total PA [52]. The sample distribution, however, did not allow for a differentiation by the amount of participation in club sports (e.g., hours, days); rather, only participants vs. non-participants could be analyzed. An additional limitation is that participants reported frequency rather than total amount of foods consumed, which provides only limited information on the total energy content of the diet. There is also an inherent risk of selective over- or under-reporting, due to social desirability and social approval with any form of diet report. Participants may have difficulties remembering all the foods, and some foods that they consumed may not have been listed on the questionnaire, which could have affected the reported dietary pattern. The cross-sectional nature of the study further does not allow for the establishment of causal relationships and temporal trends between sports participation, dietary pattern, and motor competence. In addition, the generalizability of the results may be limited, due to the small sample size and homogeneity of the study population. The objective assessment of various components of motor competence with a widely used and previously validated test, on the other hand, should be considered a strength of the study.

5. Conclusions

PA and healthy dietary habits play a crucial role in the development and general health of children and adolescents [1]. The present study also showed that both behaviors are independently associated with motor competence, which is an important component in the facilitation of an active lifestyle [36]. The facilitation of participation in sports, along with the promotion of healthy dietary choices may be particularly important during adolescence, as this is a critical time for the development of future lifestyle choices [3]. Accordingly, coaches, parents, and youth need to be educated on the importance of adequate nutrition in addition to participation in various forms of PA, including sports, for optimal motor development. Even though this may require additional efforts and resources, it may be a worthwhile investment to enhance the health and well-being of future generations.

Author Contributions: C.D. and K.G. conceptualized the study. C.D. analyzed the data and drafted the initial manuscript. K.G. organized data collection and provided critical input on the final version of the manuscript.

Funding: This research received no external funding.

Conflicts of Interest: The authors declare no conflict of interest.

References

1. Office of the Surgeon General (US and National Institutes of Health). *The Surgeon General's Call to Action to Prevent and Decrease Overweight and Obesity*; USDHHS, Office of the Surgeon General: Rockville, MD, USA, 2001.
2. Resnick, M.D.; Catalano, R.F.; Sawyer, S.M.; Viner, R.; Patton, G.C. Seizing the opportunities of adolescent health. *Lancet* **2012**, *379*, 1564–1567. [CrossRef]
3. Sawyer, S.M.; Afifi, R.A.; Bearinger, L.H.; Blakemore, S.J.; Dick, B.; Ezeh, A.C.; Patton, G.C. Adolescence: A foundation for future health. *Lancet* **2012**, *379*, 1630–1640. [CrossRef]
4. US Department of Health and Human Services. *Physical Activity Guidelines for Americans*, 2nd ed.; US Department of Health and Human Services: Washington, DC, USA, 2018. Available online: https://health.gov/paguidelines/second-edition/pdf/Physical_Activity_Guidelines_2nd_edition.pdf (accessed on 14 November 2018).
5. Institute of Medicine. *Nutrition Standards for Foods in Schools: Leading the Way Toward Healthier Youth*; The National Academic Press: Washington, DC, USA, 2007.
6. Kaiser-Jovy, S.; Scheu, A.; Greier, K. Media use, sports activities, and motor fitness in childhood and adolescence. *Wien. Klin. Wochenschr.* **2017**, *129*, 464–471. [CrossRef] [PubMed]
7. Biddle, S.J.; Marshall, S.J.; Gorely, T.; Cameron, N. Temporal and environmental patterns of sedentary and active behaviors during adolescents' leisure time. *Int. J. Behav. Med.* **2009**, *16*, 278–286. [CrossRef] [PubMed]
8. Mathers, M.; Canterford, L.; Olds, T.; Hesketh, K.; Ridley, K.; Wake, M. Electronic media use and adolescent health and well-being: Cross-sectional community study. *Acad. Pediatr.* **2009**, *9*, 307–314. [CrossRef] [PubMed]
9. Mensink, G.B.; Kleiser, C.; Richter, A. [Food consumption of children and adolescents in Germany. Results of the German Health Interview and Examination Survey for Children and Adolescents (KiGGS)]. *Bundesgesundheitsblatt Gesundheitsforschung Gesundheitsschutz* **2007**, *50*, 609–623. [CrossRef] [PubMed]
10. Krebs-Smith, S.M.; Guenther, P.M.; Subar, A.F.; Kirkpatrick, S.I.; Dodd, K.W. Americans do not meet federal dietary recommendations. *J. Nutr.* **2010**, *140*, 1832–1838. [CrossRef] [PubMed]
11. Hardy, L.L.; Barnett, L.; Espinel, P.; Okely, A.D. Thirteen-year trends in child and adolescent fundamental movement skills: 1997–2010. *Med. Sci. Sports Exerc.* **2013**, *45*, 1965–1970. [CrossRef] [PubMed]

12. Tomkinson, G.R.; Olds, T.S. Secular changes in pediatric aerobic fitness test performance: The global picture. *Med. Sport Sci.* **2007**, *50*, 46–66. [PubMed]

13. Lubans, D.R.; Morgan, P.J.; Cliff, D.P.; Barnett, L.M.; Okely, A.D. Fundamental movement skills in children and adolescents: Review of associated health benefits. *Sports Med.* **2010**, *40*, 1019–1035. [CrossRef] [PubMed]

14. Holfelder, B.; Schott, N. Relationship of fundamental movement skills in physical activity in children and adolescents: A systematic review. *Psychol. Sport Exerc.* **2014**, *15*, 382–391. [CrossRef]

15. Geidne, S.; Quennerstedt, M.; Eriksson, C. The youth sports club as a health-promoting setting: An integrative review of research. *Scand. J. Public Health* **2013**, *41*, 269–283. [CrossRef] [PubMed]

16. Badura, P.; Geckova, A.M.; Sigmundova, D.; van Dijk, J.P.; Reijneveld, S.A. When children play, they feel better: Organized activity participation and health in adolescents. *BMC Public Health* **2015**, *15*, 1090. [CrossRef] [PubMed]

17. Hardy, L.L.; O'Hara, B.J.; Rogers, K.; St George, A.; Bauman, A. Contribution of organized and nonorganized activity to children's motor skills and fitness. *J. Sch. Health* **2014**, *84*, 690–696. [CrossRef] [PubMed]

18. Hallal, P.C.; Wells, J.C.; Reichert, F.F.; Anselmi, L.; Victora, C.G. Early determinants of physical activity in adolescence: Prospective birth cohort study. *BMJ* **2006**, *332*, 1002–1007. [CrossRef] [PubMed]

19. Azevedo, M.R.; Araújo, C.L.; Cozzensa da Silva, M.; Hallal, P.C. Tracking of physical activity from adolescence to adulthood: A population-based study. *Rev. Saude Publica* **2007**, *41*, 69–75. [CrossRef] [PubMed]

20. Kokko, S.; Martin, L.; Geidne, S.; Van Hoye, A.; Lane, A.; Meganck, J.; Scheerder, J.; Seghers, J.; Villberg, J.; Kudlacek, M.; et al. Does sports club participation contribute to physical activity among children and adolescents? A comparison across six European countries. *Scand. J. Public Health* **2018**. [CrossRef] [PubMed]

21. Hebert, J.J.; Møller, N.C.; Andersen, L.B.; Wedderkopp, N. Organized Sport Participation Is Associated with Higher Levels of Overall Health-Related Physical Activity in Children (CHAMPS Study-DK). *PLoS ONE* **2015**, *10*, e0134621. [CrossRef] [PubMed]

22. Mäkelä, K.; Kokko, S.; Kannas, L.; Villberg, J.; Vasankari, T.; Heinonen, O.; Savonen, K.; Alanko, L.; Korpelainen, R.; Selänne, H.; et al. Physical activity, screen time, and sleep among youth participating and non-participating in organized sports—The Finnish health promoting Sports Club (FHPSC) Study. *Adv. Phys. Educ.* **2016**, *6*, 378–388. [CrossRef]

23. Marques, A.; Ekelund, U.; Sardinha, L. Associations between organized sports participation and objectively measured physical activity, sedentary time and weight status in youth. *J. Sci. Med. Sport* **2016**, *19*, 154–157. [CrossRef] [PubMed]

24. Hands, B. Changes in motor skill and fitness measures among children with high and low motor competence: A five-year longitudinal study. *J. Sci. Med. Sport* **2008**, *11*, 155–162. [CrossRef] [PubMed]

25. Ortega, F.B.; Ruiz, J.R.; Castillo, M.J.; Sjöström, M. Physical fitness in childhood and adolescence: A powerful marker of health. *Int. J. Obes.* **2008**, *32*, 1–11. [CrossRef] [PubMed]

26. Eime, R.M.; Young, J.A.; Harvey, J.T.; Charity, M.J.; Payne, W.R. A systematic review of the psychological and social benefits of participation in sport for adults: Informing development of a conceptual model of health through sport. *Int. J. Behav. Nutr. Phys. Act.* **2013**, *10*, 135. [CrossRef] [PubMed]

27. Badura, P.; Sigmund, E.; Geckova, A.M.; Sigmundova, D.; Sirucek, J.; van Dijk, J.P.; Reijneveld, S.A. Is Participation in Organized Leisure-Time Activities Associated with School Performance in Adolescence? *PLoS ONE* **2016**, *11*, e0153276. [CrossRef] [PubMed]

28. Morris, D. Actively closing the gap? Social class, organized activities, and academic achievement in high school. *Youth Soc.* **2015**, *47*, 267–290. [CrossRef]

29. Voráčová, J.; Badura, P.; Hamrik, Z.; Holubčíková, J.; Sigmund, E. Unhealthy eating habits and participation in organized leisure-time activities in Czech adolescents. *Eur. J. Pediatr.* **2018**, *177*, 1505–1513. [CrossRef] [PubMed]

30. Moreno, L.A.; Rodríguez, G. Dietary risk factors for development of childhood obesity. *Curr. Opin. Clin. Nutr. Metab. Care* **2007**, *10*, 336–341. [CrossRef] [PubMed]

31. Lahoz-García, N.; García-Hermoso, A.; Milla-Tobarra, M.; Díez-Fernández, A.; Soriano-Cano, A.; Martínez-Vizcaíno, V. Cardiorespiratory Fitness as a Mediator of the Influence of Diet on Obesity in Children. *Nutrients* **2018**, *10*, 358. [CrossRef] [PubMed]

32. Cuenca-García, M.; Ortega, F.B.; Huybrechts, I.; Ruiz, J.R.; González-Gross, M.; Ottevaere, C.; Sjöström, M.; Dìaz, L.E.; Ciarapica, D.; Molnar, D.; et al. Cardiorespiratory fitness and dietary intake in European adolescents: The Healthy Lifestyle in Europe by Nutrition in Adolescence study. *Br. J. Nutr.* **2012**, *107*, 1850–1859. [CrossRef] [PubMed]

33. Zaqout, M.; Vyncke, K.; Moreno, L.A.; De Miguel-Etayo, P.; Lauria, F.; Molnar, D.; Lissner, L.; Hunsberger, M.; Veidebaum, T.; Tornaritis, M.; et al. Determinant factors of physical fitness in European children. *Int. J. Public Health* **2016**, *61*, 573–582. [CrossRef] [PubMed]

34. Howe, A.S.; Skidmore, P.M.; Parnell, W.R.; Wong, J.E.; Lubransky, A.C.; Black, K.E. Cardiorespiratory fitness is positively associated with a healthy dietary pattern in New Zealand adolescents. *Public Health Nutr.* **2016**, *19*, 1279–1287. [CrossRef] [PubMed]

35. Saeedi, P.; Black, K.E.; Haszard, J.J.; Skeaff, S.; Stoner, L.; Davidson, B.; Harrex, H.A.L.; Meredith-Jones, K.; Quigg, R.; Wong, J.E.; et al. Dietary Patterns, Cardiorespiratory and Muscular Fitness in 9–11-Year-Old Children from Dunedin, New Zealand. *Nutrients* **2018**, *10*, 887. [CrossRef] [PubMed]

36. Drenowatz, C. A focus on motor competence as alternative strategy for weight management. *J. Obes. Chron. Dis.* **2017**, *1*, 31–38. [CrossRef]

37. Kromeyer-Hauschild, K.; Wabitsch, M.; Kunze, D.; Geller, F.; Geiß, H.; Hesse, V.; von Hippel, A.; Jaeger, U.; Johnson, D.; Korte, W.; et al. Perzentile für den Body-mass-Index für das Kindes- und Jugendalter unter Heranziehung verschiedener deutscher Stichproben. *Monatsschrift Kinderheilkunde* **2001**, *149*, 807–818. [CrossRef]

38. Bös, K.; Schlenker, L.; Büsch, D.; Lämmle, L.; Müller, H.; Oberger, J.; Seidl, I.; Tittlbach, S. *Deutscher Motorik-Test 6-18 (DMT6-18) [German Motor Abilities Test 6-18 (DMT6-18)]*; Czwalina: Hamburg, Germany, 2009.

39. Greier, K.; Ruedl, G.; Weber, C.; Thöni, G.; Riechelmann, H. Ernährungsverhalten und motorische Leistungsfähigkeit von 10- bis 14-jährigen Jugendlichen. *Ernährung Medizin* **2016**, *31*, 166–171. [CrossRef]

40. D'Hondt, E.; Deforche, B.; Gentier, I.; De Bourdeaudhuij, I.; Vaeyens, R.; Philippaerts, R.; Lenoir, M. A longitudinal analysis of gross motor coordination in overweight and obese children versus normal-weight peers. *Int. J. Obes.* **2013**, *37*, 61–67. [CrossRef] [PubMed]

41. Vandorpe, B.; Vandendriessche, J.; Vaeyens, R.; Pion, J.; Matthys, S.; Lefevre, J.; Philippaerts, R.; Lenoir, M. Relationship between sports participation and the level of motor coordination in childhood: A longitudinal approach. *J. Sci. Med. Sport* **2012**, *15*, 220–225. [CrossRef] [PubMed]

42. Okely, A.D.; Booth, M.L.; Patterson, J.W. Relationship of physical activity to fundamental movement skills among adolescents. *Med. Sci. Sports Exerc.* **2001**, *33*, 1899–1904. [CrossRef] [PubMed]

43. Jaakkola, T.; Kalaja, S.; Liukkonen, J.; Jutila, A.; Virtanen, P.; Watt, A. Relations among physical activity patterns, lifestyle activities, and fundamental movement skills for Finnish students in grade 7. *Percept. Mot. Skills* **2009**, *108*, 97–111. [CrossRef] [PubMed]

44. Campos, C.; Queiroz, D.; Silva, J.; Feitoza, A.; Cattuzzo, M. Relationship between organized physical activity and motor competence in teenagers. *Am. J. Sport Sci. Med.* **2017**, *5*, 82–85. [CrossRef]

45. Fransen, J.; Deprez, D.; Pion, J.; Tallir, I.B.; D'Hondt, E.; Vaeyens, R.; Lenoir, M.; Philippaerts, R.M. Changes in physical fitness and sports participation among children with different levels of motor competence: A 2-year longitudinal study. *Pediatr. Exerc. Sci.* **2014**, *26*, 11–21. [CrossRef] [PubMed]

46. Stodden, D.; Goodway, J.; Langendorfer, S.; Roberton, M.; Rudisill, M.; Garcia, C.; Garcia, L. A developmental perspective on the role of motor skill competence in physical activity: An emergent relationshihp. *Quest* **2008**, *60*, 290–306. [CrossRef]

47. Drenowatz, C.; Greier, K. Cross sectional and longitudinal assocaition between club sports participation, media consumption and motor competence in adolescents. *Scand. J. Med. Sci. Sports* **2019**, in press.

48. Khodaverdi, Z.; Bahram, A.; Stodden, D.; Kazemnejad, A. The relationship between actual motor competence and physical activity in children: Mediating roles of perceived motor competence and health-related physical fitness. *J. Sports Sci.* **2016**, *34*, 1523–1529. [CrossRef] [PubMed]

49. Torstveit, M.K.; Johansen, B.T.; Haugland, S.H.; Stea, T.H. Participation in organized sports is associated with decreased likelihood of unhealthy lifestyle habits in adolescents. *Scand. J. Med. Sci. Sports* **2018**, *28*, 2384–2396. [CrossRef] [PubMed]

50. Taliaferro, L.A.; Rienzo, B.A.; Donovan, K.A. Relationships between youth sport participation and selected health risk behaviors from 1999 to 2007. *J. Sch. Health* **2010**, *80*, 399–410. [CrossRef] [PubMed]

51. Dortch, K.S.; Gay, J.; Springer, A.; Kohl, H.W.; Sharma, S.; Saxton, D.; Wilson, K.; Hoelscher, D. The association between sport participation and dietary behaviors among fourth graders in the school physical activity and nutrition survey, 2009–2010. *Am. J. Health Promot.* **2014**, *29*, 99–106. [CrossRef] [PubMed]

52. Nelson, T.F.; Stovitz, S.D.; Thomas, M.; LaVoi, N.M.; Bauer, K.W.; Neumark-Sztainer, D. Do youth sports prevent pediatric obesity? A systematic review and commentary. *Curr. Sports Med. Rep.* **2011**, *10*, 360–370. [CrossRef] [PubMed]

53. Vella, S.A.; Cliff, D.P.; Okely, A.D.; Scully, M.L.; Morley, B.C. Associations between sports participation, adiposity and obesity-related health behaviors in Australian adolescents. *Int. J. Behav. Nutr. Phys. Act.* **2013**, *10*, 113. [CrossRef] [PubMed]

54. Bauer, K.W.; Larson, N.I.; Nelson, M.C.; Story, M.; Neumark-Sztainer, D. Socio-environmental, personal and behavioural predictors of fast-food intake among adolescents. *Public Health Nutr.* **2009**, *12*, 1767–1774. [CrossRef] [PubMed]

55. Arriscado, D.; Muros, J.J.; Zabala, M.; Dalmau, J.M. Factors associated with low adherence to a Mediterranean diet in healthy children in northern Spain. *Appetite* **2014**, *80*, 28–34. [CrossRef] [PubMed]

56. Greier, K.; Drenowatz, C. Bidirectional association between weight status and motor skills in adolescents: A 4-year longitudinal study. *Wien. Klin. Wochenschr.* **2018**, *130*, 314–320. [CrossRef] [PubMed]

57. Kremers, S.P.; van der Horst, K.; Brug, J. Adolescent screen-viewing behaviour is associated with consumption of sugar-sweetened beverages: The role of habit strength and perceived parental norms. *Appetite* **2007**, *48*, 345–350. [CrossRef] [PubMed]

58. Utter, J.; Scragg, R.; Schaaf, D. Associations between television viewing and consumption of commonly advertised foods among New Zealand children and young adolescents. *Public Health Nutr.* **2006**, *9*, 606–612. [CrossRef] [PubMed]

59. Mota, J.; Ribeiro, J.C.; Carvalho, J.; Santos, M.P.; Martins, J. Television viewing and changes in body mass index and cardiorespiratory fitness over a two-year period in schoolchildren. *Pediatr. Exerc. Sci.* **2010**, *22*, 245–253. [CrossRef] [PubMed]

60. Must, A.; Tybor, D.J. Physical activity and sedentary behavior: A review of longitudinal studies of weight and adiposity in youth. *Int. J. Obes.* **2005**, *29* (Suppl. S2), S84–S96. [CrossRef]

61. Leek, D.; Carlson, J.A.; Cain, K.L.; Henrichon, S.; Rosenberg, D.; Patrick, K.; Sallis, J.F. Physical activity during youth sports practices. *Arch. Pediatr. Adolesc. Med.* **2011**, *165*, 294–299. [CrossRef] [PubMed]

62. Wickel, E.E.; Eisenmann, J.C. Contribution of youth sport to total daily physical activity among 6- to 12-yr-old boys. *Med. Sci. Sports Exerc.* **2007**, *39*, 1493–1500. [CrossRef] [PubMed]

63. Stallmann-Jorgensen, I.S.; Gutin, B.; Hatfield-Laube, J.L.; Humphries, M.C.; Johnson, M.H.; Barbeau, P. General and visceral adiposity in black and white adolescents and their relation with reported physical activity and diet. *Int. J. Obes.* **2007**, *31*, 622–629. [CrossRef] [PubMed]

64. Telford, R.D.; Cunningham, R.B.; Telford, R.M.; Riley, M.; Abhayaratna, W.P. Determinants of childhood adiposity: Evidence from the Australian LOOK study. *PLoS ONE* **2012**, *7*, e50014. [CrossRef] [PubMed]

65. McGloin, A.F.; Livingstone, M.B.; Greene, L.C.; Webb, S.E.; Gibson, J.M.; Jebb, S.A.; Cole, T.J.; Coward, W.A.; Wright, A.; Prentice, A.M. Energy and fat intake in obese and lean children at varying risk of obesity. *Int. J. Obes. Relat. Metab. Disord.* **2002**, *26*, 200–207. [CrossRef] [PubMed]

66. Elliott, S.A.; Truby, H.; Lee, A.; Harper, C.; Abbott, R.A.; Davies, P.S. Associations of body mass index and waist circumference with: Energy intake and percentage energy from macronutrients, in a cohort of Australian children. *Nutr. J.* **2011**, *10*, 58. [CrossRef] [PubMed]

67. Skinner, A.C.; Steiner, M.J.; Perrin, E.M. Self-reported energy intake by age in overweight and healthy-weight children in NHANES, 2001–2008. *Pediatrics* **2012**, *130*, e936–e942. [CrossRef] [PubMed]

nutrients

MDPI

Article

Regular Practice of Moderate Physical Activity by Older Adults Ameliorates Their Anti-Inflammatory Status

Miguel D. Ferrer [1,2,3], Xavier Capó [1,2,3], Miquel Martorell [1,3,4], Carla Busquets-Cortés [1,2,3], Cristina Bouzas [1,2], Sandra Carreres [1,2], David Mateos [1,2], Antoni Sureda [1,2,3], Josep A. Tur [1,2] and Antoni Pons [1,2,*]

[1] Grup de Nutrició Comunitària i Estrès Oxidatiu, Departament de Biologia Fonamental i Ciències de la Salut, Universitat de les Illes Balears, 07122 Palma, Spain; miguel-david.ferrer@uib.es (M.D.F.); xavier.capo@uib.es (X.C.); martorellpons@gmail.com (M.M.); carla_busquets@hotmail.com (C.B.-C.); cristina.bouzas@uib.es (C.B.); sandra.carreres@uib.cat (S.C.); david-mateos@hotmail.com (D.M.); antoni.sureda@uib.es (A.S.); pep.tur@uib.es (J.A.T.)
[2] CIBER: CB12/03/30038 Fisiopatología de la Obesidad la Nutrición, CIBEROBN, Instituto de Salud Carlos III (ISCIII), University of Balearic Islands, 07122 Palma, Spain
[3] Laboratori de Ciències de l'Activitat Física, Departament de Biologia Fonamental i Ciències de la Salut, Universitat de les Illes Balears, 07122 Palma, Spain
[4] Departamento de Nutrición y Dietética, Facultad de Farmacia, Universidad de Concepción, 4070386 Concepción, Chile
* Correspondence: antonipons@uib.es; Tel.: +34-97117-3171

Received: 10 October 2018; Accepted: 14 November 2018; Published: 16 November 2018

Abstract: A chronic inflammatory state is a major characteristic of the aging process, and physical activity is proposed as a key component for healthy aging. Our aim was to evaluate the body composition, hypertension, lipid profile, and inflammatory status of older adults, and these factors' association with physical activity. A total of 116 elderly volunteers were categorized into terciles of quantitative metabolic equivalents of task (MET). Subjects in the first and third terciles were defined as sedentary and active subjects, respectively. Anthropometric and biochemical parameters, hemograms, and inflammatory markers were measured in plasma or peripheral mononuclear blood cells (PBMCs). The active groups exercised more than their sedentary counterparts. The practice of physical activity was accompanied by lower weight, fat mass, body mass index, and diastolic blood pressure when compared to a more sedentary life-style. Physical activity also lowered the haematocrit and total leukocyte, neutrophil, and lymphocyte counts. The practice of exercise induced a decrease in the IL-6 circulating levels and the TLR2 protein levels in PBMCs, while the expression of the anti-inflammatory IL-10 was activated in active subjects. The regular practice of physical activity exerts beneficial effects on body composition and the anti-inflammatory status of old people.

Keywords: immunity; inflammation; metabolism; physical activity

1. Introduction

Ageing is an unavoidable process in all animals and is characterized by progressive accumulation of cell and organ damage, which result in organism malfunction. In the past few centuries, the proportion of elderly people has been continuously increasing worldwide and is projected to reach 19.3% of total population by 2050 [1]. Although ageing has an unavoidable and intrinsic component, it is also importantly modulated by several external factors, such as exposure to chemicals, lifestyle, or nutrition [2]. Therefore, it seems clear that the progression of ageing can be, at least in part, counteracted by the combination of adequate nutritional intake and a healthy lifestyle, the latter including the regular practice of physical activity [3].

Several studies have evidenced the beneficial effects of physical activity on longevity, showing that regular physical activity is associated with a 30% reduction in the risk of mortality in subjects without CV disease [4], which might correspond to one to two years of additional life [5]. On the other hand, physical inactivity causes 6–10% of the burden of several diseases (including coronary heart disease, diabetes, and cancer) and 9% of premature mortality [6]. In fact, regular physical activity prescription for healthy ageing is a key point for chronic disease management and prevention [4,7]. These positive effects of physical activity might be related to greater conservation of lean tissue [8], lower body mass, and less relative body fat [9] in old adults engaging in high levels of physical activity in comparison to individuals who are more sedentary.

Regular physical activity has also been shown to reduce the risk of several diseases, such as cardiovascular disease, stroke, hypertension, type 2 diabetes, osteoporosis, obesity, colon cancer, breast cancer, anxiety, and depression [10]. Most of these diseases are directly or indirectly related to inflammation processes. In this instance, the benefits of exercise on life-span have been related to different cardioprotective mechanisms, including effects on endothelial function and inflammation [4]. Actually, it has been shown that exercise training exerts anti-inflammatory effects in aged or diseased populations [11], and these effects might be mediated by decreases in TNF-α expression in skeletal muscle, among other effectors [12]. Increased risk of chronic diseases has been associated with elevated inflammation markers [13,14], while the practice of physical activity has shown to reduce pro-inflammatory biomarkers such as C-reactive protein [15], TNF-α [16], or interleukin (IL) 6 [17,18].

Old people usually face a situation of chronic low-grade inflammation. It has been stated that inflammatory cytokines are elevated, and anti-inflammatory cytokine concentrations are lowered, in healthy adults over 50 years of age [19]. This behaviour has been associated with redistribution of body fat and concomitant increases in circulating fatty acids that lead to the activation of proinflammatory macrophages [20]. This chronic low-grade inflammatory status, termed inflamm-aging by some authors [14,21], appears to be a major component of the most common age-related diseases, such as diabetes, osteoporosis, cardiovascular diseases, and cancer.

Therefore, the aim of this study was to evaluate the body composition, hypertension, and lipid metabolic profile, as well as the inflammatory status, of older adults, as well as its association with the regular practice of physical activity.

2. Materials and Methods

2.1. Subjects and Study Design

A total of 116 elderly volunteers (58 men aged between 55 and 80 years and 58 women aged between 60 and 80 years) participated in the study. These volunteers were selected from a larger study population conforming the PHYSMED project (with a total of 380 participants), a multi-centre, cross-sectional study aiming at identifying cardiovascular risk factors in sedentary and active elderly subjects. The 116 volunteers included in this study were recruited in social and municipal clubs, health centres, and sport clubs in different villages and cities of Mallorca, Spain. Exclusion criteria included being institutionalized, suffering from a physical or mental illness that would have limited their participation in physical fitness or their ability to respond to questionnaires, chronic alcoholism or drug addiction, and intake of drugs for clinical research over the past year.

The physical activity performed by the participants was measured using the Minnesota Leisure-time Physical Activity Questionnaire previously validated for the Spanish old adult population [22,23]. This questionnaire included a list of physical activities, and the participants were asked about what type of leisure-time physical activities (LTPA) they had performed during the last year. The participants estimated the duration of the activities performed each hour/week by using metabolic equivalents of task (MET, defined as 1 kcal/kg/hour and equivalent to the energy cost of sitting quietly) [24]. The resulting quantitative MET for each participant were categorized into terciles [25], and subjects in the first and the third terciles were selected to take part in this study and defined as sedentary and active subjects, respectively.

The study was conducted according to the guidelines laid down in the Declaration of Helsinki, and all procedures were approved by the Ethics Committee of Clinical Research of the Balearic Islands (CEIC-IB, ref. 1295/09 PI). All the subjects were informed of the purpose and demands of the study before giving their written consent to participate.

Venous blood samples were obtained from the antecubital vein of participants in resting conditions after overnight fasting. The peripheral blood mononuclear cell (PBMC) fraction was purified from whole blood following an adaptation of the method described by Boyum [26] using Ficoll-Paque PLUS reagent (GE Healthcare). This procedure ensures a PBMC purity and viability of $95 \pm 5\%$.

2.2. Anthropometric Characteristics

Anthropometric measurements were performed by well-trained dieticians who underwent identical and rigorous training as an effort to minimize the effects of inter-observer variation. Height was determined using a mobile anthropometer (Seca 213, SECA Deutchland, Hamburg, Germany) to the nearest millimetre, with the subject's head in the Frankfurt plane. Body weight, body fat, and muscle mass were determined using a Segmental Body Composition Analyzer (Tanita BC-418, Tanita, Tokyo, Japan). The participants were weighed in bare feet and light clothes, subtracting 0.6 g for their clothes. Body mass index (BMI) was calculated using the following formula: BMI = mass (kg)/squared height (m).

2.3. Biochemical Parameters and Hemogram

Glucose, triglycerides, total, high-density lipoprotein (HDL), low-density lipoprotein (LDL) and very low-density lipoprotein (VLDL) cholesterol, urea, uric acid, and creatinine were determined by standard procedures using commercial clinical kits in an autoanalyzer system (Technicon DAX System).

Haematological parameters and hemogram were determined in an automatic flow cytometer analyser Technicon H2 (Bayer, Leverkusen, Germany) VCS system. Haemoglobin concentration was determined using Drabkin reagent (Sigma Aldrich, St. Louis, MO, USA).

2.4. Circulating Inflammatory Parameters

IL-6, sCD62L, and sICAM3 plasma levels were determined using individual ELISA kits from Diaclone (Besançon, France). TNFα was determined using the RayBiotech (Norcross, GA, USA) ELISA kit. All procedures were performed following the supplier instructions for use.

MPO activity in plasma was measured by guaiacol oxidation, under identical conditions to those previously described [27].

2.5. mRNA Gene Expression

mRNA expressions were determined by real time-polymerase chain reaction (RT-PCR). For this purpose, mRNA was isolated from PBMC by extraction with Tripure Isolation Reagent (Roche, Basel, Switzerland). cDNA was synthesized from 1 μg total RNA using reverse transcriptase with oligo-dT primers. Quantitative PCR was performed using the LightCycler instrument (Roche Diagnostics, Basel, Switzerland) with DNA-master SYBR Green I. The primers used are shown in Table 1. For all PCRs, there was one cycle at 95 °C for 10 min, followed by 40 cycles at the conditions shown in Table 1.

Table 1. Primers and conditions used in the PCRs.

Gene	Primer		Conditions	
18S	Fw:	5′-ATG TGA AGT CAC TGT GCC AG-3′	95 °C	10 s
	Rv:	5′-GTG TAA TCC GTC TCC ACA GA-3′	60 °C	10 s
			72 °C	12 s
IL-1ra	Fw:	5′-GAA GAT GTG CCT GTC CTG TGT-3′	95 °C	10 s
	Rv:	5′-CGC TCA GGT CAG TGA TGT TAA-3′	60 °C	10 s
			72 °C	15 s
IL10	Fw:	5′-AGA ACC TGA AGA CCC TCA GGC-3′	95 °C	10 s
	Rv:	5′-CCA CGG CCT TGC TCT TGT T-3′	60 °C	10 s
			72 °C	15 s
IL1β	Fw:	5′-GGA CAG GAT ATG GAG CAA CA-3′	95 °C	10 s
	Rv:	5′-GGC AGA CTC AAA TTC CAG CT-3′	58 °C	10 s
			72 °C	15 s
NFκB	Fw:	5′-AAA CAC TGT GAG GAT GGG ATC TG-3′	95 °C	10 s
	Rv:	5′-CGA AGC CGA CCA CCA TGT-3′	60 °C	10 s
			72 °C	15 s
TLR4	Fw:	5′-GGT CAC CTT TTC TTG ATT CCA-3′	95 °C	10 s
	Rv:	5′-TCA GAG GTC CAT CAA ACA TCA C-3′	60 °C	10 s
			72 °C	15 s
TNFα	Fw:	5′-CCC AGG CAG TCA GAT CAT CTT CTC GGA A-3′	94 °C	10 s
	Rv:	5′-CTG GTT ATC TCT CAG CTC CAC GCC ATT-3′	63 °C	10 s
			72 °C	15 s
IL6	Fw:	5′-ACC TGA ACC TTC CAA AGA TGG C-3′	95 °C	10 s
	Rv:	5′-TCA CCA GGC AAG TCT CCT CAT TG-3′	63 °C	10 s
			72 °C	15 s

The relative quantification was performed by standard calculations considering $2^{(-\Delta\Delta Ct)}$. mRNA levels of sedentary males were arbitrarily referred to as 1. The expression of the target gene was normalized with respect to ribosomal 18S.

2.6. Western Blot Analysis in PBMCs

Toll-Like Receptor (TLR) 2 and 4 protein levels were determined in PBMCs by Western blot. Protein extracts were analysed by SDS–polyacrylamide gel electrophoresis (SDS–PAGE). Total protein concentrations were measured by the method of Bradford [28]. 80 µg of total protein was loaded on a 12% agarose gel. Following electrophoresis, samples were transferred onto a nitrocellulose membrane and incubated with a primary monoclonal anti-TLR2 or anti-TLR4 antibody (Santa Cruz Biotechnology, Dallas, TX, USA) and a secondary anti-mouse IgG peroxidase-conjugated antibody. Protein bands were visualized by Immun-Star® Western C® Kit reagent (Bio-Rad Laboratories, Hercules, CA, USA) Western blotting detection systems. The chemiluminiscence signal was captured with a Chemidoc XRS densitometer (Bio-Rad Laboratories) and analyzed with Quantity One-1D Software (Bio-Rad Laboratories).

2.7. Statistical Analysis

Statistical analysis was carried out using a statistical package for social sciences (SPSS 22 for Windows, SPSS Inc., Chicago, IL, USA). Results are expressed as mean ± standard error of the mean (SEM) and $p < 0.05$ was considered statistically significant. The statistical significance of the data was assessed by a two-way analysis of variance (ANOVA). The statistical factors analysed were (S) sex and (E) exercise. When significant effects were found, one-way ANOVA was used to determine the differences between the groups involved.

3. Results

The anthropometric characteristics of the participants are shown in Table 2. The active groups (both male and female) exercised more than their sedentary counterparts, as evidenced by the significantly higher degree of physical activity measured in MET-hours/week. No differences in the degree of physical activity performed were evidenced between males and females. However, males were taller, weighed more, and presented higher fat-free mass and body mass index than females. The practice of regular physical activity was accompanied by significantly lower total weight, fat mass, body mass index, and diastolic blood pressure when compared to a more sedentary life-style.

Table 2. Anthropometric characteristics of the participants.

		Sedentary	Active	ANOVA		
				Sex	Exercise	SxE
Age (years)	Male	64.6 ± 1.1	62.5 ± 0.9	0.000	0.339	0.281
	Female	67.3 ± 1.1	67.4 ± 1.0 *			
Physical activity (MET-hours/week)	Male	40.4 ± 4.4	141 ± 9 #	0.602	0.000	0.071
	Female	48.4 ± 3.3	126 ± 6 #			
Weight (kg)	Male	86.1 ± 1.9	78.2 ± 2.0 #	0.000	0.000	0.875
	Female	69.3 ± 2.2 *	62.0 ± 1.7 *			
Height (cm)	Male	170 ± 1	171 ± 1	0.000	0.808	0.624
	Female	157 ± 1 *	156 ± 1 *			
Fat-free mass (kg)	Male	61.1 ± 1.1	58.8 ± 1.4	0.000	0.142	0.531
	Female	41.8 ± 0.9 *	40.9 ± 0.6 *			
Fat mass (kg)	Male	25.0 ± 1.1	19.4 ± 0.9 #	0.090	0.000	0.765
	Female	27.5 ± 1.6	21.2 ± 1.2 #			
Body Mass Index (kg/m^2)	Male	29.6 ± 0.6	26.8 ± 0.5 #	0.038	0.000	0.874
	Female	28.1 ± 0.8	25.5 ± 0.7 #			
Systolic blood pressure (mm Hg)	Male	141 ± 3	138 ± 4	0.312	0.312	0.796
	Female	138 ± 4	133 ± 3			
Diastolic blood pressure (mm Hg)	Male	84.8 ± 1.4	81.2 ± 1.8	0.099	0.039	0.883
	Female	82.0 ± 2.2	77.8 ± 1.8			

Mean ± SEM. Statistical analysis: two-way ANOVA, $p < 0.05$. (S) effect of sex, (E) effect of exercise, and (SxE) interaction between the two factors. (*) significant differences between sexes; (#) significant differences between sedentary and active groups.

Neither sex or exercise influenced glucose or triglyceride circulating levels (Table 3). Total circulating cholesterol was, however, significantly affected by the sex of the participants, with higher levels observed in women when compared to men. These higher levels of total cholesterol found in females seem attributable to higher HDL-cholesterol levels, which were also higher in women compared to men. HDL-cholesterol circulating levels were also significantly affected by exercise, with those groups of active participants presenting higher levels than their sedentary counterparts. A significant effect of sex was also evidenced in the circulating levels of uric acid and creatinine: females presented significantly lower levels of uric acid and creatinine than their respective male counterparts.

Sex also affected several hemogram parameters, as shown in Table 3. In this instance, females presented significantly lower counts of red blood cells (which resulted in a lower haematocrit and lower haemoglobin content) and eosinophils, as well as a higher platelet count. On the other side, the practice of physical activity lowered the haematocrit, through the significant decrease on total leukocyte count, as well as neutrophil and lymphocyte count.

The circulating levels of key pro-inflammatory proteins are shown in Table 4. No effects of sex were evidenced in any of the circulating pro inflammatory proteins measured. A significant effect of exercise was observed only in IL-6 levels. The practice of exercise induced a decrease in the circulating levels of IL-6, although this decrease was only significant in the group of females.

Table 3. Biochemical parameters and hemogram of the participants.

		Sedentary	Active	ANOVA Sex	ANOVA Exercise	ANOVA SxE
Glucose (mg/dL)	Male	100 ± 2	98.8 ± 2.8	0.636	0.153	0.217
	Female	105 ± 12	87.8 ± 1.8			
Triglycerides (mg/dL)	Male	111 ± 7	100 ± 6	0.360	0.564	0.289
	Female	97.2 ± 6.3	100 ± 6			
Total cholesterol (mg/dL)	Male	197 ± 5	199 + 5	0.016	0.919	0.732
	Female	214 ± 7	211 ± 6			
HDL (mg/dL)	Male	44.7 ± 1.6	51.6 ± 2.2	0.000	0.011	0.520
	Female	57.2 ± 2.0 *	61.3 ± 2.5 *			
LDL (mg/dL)	Male	130 + 5	127 ± 5	0.360	0.317	0.676
	Female	137 ± 6	130 ± 5			
VLDL (mg/dL)	Male	22.3 ± 1.4	19.9 ± 1.3	0.364	0.532	0.285
	Female	19.5 ± 1.3	20.1 ± 1.8			
Urea (mg/dL)	Male	36.3 ± 1.5	36.2 ± 1.7	0.754	0.883	0.909
	Female	36.0 ± 1.5	35.6 ± 1.2			
Uric acid (mg/dL)	Male	6.25 ± 0.21	6.03 ± 0.18	0.000	0.087	0.515
	Female	5.04 ± 0.22 *	4.56 ± 0.20 *			
Creatinine (mg/dL)	Male	0.829 ± 0.018	0.841 ± 0.016	0.000	0.978	0.460
	Female	0.728 ± 0.013 *	0.716 ± 0.018 *			
Red blood cells (10^6/mm^3)	Male	5.03 ± 0.08	4.90 ± 0.07	0.000	0.103	0.860
	Female	4.62 ± 0.06 *	4.52 ± 0.08 *			
Haemoglobin (g/dL)	Male	15.5 ± 0.2	15.3 ± 0.2	0.000	0.056	0.798
	Female	14.2 ± 0.2 *	13.8 ± 0.1 *			
Haematocrit (%)	Male	46.0 ± 0.5	45.0 ± 0.6	0.000	0.048	0.971
	Female	42.1 ± 0.5 *	41.1 ± 0.4 *			
Mean corpuscular volume (fL)	Male	91.7 ± 0.9	91.9 ± 0.6	0.480	0.916	0.857
	Female	91.3 ± 0.7	91.2 ± 0.9			
Platelets (10^3/mm^3)	Male	222 ± 10	214 ± 8	0.018	0.195	0.743
	Female	246 ± 9	232 ± 8			
Leucocytes (10^3/mm^3)	Male	6.39 ± 0.29	5.85 ± 0.21	0.423	0.002	0.244
	Female	6.49 ± 0.32	5.33 ± 0.21 #			
Neutrophils (10^3/mm^3)	Male	3.43 ± 0.21	3.15 ± 0.18	0.074	0.006	0.185
	Female	3.35 ± 0.22	2.56 ± 0.13 #			
Lymphocytes (10^3/mm^3)	Male	2.20 ± 0.13	1.94 ± 0.09	0.084	0.039	0.997
	Female	2.41 ± 0.14	2.15 ± 0.12			
Monocytes (10^3/mm^3)	Male	0.512 ± 0.025	0.526 ± 0.027	0.114	0.306	0.116
	Female	0.511 ± 0.024	0.446 ± 0.022			
Eosinophils (10^3/mm^3)	Male	0.220 ± 0.025	0.196 ± 0.020	0.037	0.088	0.586
	Female	0.187 ± 0.023	0.140 ± 0.013			
Basophils (10^3/mm^3)	Male	0.035 ± 0.004	0.037 ± 0.004	0.721	0.855	0.775
	Female	0.034 ± 0.004	0.034 ± 0.003			

Mean ± SEM. Statistical analysis: two-way ANOVA, $p < 0.05$. (S) effect of sex, (E) effect of exercise, and (SxE) interaction between the two factors. (*) significant differences between sexes; (#) significant differences between sedentary and active groups.

Table 4. Plasma markers of inflammation.

		Sedentary	Active	ANOVA Sex	ANOVA Exercise	ANOVA SxE
IL6 (pg/mL)	Male	3.33 ± 0.27	2.55 ± 0.18	0.599	0.001	0.478
	Female	3.39 ± 0.31	2.19 ± 0.35 #			
TNFα (pg/mL)	Male	26.6 ± 3.0	21.5 ± 1.8	0.480	0.148	0.997
	Female	29.1 ± 3.2	23.9 ± 5.6			
sCD62L (ng/mL)	Male	1507 ± 61	1298 ± 77	0.584	0.132	0.825
	Female	1631 ± 239	1351 ± 110			
sICAM3 (ng/mL)	Male	523 ± 24	496 ± 18	0.549	0.318	0.832
	Female	531 ± 20	514 ± 25			
Myeloperoxidase (μkat/mL)	Male	139 ± 43	179 ± 45	0.155	0.799	0.514
	Female	105 ± 36	87 ± 29			

Mean ± SEM. Statistical analysis: two-way ANOVA, $p < 0.05$. (S) effect of sex, (E) effect of exercise, and (SxE) interaction between the two factors. (#) significant differences between sedentary and active groups.

The inflammatory status of the organism was additionally studied through the gene (Figure 1) and protein (Figure 2) expression of pro- and anti-inflammatory cytokines in PBMC. The regular practice of physical activity influenced the expression of the anti-inflammatory IL-10, with significantly higher expression levels in active males compared to sedentary males. A similar pattern of response, although non-significant, was also observed in females. Similarly, exercise also significantly influenced the gene expression of NF-κB, tending to higher expressions in active participants when compared to their sedentary counterparts. A significant effect of sex was observed regarding TLR4 gene expression: significantly higher expression of this gene was observed in active females when compared to active males.

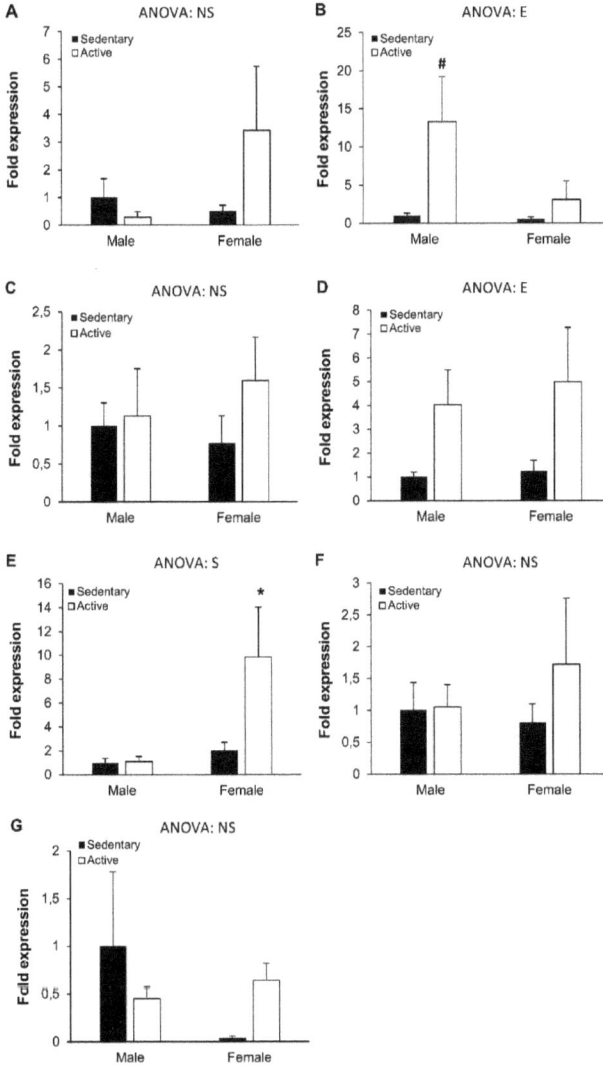

Figure 1. Peripheral blood mononuclear cells gene expression. (**A**) Interleukin(IL)-1 receptor antagonist, (**B**) IL-10, (**C**) IL-1β, (**D**) NF-κB, (**E**) TLR4, (**F**) TNFα, (**G**) IL-6. Results represent mean ± SEM. Statistical analysis: two-way ANOVA, $p < 0.05$. (S) effect of sex, (E) effect of exercise, (NS) non-significant. (*) Significant differences between sexes, (#) significant differences between sedentary and active groups.

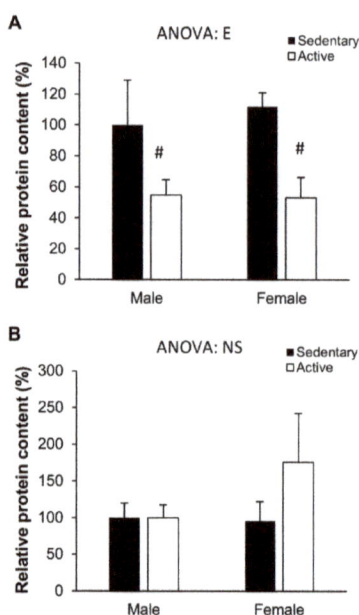

Figure 2. Peripheral blood mononuclear cells protein levels. (**A**) Tol-like receptor (TLR)2 and (**B**) TLR4. Results represent mean ± SEM. Statistical analysis: two-way ANOVA, $p < 0.05$. (E) effect of exercise. (#) Significant differences between sedentary and active groups.

A similar (but non-significant) tendency was also observed in TLR4 protein levels (Figure 2). Finally, TLR2 protein levels were affected by exercise, as evidenced by the significantly lower TLR2 protein levels in the PBMC of active vs sedentary participants (both in males and females).

4. Discussion

Aging has been associated with the functioning of the immune system, and more concretely with inflammatory responses. A chronic, low-grade inflammatory state, called inflamm-aging by some authors [14,21], has been proposed as being responsible for a progressive pro-inflammatory status, which appears to be a major characteristic of the aging process and age-related disease [29]. Therefore, the modulation of the inflammatory status throughout one's life might be an adequate strategy to attain healthy ageing. In this instance, the practice of physical activity has been proposed as a key component of healthy aging [4,30], and the benefits exerted by exercise might be attributable to the acquisition of an anti-inflammatory status. In the present study, we demonstrate that regular practice of physical activity exerts beneficial effects on body composition and the anti-inflammatory status of old people.

The physical activity performed by the participants was measured in the current study using the Minnesota Leisure-time Physical Activity Questionnaire, which had been previously validated for the Spanish old adult population [22,23], and the participants estimated the duration of the activities performed in hour/week by using metabolic equivalents of task (MET, defined as 1 kcal/kg/hour and equivalent to the energy cost of sitting quietly) [24]. The subjects in the first tercile (<82 MET-hours/week) were defined as sedentary, while the subjects in the third tercile (>84 MET-hours/week) were defined as active subjects. As expected by this classification, the active subjects performed around three-fold more physical activity than the sedentary subjects, both in the male and female groups. This regular practice of physical activity translated into a lower weight, a lower fat mass content, and a lower BMI. These results are in accordance with previous reports that evidenced that the regular practice of physical activity by old people reduces fat mass and BMI and increases fat-free mass [31,32]. These effects on body composition were also accompanied by reduced diastolic blood pressure in the physically active subjects, as has been extensively reported in subjects performing aerobic exercise [33]. The effects of sex on the body composition were also evidenced, as women presented lower fat-free mass and higher fat mass than their male counterparts.

The practice of physical activity also had positive effects on the levels of HDL-cholesterol, which is in accordance with previous reports [34,35]. Lipid parameters were also affected by sex, with women presenting higher total cholesterol circulating levels, which were attributable to higher HDL-cholesterol. These results are in accordance with previous reports on different European populations showing that HDL-cholesterol circulating levels are higher in women than in men [36,37].

The active subjects presented a lower haematocrit than their sedentary counterparts, but similar values of red blood cells counts, haemoglobin levels, and mean corpuscular erythrocyte volume. The lower haematocrit was accompanied by a certain degree of leucopoenia. Although decreases in the number of circulant erythrocytes have been reported in response to acute bouts of physical activity [38], these changes are not always found in well-trained subjects [39], and even increases in the haematocrit and erythrocyte number have been reported in both amateur and professional sportsmen after maximal and submaximal tests and a cycling stage [40]. The leucopoenia found in the active participants of the current study was explained by lower counts of both neutrophils and lymphocytes and is in accordance with previous studies reporting a certain degree of leucopoenia in response to the regular practice of physical activity, which in turn is interpreted as part of an anti-inflammatory response [41].

The systemic inflammatory status of the participants in the study was evaluated through the measurement of circulating pro- and anti-inflammatory proteins and gene and protein expression of different cytokines in PBMCs. A decrease in the circulating levels of IL6 was observed in the active groups. Although IL6 can also exert anti-inflammatory activity (after an acute bout of exercise, IL-6 may induce the anti-inflammatory cytokines IL-10 and IL-1ra [42]), the presence of chronic circulating concentrations of IL6 can induce an acute phase immune response [43], and the regular practice of physical activity induces lower basal concentrations of IL-6 when compared to a sedentary lifestyle [44]. The plasma concentration of this interleukin has been associated with lower muscle mass [45] and higher adiposity [46], although in the present study we have not evidenced differences in fat-free mass. Although a reduction in IL6 levels was observed in the group of active volunteers, no changes in circulating TNF-α were evidenced. Gene expression of pro-inflammatory cytokines such as IL1β or TNF-α or the pro-inflammatory receptor TLR4 in PBMCs were not significantly affected by the practice of regular exercise. These results are in accordance with previous studies reporting that IL1β does not respond to different degrees of exercise, including low intensity aerobic exercise, high intensity aerobic exercise, or a combination of high intensity aerobic and resistance exercise [34]. Although no effects of physical activity were observed either on the gene expression of IL1ra, a significant activation of the anti-inflammatory cytokine IL10 gene expression was observed. Higher levels of IL10 in response to physical activity have been previously reported [34,35], and these increases have been actually related to a decrease in fat mass. In fact, the chronic inflammatory state has been related to the adiposity, and the influence of physical activity in body composition may therefore influence the inflammatory state [31,45,47,48]. The fact that anti-inflammatory cytokine concentrations are lowered in healthy adults over 50 years of age has also been associated with redistribution of body fat [20]. Our current results (lower fat mass and higher IL10 expression in active subjects) are in accordance with previous data and reinforce the anti-inflammatory effect of the regular practice of physical activity in old people.

A significant overexpression of the transcription factor NFκB was also observed in the physically active groups. This nuclear factor can be activated through the action of pro-inflammatory cytokines (such as TNFα), but it can also be activated by ROS and/or RNS [49]. Once activated, the nuclear factor migrates to the nucleus and may induce the expression of a wide variety of genes, including inflammatory cytokines such as TNFα, IL-6, and IL-1β [50,51], but also antioxidant enzymes such as superoxide dismutase and nitric oxide synthase [52,53]. As the PBMC gene expression profile shows no evidence of a pro-inflammatory phenotype, we might interpret the activation of NFκB through the ROS/RNS route rather than a proinflammatory response. In this instance, the regular practice of physical activity exposes the organism to a sustained and continuous production of low levels of ROS, and these low levels of ROS have been shown to act as second messengers leading an antioxidant and anti-inflammatory response through the activation of NFκB and other genes [52,53].

The protein levels of the inflammation-related receptors TLR2 and TLR4 were also measured. While no effects were observed regarding TLR4 levels, a decrease in both sexes in the protein levels of TLR2 was observed. Activation of the TLR4 signalling pathway stimulates an increase in pro-inflammatory cytokines such as TNFα, IL1β, or IL6, and it has been previously described that physical activity may downregulate TLR4 expression in the immune cells [54,55], together with downstream cytokines such as TNFα [16], IL1β [34], and IL6 [18]. However, a recent study reported that neither TLR4 nor TNFα responded to resistance training with or without weight loss [32], which is in accordance with our own results. Although TLR4 is usually more sensitive TLR in response to physical activity, we observed a down-regulation of TLR2 in active subjects but not of TLR4. TLR2 is another member of the TLRs family that is also involved in the cell response to immune stimuli, and shares with TLR4 its downstream signalling cascade. A recent systematic review showed that chronic exercise has anti-inflammatory effects on the organism through the downregulation of both TLR2 and TLR4 at the protein and gene expression levels [56], which is in accordance with our results.

Taken together, our results show that the regular practice of physical activity by older adults ameliorates their anthropometric characteristics by reducing their weight, fat mass, and body mass index. This effect in their body composition is accompanied by a healthier status, with lower diastolic blood pressure and higher levels of circulating HDL-cholesterol. The changes in the fat body composition and lipid profile might be responsible for the observed attenuation of pro-inflammatory parameters, such as the reduced count of lymphocytes and neutrophils, reduced IL6 circulating levels, and the changes in the expression of pro- and anti-inflammatory proteins in PMBCs. In conclusion, the regular practice of physical activity (> 84 MET-hours/week) by older adults ameliorates their anti-inflammatory status.

Author Contributions: Data curation, C.B., S.C., and D.M.; Formal analysis, C.B., S.C., and D.M.; Investigation, M.D.F., X.C., M.M., C.B.-C., and A.S.; Methodology, A.S. and A.P.; Project administration, J.A.T. and A.P.; Resources, J.A.T. and A.P.; Supervision, J.A.T.; Writing—original draft, M.D.F. and A.P.; Writing—review & editing, X.C., M.M., C.B.-C., C.B., S.C., D.M., A.S., and J.A.T.

Funding: M.D.F. receives funding from Laboratoris Sanifit for an independent project not related to the content of this manuscript. The founding sponsors had no role in the design of the study; in the collection, analyses, or interpretation of data; in the writing of the manuscript; or in the decision to publish the results.

Acknowledgments: This work was supported by the Spanish Ministry of Economy and Competitiveness, Instituto de Salud Carlos III (Projects 11/01791 and 14/00636, Red Predimed-RETIC RD06/0045/1004, and CIBEROBN CB12/03/30038); Government of the Balearic Islands (grant numbers 35/2011 and AAEE26/2017), and European Union FEDER funds. The funders had no role in study design, data collection and analysis, decision to publish, or preparation of the manuscript.

Conflicts of Interest: The authors declare no conflict of interests.

References

1. Tanaka, H.; Seals, D.R. Endurance exercise performance in Masters athletes: Age-associated changes and underlying physiological mechanisms. *J. Physiol.* **2008**, *586*, 55–63. [CrossRef] [PubMed]
2. Tosato, M.; Zamboni, V.; Ferrini, A.; Cesari, M. The aging process and potential interventions to extend life expectancy. *Clin. Interv. Aging* **2007**, *2*, 401–412. [PubMed]
3. Boirie, Y. Physiopathological mechanism of sarcopenia. *J. Nutr. Health Aging* **2009**, *13*, 717–723. [CrossRef] [PubMed]
4. Gremeaux, V.; Gayda, M.; Lepers, R.; Sosner, P.; Juneau, M.; Nigam, A. Exercise and longevity. *Maturitas* **2012**, *73*, 312–317. [CrossRef] [PubMed]

5. Franco, O.H.; de Laet, C.; Peeters, A.; Jonker, J.; Mackenbach, J.; Nusselder, W. Effects of physical activity on life expectancy with cardiovascular disease. *Arch. Intern. Med.* **2005**, *165*, 2355–2360. [CrossRef] [PubMed]
6. Lee, I.M.; Shiroma, E.J.; Lobelo, F.; Puska, P.; Blair, S.N.; Katzmarzyk, P.T. Effect of physical inactivity on major non-communicable diseases worldwide: An analysis of burden of disease and life expectancy. *Lancet* **2012**, *380*, 219–229. [CrossRef]
7. Wen, C.P.; Wai, J.P.; Tsai, M.K.; Yang, Y.C.; Cheng, T.Y.; Lee, M.C.; Chan, H.T.; Tsao, C.K.; Tsai, S.P.; Wu, X. Minimum amount of physical activity for reduced mortality and extended life expectancy: A prospective cohort study. *Lancet* **2011**, *378*, 1244–1253. [CrossRef]
8. Shephard, R.J.; Park, H.; Park, S.; Aoyagi, Y. Objectively measured physical activity and progressive loss of lean tissue in older Japanese adults: Longitudinal data from the Nakanojo study. *J. Am. Geriatr. Soc.* **2013**, *61*, 1887–1893. [CrossRef] [PubMed]
9. Timmerman, K.L.; Flynn, M.G.; Coen, P.M.; Markofski, M.M.; Pence, B.D. Exercise training-induced lowering of inflammatory (CD14+CD16+) monocytes: A role in the anti-inflammatory influence of exercise? *J. Leukoc. Biol.* **2008**, *84*, 1271–1278. [CrossRef] [PubMed]
10. Nelson, M.E.; Rejeski, W.J.; Blair, S.N.; Duncan, P.W.; Judge, J.O.; King, A.C.; Macera, C.A.; Castaneda-Sceppa, C. Physical activity and public health in older adults: Recommendation from the American College of Sports Medicine and the American Heart Association. *Circulation* **2007**, *116*, 1094–1105. [CrossRef] [PubMed]
11. Woods, J.A.; Wilund, K.R.; Martin, S.A.; Kistler, B.M. Exercise, inflammation and aging. *Aging Dis.* **2012**, *3*, 130–140. [PubMed]
12. Olesen, J.; Gliemann, L.; Bienso, R.; Schmidt, J.; Hellsten, Y.; Pilegaard, H. Exercise training, but not resveratrol, improves metabolic and inflammatory status in skeletal muscle of aged men. *J. Physiol.* **2014**, *592*, 1873–1886. [CrossRef] [PubMed]
13. Bastard, J.P.; Maachi, M.; Lagathu, C.; Kim, M.J.; Caron, M.; Vidal, H.; Capeau, J.; Feve, B. Recent advances in the relationship between obesity, inflammation, and insulin resistance. *Eur. Cytokine Netw.* **2006**, *17*, 4–12. [PubMed]
14. Ferrucci, L.; Fabbri, E. Inflammageing: Chronic inflammation in ageing, cardiovascular disease, and frailty. *Nat. Rev. Cardiol.* **2018**, *15*, 505–522. [CrossRef] [PubMed]
15. Castaneda, C.; Gordon, P.L.; Parker, R.C.; Uhlin, K.L.; Roubenoff, R.; Levey, A.S. Resistance training to reduce the malnutrition-inflammation complex syndrome of chronic kidney disease. *Am. J. Kidney Dis.* **2004**, *43*, 607–616. [CrossRef] [PubMed]
16. Adamopoulos, S.; Parissis, J.; Karatzas, D.; Kroupis, C.; Georgiadis, M.; Karavolias, G.; Paraskevaidis, J.; Koniavitou, K.; Coats, A.J.; Kremastinos, D.T. Physical training modulates proinflammatory cytokines and the soluble Fas/soluble Fas ligand system in patients with chronic heart failure. *J. Am. Coll. Cardiol.* **2002**, *39*, 653–663. [CrossRef]
17. Kohut, M.L.; McCann, D.A.; Russell, D.W.; Konopka, D.N.; Cunnick, J.E.; Franke, W.D.; Castillo, M.C.; Reighard, A.E.; Vanderah, E. Aerobic exercise, but not flexibility/resistance exercise, reduces serum IL-18, CRP, and IL-6 independent of beta-blockers, BMI, and psychosocial factors in older adults. *Brain Behav. Immun.* **2006**, *20*, 201–209. [CrossRef] [PubMed]
18. Phillips, C.M.; Dillon, C.B.; Perry, I.J. Does replacing sedentary behaviour with light or moderate to vigorous physical activity modulate inflammatory status in adults? *Int. J. Behav. Nutr. Phys. Act.* **2017**, *14*, 138. [CrossRef] [PubMed]
19. Pararasa, C.; Ikwuobe, J.; Shigdar, S.; Boukouvalas, A.; Nabney, I.T.; Brown, J.E.; Devitt, A.; Bailey, C.J.; Bennett, S.J.; Griffiths, H.R. Age-associated changes in long-chain fatty acid profile during healthy aging promote pro-inflammatory monocyte polarization via PPARgamma. *Aging Cell* **2016**, *15*, 128–139. [CrossRef] [PubMed]
20. Pararasa, C.; Bailey, C.J.; Griffiths, H.R. Ageing, adipose tissue, fatty acids and inflammation. *Biogerontology* **2015**, *16*, 235–248. [CrossRef] [PubMed]
21. Franceschi, C.; Bonafe, M.; Valensin, S.; Olivieri, F.; De Luca, M.; Ottaviani, E.; De Benedictis, G. Inflamm-aging. An evolutionary perspective on immunosenescence. *Ann. N. Y. Acad. Sci.* **2000**, *908*, 244–254. [CrossRef] [PubMed]

22. Elosua, R.; Marrugat, J.; Molina, L.; Pons, S.; Pujol, E. Validation of the Minnesota Leisure Time Physical Activity Questionnaire in Spanish men. The MARATHOM Investigators. *Am. J. Epidemiol.* **1994**, *139*, 1197–1209. [CrossRef] [PubMed]

23. Elosua, R.; Garcia, M.; Aguilar, A.; Molina, L.; Covas, M.I.; Marrugat, J. Validation of the Minnesota leisure time physical activity questionnaire in Spanish women. Investigators of the MARATDON group. *Med. Sci. Sports. Exerc.* **2000**, *32*, 1431–1437. [CrossRef] [PubMed]

24. Conway, J.M.; Irwin, M.L.; Ainsworth, B.E. Estimating energy expenditure from the Minnesota leisure time physical activity and Tecumseh occupational activity questionnaires—A doubly labeled water validation. *J. Clin. Epidemiol.* **2002**, *55*, 392–399. [CrossRef]

25. Bibiloni, M.D.M.; Julibert, A.; Argelich, E.; Aparicio-Ugarriza, R.; Palacios, G.; Pons, A.; Gonzalez-Gross, M.; Tur, J.A. Western and Mediterranean dietary patterns and physical activity and fitness among Spanish older adults. *Nutrients* **2017**, *9*. [CrossRef] [PubMed]

26. Ferrer, M.D.; Sureda, A.; Batle, J.M.; Tauler, P.; Tur, J.A.; Pons, A. Scuba diving enhances endogenous antioxidant defenses in lymphocytes and neutrophils. *Free Radic. Res.* **2007**, *41*, 274–281. [CrossRef] [PubMed]

27. Sureda, A.; Batle, J.M.; Tauler, P.; Aguilo, A.; Cases, N.; Tur, J.A.; Pons, A. Hypoxia/reoxygenation and vitamin C intake influence NO synthesis and antioxidant defenses of neutrophils. *Free Radic. Biol. Med.* **2004**, *37*, 1744–1755. [CrossRef] [PubMed]

28. Bradford, M.M. A rapid and sensitive method for the quantitation of microgram quantities of protein utilizing the principle of protein-dye binding. *Anal. Biochem.* **1976**, *72*, 248–254. [CrossRef]

29. Franceschi, C.; Olivieri, F.; Marchegiani, F.; Cardelli, M.; Cavallone, L.; Capri, M.; Salvioli, S.; Valensin, S.; De Benedictis, G.; Di Iorio, A.; et al. Genes involved in immune response/inflammation, IGF1/insulin pathway and response to oxidative stress play a major role in the genetics of human longevity: The lesson of centenarians. *Mech. Ageing Dev.* **2005**, *126*, 351–361. [CrossRef] [PubMed]

30. Knight, E.; Petrella, R.J. Prescribing physical activity for healthy aging: Longitudinal follow-up and mixed method analysis of a primary care intervention. *Phys. Sportsmed.* **2014**, *42*, 30–38. [CrossRef] [PubMed]

31. Mendham, A.E.; Duffield, R.; Marino, F.; Coutts, A.J. Small-sided games training reduces CRP, IL-6 and leptin in sedentary, middle-aged men. *Eur. J. Appl. Physiol.* **2014**, *114*, 2289–2297. [CrossRef] [PubMed]

32. Markofski, M.M.; Flynn, M.G.; Carrillo, A.E.; Armstrong, C.L.; Campbell, W.W.; Sedlock, D.A. Resistance exercise training-induced decrease in circulating inflammatory CD14+CD16+ monocyte percentage without weight loss in older adults. *Eur. J. Appl. Physiol.* **2014**, *114*, 1737–1748. [CrossRef] [PubMed]

33. Pagonas, N.; Dimeo, F.; Bauer, F.; Seibert, F.; Kiziler, F.; Zidek, W.; Westhoff, T.H. The impact of aerobic exercise on blood pressure variability. *J. Hum. Hypertens.* **2014**, *28*, 367–371. [CrossRef] [PubMed]

34. Balducci, S.; Zanuso, S.; Nicolucci, A.; Fernando, F.; Cavallo, S.; Cardelli, P.; Fallucca, S.; Alessi, E.; Letizia, C.; Jimenez, A.; et al. Anti-inflammatory effect of exercise training in subjects with type 2 diabetes and the metabolic syndrome is dependent on exercise modalities and independent of weight loss. *Nutr. Metab. Cardiovasc. Dis.* **2010**, *20*, 608–617. [CrossRef] [PubMed]

35. Kadoglou, N.P.; Iliadis, F.; Angelopoulou, N.; Perrea, D.; Ampatzidis, G.; Liapis, C.D.; Alevizos, M. The anti-inflammatory effects of exercise training in patients with type 2 diabetes mellitus. *Eur. J. Cardiovasc. Prev. Rehabil.* **2007**, *14*, 837–843. [CrossRef] [PubMed]

36. Serra-Majem, L.; Pastor-Ferrer, M.C.; Castell, C.; Ribas-Barba, L.; Roman-Vinas, B.; Ribera, L.F.; Plasencia, A.; Salleras, L. Trends in blood lipids and fat soluble vitamins in Catalonia, Spain (1992–2003). *Public Health Nutr.* **2007**, *10*, 1379–1388. [CrossRef] [PubMed]

37. Soriano-Maldonado, A.; Aparicio, V.A.; Felix-Redondo, F.J.; Fernandez-Berges, D. Severity of obesity and cardiometabolic risk factors in adults: Sex differences and role of physical activity. The HERMEX study. *Int. J. Cardiol.* **2016**, *223*, 352–359. [CrossRef] [PubMed]

38. Sureda, A.; Tauler, P.; Aguilo, A.; Cases, N.; Fuentespina, E.; Cordova, A.; Tur, J.A.; Pons, A. Relation between oxidative stress markers and antioxidant endogenous defences during exhaustive exercise. *Free Radic. Res.* **2005**, *39*, 1317–1324. [CrossRef] [PubMed]

39. Tauler, P.; Aguilo, A.; Gimeno, I.; Fuentespina, E.; Tur, J.A.; Pons, A. Response of blood cell antioxidant enzyme defences to antioxidant diet supplementation and to intense exercise. *Eur. J. Nutr.* **2006**, *45*, 187–195. [CrossRef] [PubMed]

40. Tauler, P.; Aguilo, A.; Guix, P.; Jimenez, F.; Villa, G.; Tur, J.A.; Cordova, A.; Pons, A. Pre-exercise antioxidant enzyme activities determine the antioxidant enzyme erythrocyte response to exercise. *J. Sports Sci.* **2005**, *23*, 5–13. [CrossRef] [PubMed]

41. Moro-García, A.; Fernández-García, B.; Alonso-Arias, R.; Rodriguez-Alonso, M.; Suárez, M.; López-Larrea, C. Effects of maintained intense exercise throughout the life on the adaptive immune response in elderly and young athletes. *Br. J. Sports Med.* **2013**, *47*, e3. [CrossRef]

42. Steensberg, A.; Fischer, C.P.; Keller, C.; Moller, K.; Pedersen, B.K. IL-6 enhances plasma IL-1ra, IL-10, and cortisol in humans. *Am. J. Physiol. Endocrinol. Metab.* **2003**, *285*, E433–E437. [CrossRef] [PubMed]

43. Gleeson, M.; Bishop, N.C.; Stensel, D.J.; Lindley, M.R.; Mastana, S.S.; Nimmo, M.A. The anti-inflammatory effects of exercise: Mechanisms and implications for the prevention and treatment of disease. *Nat. Rev. Immunol.* **2011**, *11*, 607–615. [CrossRef] [PubMed]

44. Panagiotakos, D.B.; Pitsavos, C.; Chrysohoou, C.; Kavouras, S.; Stefanadis, C. The associations between leisure-time physical activity and inflammatory and coagulation markers related to cardiovascular disease: The ATTICA Study. *Prev. Med.* **2005**, *40*, 432–437. [CrossRef] [PubMed]

45. Visser, M.; Pahor, M.; Taaffe, D.R.; Goodpaster, B.H.; Simonsick, E.M.; Newman, A.B.; Nevitt, M.; Harris, T.B. Relationship of interleukin-6 and tumor necrosis factor-alpha with muscle mass and muscle strength in elderly men and women: The Health ABC Study. *J. Gerontol. A Biol. Sci. Med. Sci.* **2002**, *57*, M326–332. [CrossRef] [PubMed]

46. Nikseresht, M. Comparison of Serum Cytokine Levels in Men Who are Obese or Men Who are Lean: Effects of Nonlinear Periodized Resistance Training and Obesity. *J. Strength Cond. Res.* **2018**, *32*, 1787–1795. [CrossRef] [PubMed]

47. Fried, S.K.; Bunkin, D.A.; Greenberg, A.S. Omental and subcutaneous adipose tissues of obese subjects release interleukin-6: Depot difference and regulation by glucocorticoid. *J. Clin. Endocrinol. Metab.* **1998**, *83*, 847–850. [CrossRef] [PubMed]

48. Manna, P.; Jain, S.K. Obesity, Oxidative Stress, Adipose Tissue Dysfunction, and the Associated Health Risks: Causes and Therapeutic Strategies. *Metab. Syndr. Relat. Disord.* **2015**, *13*, 423–444. [CrossRef] [PubMed]

49. Hoesel, B.; Schmid, J.A. The complexity of NF-κB signaling in inflammation and cancer. *Mol. Cancer* **2013**, *12*, 86. [CrossRef] [PubMed]

50. Khansari, N.; Shakiba, Y.; Mahmoudi, M. Chronic inflammation and oxidative stress as a major cause of age-related diseases and cancer. *Recent Pat. Inflamm. Allergy Drug Discov.* **2009**, *3*, 73–80. [CrossRef] [PubMed]

51. Mendes, K.L.; Lelis, D.F.; Santos, S.H.S. Nuclear sirtuins and inflammatory signaling pathways. *Cytokine Growth Factor Rev.* **2017**, *38*, 98–105. [CrossRef] [PubMed]

52. Gomez-Cabrera, M.C.; Domenech, E.; Vina, J. Moderate exercise is an antioxidant: Upregulation of antioxidant genes by training. *Free Radic. Biol. Med.* **2008**, *44*, 126–131. [CrossRef] [PubMed]

53. Ji, L.L.; Gomez-Cabrera, M.C.; Vina, J. Role of nuclear factor κB and mitogen-activated protein kinase signaling in exercise-induced antioxidant enzyme adaptation. *Appl. Physiol. Nutr. Metab.* **2007**, *32*, 930–935. [CrossRef] [PubMed]

54. Flynn, M.G.; McFarlin, B.K.; Phillips, M.D.; Stewart, L.K.; Timmerman, K.L. Toll-like receptor 4 and CD14 mRNA expression are lower in resistive exercise-trained elderly women. *J. Appl. Physiol.* **2003**, *95*, 1833–1842. [CrossRef] [PubMed]

55. Stewart, L.K.; Flynn, M.G.; Campbell, W.W.; Craig, B.A.; Robinson, J.P.; McFarlin, B.K.; Timmerman, K.L.; Coen, P.M.; Felker, J.; Talbert, E. Influence of exercise training and age on CD14+ cell-surface expression of toll-like receptor 2 and 4. *Brain Behav. Immun.* **2005**, *19*, 389–397. [CrossRef] [PubMed]

56. Cavalcante, P.A.M.; Gregnani, M.F.; Henrique, J.S.; Ornellas, F.H.; Araujo, R.C. Aerobic but not Resistance Exercise Can Induce Inflammatory Pathways via Toll-Like 2 and 4: A Systematic Review. *Sports Med. Open* **2017**, *3*, 42. [CrossRef] [PubMed]

nutrients

MDPI

Article

Childhood Experiences and Sporting Event Visitors' Preference for Unhealthy versus Healthy Foods: Priming the Route to Obesity?

Joerg Koenigstorfer

Department of Sport & Health Management, Technical University of Munich, Uptown Munich—Campus D, Georg Brauchle Ring 60/62, 80992 Munich, Germany; joerg.koenigstorfer@tum.de; Tel.: +49-89-289-24558

Received: 30 August 2018; Accepted: 29 October 2018; Published: 5 November 2018

Abstract: To date, there is little knowledge about how experiences in childhood frame adults' food and drink consumption patterns in the context of attending sporting events as spectators. Therefore, the goal of this study was to explore the childhood memories of adults when they visited sporting events and find out whether and why this particular setting makes individuals indulge in unhealthy food. The study comprises two components: Study 1 and Study 2. In Study 1, 30 individuals recalled their childhood experiences of sport stadium visits at the age of ten years or younger. Inductive coding of the stories revealed that on-site enjoyment is an important factor that may lead to unhealthy food consumption. In Study 2 ($n = 240$), the effect of enjoyment on the intentions to eat unhealthy versus healthy food at sporting events was tested empirically and contrasted with two other leisure-time activities. The results of the experiment revealed that it is not enjoyment, but the visit to sporting or music events (versus a flea market) that increased the preference for unhealthy versus healthy foods. Implications to decrease (increase) the preference for unhealthy (healthy) food in these particular settings against the background of childhood experiences can be drawn.

Keywords: childhood memories; sport spectators; sport games; sporting events; music events; food consumption; out-of-home eating; food healthiness

1. Introduction

What food and drinks do you associate with your family's most recent visit to a baseball game? It would be of little surprise if you mentioned that your children had hotdogs and soft drinks. It is noteworthy that these foods are in conflict with the diets of athletes whom you and your family members were following during the course of the game (assuming that athletes stick to the recommendations of sport nutritionists, e.g., [1]).

In consumer behavior and food research, it is well known that childhood experiences influence both purchase behaviors and consumption behaviors in adulthood [2,3]. The attendance of sporting events as a spectator, however, has not been researched extensively yet, particularly with regard to people's eating and drinking behaviors that are potentially influenced by childhood experiences. This is despite the fact that the food that is provided to children (and adults) when attending sporting events has been subjected to criticism, particularly because of its low nutritional value and high calorie density [4,5]. If childhood memories about the association between food and sporting event attendance as spectators are important, it is crucial that children are exposed to healthy, and not to unhealthy, food environments today (arguing from the perspective of public policy), because these contextual factors influence behavioral patterns later in life.

1.1. Food Provision to Sporting Event Spectators

At sporting events, concession stands offer various food and drink options to spectators, and many spectators consume food and drinks before or while following a sporting event. Often, sponsorship and the exclusiveness rights that go along with the sponsorship determine the kind of food and drink provision [6,7], and sport spectators can choose from those options that are made available to them [8].

Within the context of the attendance of sporting events, four factors characterize the market for food and drinks, including the physical facility (i.e., the built environment for food and drink provision/availability as well as for eating and drinking), group experience (i.e., the influence of peer groups, such as spouses, friends, siblings, and strangers, on food and drink choices as well as eating and drinking), history and tradition factors (e.g., having hotdogs at half time for nostalgic reasons), and rituals (e.g., scripts that sporting event visitors follow when attending a game, such as drinking beer when tailgating before the game) [9]. These characteristics affect what and how much people eat and drink.

Marketers make use of the connection between the various food and drink options and the consumers following sporting events in their leisure time, particularly for the promotion of unhealthy food and drinks, targeting children and adolescents [6], generation Y consumers [10], sport gamblers [11], and sport stadium visitors [4], for example. In sport stadiums, concession stands contribute to the perception and appeal of the 'sportscape', that is, the stadium as an environment in which services are provided and value is (co-)created [12], and the concession stands may become part of the psychological associations within this environment, a 'home ground', a beloved place to sporting event spectators (sometimes called 'topophilia' [13,14]).

1.2. Childhood Memories about Eating and Drinking When Following Sporting Events On-Site

Attending sporting events is a common leisure time activity for families with children around the world [15,16]. The sensory experience with certain foods and drinks at sporting events can then influence children's preferences for food and drinks (particularly sweet and salty food and sweet drinks) or their avoidance of food and drinks that they do not like or know of [17,18]. As dietary habits and the acceptance of certain foods and drinks and amounts influence individuals' health and important health determinants later in life [17], the sporting event setting may be of relevance in this context [5–8].

The emotions associated with the food and drinks provided and consumed at sporting events relate to the discrete emotions that have been generally studied in food decision-making models (e.g., enjoyment), such as in the goal conflict model of eating [19]. According to the model, there is a conflict between the enjoyment goals of eating and the cognitive representations of weight control (or healthy eating). At sporting events, spectators have been reported to enjoy foods and drinks such as meat pies, burgers, fries, popcorn, and soft drinks [4]—options that are considered to contribute to overweight and obesity because of their low nutritional value, high calorie density, and large portion sizes, which often go along with relatively high consumption volumes (and calorie intake). Yet, some stadiums also have healthy food items on offer, such as salad (e.g., from the stadium's rooftop garden at Fenway Park, Boston, MA, USA) and cauliflower sandwiches (Wrigley Field, Chicago, IL, USA). The memories about eating and drinking may then influence consumers' preferences for certain food and drinks at sporting events.

1.3. Aims and Research Goals

The present study aimed to capture and categorize the food- and drinks-related memories of adults when they visited sporting events in their childhood. Since enjoyment might be a central emotion (be it related to the game itself, such as when a goal is scored, or the experience in the stands, such as when cheering for the team, being with family and friends, as well as eating and drinking), the present study also addressed the relevance of enjoyment in this context. In particular, it aimed to find out whether the attraction to unhealthy food and the avoidance of healthy food observed in sporting event visitors is merely due to the activation of enjoyment goals (and thus should be replicable across contexts if enjoyment goals are activated) or whether sporting event visits can lead spectators to indulge or to control the effect of enjoying the leisure time activity on the intentions to indulge. Thus, the research questions that guided the research are as follows:

RQ 1   What are the childhood memories of adults when they visited sporting events, particularly in relation to food and drinks?

RQ 2   What is the influence of the particular sporting event setting and the enjoyable experience of the event visit on individuals' intentions to indulge in unhealthy foods versus healthy foods?

In what follows, two studies are presented: Study 1, which aimed to answer RQ 1, and Study 2, which aimed to answer RQ 2.

2. Study 1

2.1. Materials and Methods

Thirty informants (15 women) were recruited via Amazon's Mechanical Turk and took part in the study in exchange for monetary compensation (M [mean] = 34.8 years, SD [standard deviation] = 8.9). Informants were only interviewed when they could recall a professional sporting event that they had visited at the age of 10 years or younger [20,21]. They were US residents and recalled the first time they visited a sporting event, at the age between four and 10 years, according to the written interviews with the informants. The following sports were mentioned: baseball, basketball, football, and ice hockey.

Similar to the procedure in Braun–Latour et al.'s study [3], informants were asked to write down a memory story of their earliest childhood memory. MAXQDA software (VERBI, Berlin, Germany) was used to inductively code the response, based on content analysis procedures [22]. Categories were withdrawn directly from the raw data. The categories that were extracted were treated as content units. The coded content units consisted of three categories (level a) and eleven subcategories (level b). Two coders performed the coding; inter-rater reliability was satisfied with Cohen's $\kappa = 0.82$.

2.2. Results and Discussion

In the memory stories, a number of different themes were mentioned. The inductive coding revealed three categories (level a): individual experiences of the visit; external factors related to the visit; social factors related to the visit. Table 1 shows the eleven subcategories (level b) of the three categories and some example statements.

Every informant referred to food and drinks in the memory task, an indicator that there is a close association between food and drinks and sporting event visits in childhood. Within the emotions subcategory, enjoyment and enjoyment-related facets, such as happiness and excitement, were dominant themes in the informants' memory stories. This supports the assumption that enjoyment is central to the attendance of a sporting event in childhood. The data further revealed that enjoyment relates to many aspects: the game itself, the players, the audience, and the stadium, for example. Most importantly to the present study, informants also made connections between enjoyment and the food experience. For example, informant 11 made the following statement: "The hotdogs there smelled and tasted much better than the ones my mom would make at home. I was not a big hotdog fan, but I did really enjoy the ones there." Informant 26 highlighted that her food preferences are still the same today: "The hotdog was definitely the classic, made up the whole experience, and I still love pretzels today." The foods that were mentioned in the stories were the following: burger, chili, french fries, hotdogs, nachos, pizza, and snacks (cotton candy, crackers, peanuts, and popcorn were mentioned explicitly). The drinks that were mentioned were the following: coke, soda, water, and hot chocolate. Table 1 shows some further example statements taken from the memory stories.

Table 1. Categories of childhood memories about the attendance of sporting events.

(Sub-)Category	Example Statement
Individual experiences	
Food and drink consumption	"I didn't really like hotdogs that much, but for some reason they smelled so much better there." (informant 11) "I was very happy and enjoying myself as I got a new hat, eating junk food, and learning about baseball. It seemed like a dream because everything was perfect." (informant 20) "I remember vividly the taste of the hotdogs and excitedly watching the game in front of us. It is a highlight of my childhood." (informant 26)
Emotions	"From the time the kick-off started until the final whistle, I did not sit down once. We thoroughly enjoyed the game." (informant 4) "I enjoyed myself very much. It is a great day I loved to remember." (informant 25)
Clothing	"We all wore green t-shirts." (informant 2)
Merchandise	"I can almost see it now, sitting with my baseball cards from the gift shop (…) watching the game." (informant 24)
Inspiration for life	"It was amazing to watch and experience as it taught me never give up even when things seem the bleakest." (informant 29)
Nostalgia	"I still reflect back to the first time I stepped onto that hallowed ground, many, many years ago." (informant 14)
External factors	
Game/show performance	"I was all excited to see my favorite player, Aurelio Rodriguez (third baseman)." (informant 18)
Built environment	"I was astonished by (…) the bowl-shaped facade." (informant 14)
Weather	"It was so hot and muggy that day with no breeze or shade at all." (informant 11)
Social factors	
Belongingness to family and friends	"It was a special day for me, and I'll always remember seeing my first baseball game with my father and grandfather." (informant 20)
Spectator-generated atmosphere	"The thrill of the audience was the best part." (informant 25)

To find out whether enjoyment generally predicts intentions to eat unhealthy foods (e.g., the types of foods mentioned in the memory stories) versus healthy foods and whether the sporting event context provides a unique setting to spectators in terms of the preference for unhealthy versus healthy food consumption, Study 2 was conducted. In the study, the visit to a sporting event was contrasted with other leisure time activities, that is, the visit to a music event and the visit to a flea market. This allowed drawing conclusions about the peculiarities of the different settings in which food is consumed. The study exclusively looked at intended food (but not drink) consumption.

3. Study 2

3.1. Materials and Methods

Two hundred forty students (149 women) were recruited on a university campus and took part in the study in exchange for monetary compensation (M = 26.6 years, SD = 8.5). The experimental study applied a 2 × 2 design, manipulating enjoyment (high versus low) and event type (sporting versus music event) between participants, and a control group was added as a fifth experimental condition (i.e., the visit to a flea market with low enjoyment). In the study, the participants were randomly assigned to one of the five conditions.

The participants first read a description of an event visit (Appendix A) and they were asked to imagine that they would visit the event. After they read the description, they were asked to rate the likeliness to consume 15 different foods on a scale anchored at 1 ("I would not eat this food at all") and 10 ("I would definitely eat this food"), including both healthy and unhealthy foods ([23,24]; Appendix B). They were pretested to represent healthy and unhealthy foods, which were presented in a random order. An overall score for the preference of unhealthy foods versus healthy foods was computed (with reverse-coded items for healthy options) (α = 0.76).

Beside these intentions, the survey assessed sociodemographics and an item that was used to assess whether the experimental manipulation worked or not ("How enjoyable do you rate the visit to the sporting event (or music event or flea market)?", anchored at 1 = "Not enjoyable at all" and 7 = "Very enjoyable"). A funneled debriefing was applied at the end of the study, which revealed that none of the participants guessed the research questions of the study.

3.2. Results and Discussion

The experimental manipulation worked as intended: when the event visit was described as highly enjoyable, enjoyment was rated higher (M = 5.69, SD = 1.14) compared to when the visit was not described as highly enjoyable (M = 4.24, SD = 1.55; t (238) = 8.35, $p < 0.001$).

Figure 1 displays the participants' intention to indulge in unhealthy foods versus healthy foods depending on the experimental conditions.

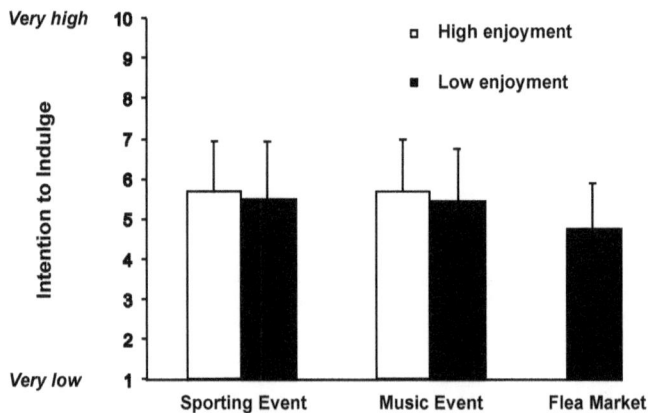

Figure 1. Intention to indulge in unhealthy foods versus healthy foods depending on the five experimental conditions: visit to a sporting event (low and high enjoyment), visit to a music event (low and high enjoyment), and visit to a flea market (low enjoyment only).

A linear regression analysis was performed to assess the influence of enjoyment (coded 1 for high and 0 for low) as well as the event (dummy 1, coded 1 for sporting event and 0 for other events; dummy 2, coded 1 for music event and 0 for other events) on the intention to indulge in unhealthy foods versus healthy foods. The variables explained 6% of the variance in the participants' intention. While the influence of enjoyment was not significant (b [beta coefficient] = 0.23, SE [standard error] = 0.19, $p = 0.22$), both the visit to the sporting event (b = 0.67, SE = 0.26, $p < 0.01$) and the visit to the music event (b = 0.65, SE = 0.25, $p < 0.01$) increased the intention to indulge in unhealthy foods versus healthy foods. The contrast between sporting event and music event visits was not significant (b = 0.02, SE = 0.19, $p = 0.93$).

To conclude, it can be stated that, controlling for the influence of primed enjoyment of the visit, the attendance at a sporting event (versus a flea market) increased the intention to choose unhealthy foods versus healthy foods. The effect was similar for the attendance at a music event (versus a flea market), while the sporting event context and the music event context did not differ in their effects. We discuss the general implications below.

4. Discussion

The purpose of the study was to explore food- and drinks-related childhood memories about the attendance at sporting events and to find out whether sporting event attendance influences people's preference for unhealthy foods versus healthy foods. The study contributes to the existing research in three ways.

First, the study revealed that eating and drinking contributed to an enjoyable stadium visit when adults reflected on their childhood experiences of spectator sports. Mostly unhealthy foods were recalled. While the provision of unhealthy foods at sporting events has been criticized [4,5,8], none of the previous studies has shown that childhood experiences with unhealthy foods are recalled even decades later nor how they shaped children's food preferences. We note that any type of food can be part of a healthy, nutritious diet. However, when certain types of foods are consumed too often as well as in high amounts and when the consumption of high volumes of low-nutritious and high-caloric foods turns into a habit when eating at home or in other out-of-home contexts, the risk of children becoming overweight and obese may increase. The lived experience at sporting events may contribute to this, similarly to sponsorship-linked marketing activities of unhealthy food and drinks with role-model athletes as endorsers [7,9].

Second, the study revealed that both the sporting event context and the music event context are leisure-time activities that, controlling for enjoyment effects, increased the likelihood that individuals prefer unhealthy foods over healthy foods compared to a control group (here: people who imagined a visit to a flea market). The model explained 6% of the variance in the dependent variable. It is plausible that other factors than the venue influenced the preference for unhealthy versus healthy foods, such as individual taste preferences, social norms, and people's general attitudes and values. Because of the random assignment of the participants to the various experimental conditions, however, the results should be unaffected by these individual differences. From a theoretical perspective, it is interesting that the two leisure-time activities—sporting event visit and music event visit—primed the intended food preferences of individuals. The potential reasons for this include that implicitly learned associations contribute to people preferring certain foods over others (such as eating hotdogs when watching a baseball game and eating popcorn when watching a movie in a cinema; [2]). Also, the anticipated convenience of certain foods may matter to food consumption intentions, depending on whether individuals assumed that seating was available or not, how much time was available, and how easy it was to dispose of left overs. Lastly, anticipated differences in social factors between contexts, such as crowding, which might be higher for music and sporting events than for a flea market, may have influenced the results.

Third, the finding that high (versus low) enjoyment of the event experience did not increase the preference for unhealthy and the avoidance of healthy food is noteworthy. On the one hand, one may have expected that individuals switch goals: when one goal (enjoyment) is achieved in the context of following the favorite sports team or music band, they may be directed to pursuing opposite goals (health goals in the context of eating) [25]. This was not the case in the present study. On the other hand, one may have expected that individuals want to nourish their consumption episode further: when they enjoy following their favorite sports team or music band, they want to enhance the enjoyment experience and continue to enjoy to have a 'perfect evening' (highlighting in the context of consumption episodes [26]). Although, in the present study, the means pointed in this direction (Figure 1), there were no significant effects in the analyses. Thus, the results did not support any of the two assumptions.

This study has important implications for public health policy and practice. There is a need for healthier food availability and food policy at sport settings in general and at sporting venues in particular. In the present study, all of the childhood memories about the association between food and sporting event attendance related to unhealthy foods (Study 1). Furthermore, the sporting context increased the preference for unhealthy over healthy foods (Study 2). To fight these links that promote unhealthy eating behaviors, public policy makers have several possibilities: (1) incentivize stakeholders to offer relatively more healthy food and drinks (such as providing funding to them if they offer sustainable, healthy options at sporting events); (2) make unhealthy (healthy) food and drinks less (more) accessible to sporting event visitors (e.g., via an increase in prices for unhealthy foods and drinks due to higher taxes); (3) make unhealthy (healthy) food and drinks less (more) attractive to sporting event visitors (e.g., via menu labeling or sponsorship of healthy food); (4) make unhealthy (healthy) food and drinks less (more) available by forced policy changes (this can range from forcing stakeholders to implement corporate social responsibility policies in relation to food and drinks to forbidding certain foods and drinks (e.g., high-sugar and high-fat options) per se in certain contexts, similarly to banning alcohol and smoking in sport stadiums). These measures may be most effective when both parents and children commit to the goal of healthy eating and drinking in sport settings [27] and when public policy-makers and practitioners include other stakeholders as well [28]. Consensus-based approaches are needed, because managers are afraid of loosing profits when strict policies are introduced [29].

The present study is not free of limitations. Study 1's memorization task related to food and drinks that were available at the time that the informants recalled. Therefore, if healthy food and drink items were not offered to them, it is no surprise that they were not mentioned. Future studies should replicate the findings for sport fans that attended stadiums that offered about the same number of healthy and unhealthy food and drink options (so that there are equal chances to enter the recalled set of items), for example. Study 2's experimental design controlled for enjoyment, but not for other contextual variables that may be relevant, such as convenience to eat food in relation to time and space. Also, the sample was skewed towards well-educated individuals, younger age, and females (versus males). Future studies may consider samples that are representative for stadium visitors. Future research may also look at the associations between (un)healthy food consumption and sporting event visits over time, as well as at choices depending on situational factors (such as an increase in the accessibility, attractiveness, and availability of healthy options; in particular: the introduction of menu-labeling schemes or when changes in family rituals in relation to food and drink consumption occur). While this study focused on memories in relation to healthy versus unhealthy food, future studies may focus on alcohol and tobacco consumption, factors that have also been identified as health threats to children and adults in sports settings [27,28,30].

5. Conclusions

To conclude, we can state that children should be exposed to healthy, but not unhealthy, food and drink environments when they attend sporting events as spectators, because these contextual factors and the memories related to the visits influence behavioral patterns later in life. The present study highlights the need for further action and further research in this important area.

Funding: This research received no external funding.

Acknowledgments: The author thanks Julia Over as well as the research assistants for their help with the data collection (Study 1 and Study 2) and the coding (Study 1).

Conflicts of Interest: The author declares no conflict of interest.

Appendix A. Experimental Manipulations in Study 2

Appendix A.1. Visit to a Sporting Event (High Enjoyment)

Please imagine that you attend a game of your favorite sports team. You have been looking forward to attending the game for months, and the media considers the game as the game of the year. Please read the brief description of how you feel on the day of the game. Please imagine how it would feel like to be there and to experience the game with all senses.

Before the game, you have met some of your family and friends and you drove to the stadium with them. You are in the stadium now and you cannot wait until the game begins. Your favorite team will play the biggest rival, and the game will have great influence on the league table. You can hear the fans already: they cheer for the team and there are amazing vibes in the stadium.

When you take your seat, the stadium, the fans, and the colorful flags that are all over the stadium overwhelm you. Your favorite players are on the pitch, and there is a tension in the air, anticipating that something special could occur today. You feel great, and everyone is ready to have a great evening, full of excitement and enjoyment. Nothing compares to this, and you have your ticket and are right there.

Before the game starts, you still have some time. Therefore, you want to get some food and drinks. The concession stands offer various food options. Please look at the following options and indicate the degree to which you would want to eat the respective food.

Appendix A.2. Visit to a Sporting Event (Low Enjoyment)

Please imagine that you attend a game of your favorite sports team. You have been looking forward to attending the game, and the media broadcast the game. Please read the brief description of how you feel on the day of the game. Please imagine how it would feel like to be there and to experience the game with all senses.

Before the game, you have met some of your family and friends and you drove to the stadium with them. You are in the stadium now and you wait until the game begins. Your favorite team will play another team that is desperate to win the game. You cannot hear a lot of fans yet, because there are not many spectators inside the stadium. You hope that the stadium atmosphere will improve later.

When you take your seat, you try to get an overview of the stadium and look around in the visitor stands and on the pitch. You don't see any of the players yet. You have to kill some time and think of what you could have done otherwise rather than attending this game. You still have to wait for the teams to enter the stadium and the game to begin.

Before the game starts, you still have some time. Therefore, you want to get some food and drinks. The concession stands offer various food options. Please look at the following options and indicate the degree to which you would want to eat the respective food.

Appendix A.3. Visit to a Music Event (High Enjoyment)

Please imagine that you attend a concert of your favorite music band. You have been looking forward to attending the concert for months, and the media considers the concert as the concert of the year. Please read the brief description of how you feel on the day of the concert. Please imagine how it would feel like to be there and to experience the concert with all senses.

Before the concert, you have met some of your family and friends and you drove to the stadium with them. You are in the stadium now and you cannot wait until the concert begins. Your favorite band will host one of their greatest concerts, and the concert will feature songs from the new album. You can hear the fans already: they sing famous songs, and there are amazing vibes in the stadium.

When you take your seat, the stadium, the fans, and the colorful flags that are all over the stadium overwhelm you. Your favorite musicians are on the stage, and there is a tension in the air, anticipating that something special could occur today. You feel great, and everyone is ready to have a great evening, full of excitement and enjoyment. Nothing compares to this, and you have your ticket and are right there.

Before the concert starts, you still have some time. Therefore, you want to get some food and drinks. The concession stands offer various food options. Please look at the following options and indicate the degree to which you would want to eat the respective food.

Appendix A.4. Visit to a Music Event (Low Enjoyment)

Please imagine that you attend a concert of your favorite music band. You have been looking forward to attending the concert, and the media broadcast the concert. Please read the brief description of how you feel on the day of the concert. Please imagine how it would feel like to be there and to experience the concert with all senses.

Before the concert, you have met some of your family and friends and you drove to the stadium with them. You are in the stadium now and you wait until the concert begins. Your favorite band will host a concert, and the concert will feature well-known and often-played songs. You cannot hear a lot of fans yet, because there are not many spectators inside the stadium. You hope that the stadium atmosphere will improve later.

When you take your seat, you try to get an overview of the stadium and look around in the visitor stands and on the stage. You don't see any of the band members yet. You have to kill some time and think of what you could have done otherwise rather than attending this concert. You still have to wait for the band to enter the stadium and the concert to begin.

Before the concert starts, you still have some time. Therefore, you want to get some food and drinks. The concession stands offer various food options. Please look at the following options and indicate the degree to which you would want to eat the respective food.

Appendix A.5. Visit to a Flea Market (Low Enjoyment)

Please imagine that you attend your favorite flea market. You have been looking forward to shop some items at the flea market. Many people visit this local flea market. Please read the brief description of how you feel on the day of the flea market. Please imagine how it would feel like to be there and to experience the flea market with all senses.

Before going to the flea market, you have met some of your family and friends and you drove to the place where it is held with them. You are at the market now and you wait until you find a good bargain or an item that you like. The market features some well-known stands, and people regularly purchase from these stands. There are not a lot of visitors at the flea market yet, and you hope that the atmosphere will improve later.

When you stroll around, you try to get an overview of the market and look around among people and what the stands have to offer. You don't see anything exciting yet. You have to kill some time and think of what you could have done otherwise rather than attending this flea market. You still want to find one or two items that you can bring home with you.

Before this and before you go home, you still have some time. Therefore, you want to get some food and drinks. The concession stands offer various food options. Please look at the following options and indicate the degree to which you would want to eat the respective food.

Appendix B. Foods Used in Study 2

The following foods were used (R indicates reverse coding): Asia soup (R), burger, chicken wings, chocolate muffin, french fries, fried onion rings, fruit salad (R), hotdog, meat skewers, pizza, salad (R), vegetable quiche (R), vegetable soup (R), vegetable wrap (R), and yogurt (R). The foods were displayed in a typical serving size and in a way so that they could be readily eaten by individuals while sitting or standing, on a disposable plate or in a cup/bowl.

References

1. USADA [United States Anti-Doping Agency], TrueSport. *Nutrition Guide*; USADA, TrueSport: Colorado Springs, CO, USA, 2016.
2. Birch, L.L. Development of food preferences. *Annu. Rev. Nutr.* **1999**, *19*, 41–62. [CrossRef] [PubMed]
3. Braun-LaTour, K.A.; LaTour, M.S.; Zinkhan, G.M. Using childhood memories to gain insight into brand meaning. *J. Mark.* **2007**, *71*, 45–60. [CrossRef]
4. Parry, K.D.; Hall, T.; Baxter, A. Who ate all the pies? The importance of food in the Australian sporting experience. *Sport Soc.* **2017**, *20*, 202–218. [CrossRef]
5. Smith, M.; Signal, L.; Edwards, R.; Hoek, J. Children's and parents' opinions on the sport-related food environment: A systematic review. *Obes. Rev.* **2017**, *18*, 1018–1039. [CrossRef] [PubMed]
6. Bragg, M.A.; Roberto, C.A.; Harris, J.L.; Brownell, K.D.; Elbel, B. Marketing food and beverages to youth through sports. *J. Adolesc. Health* **2018**, *62*, 5–13. [CrossRef] [PubMed]
7. Carter, M.-A.; Signal, L.; Edwards, R.; Hoek, J.; Maher, A. Food, fizzy, and football: Promoting unhealthy food and beverages through sport—A New Zealand case study. *BMC Public Health* **2013**, *13*, 126. [CrossRef] [PubMed]

8. McIsaac, J.-L.D.; Jarvis, S.L.; Spencer, R.; Kirk, S.F.L. "A tough sell": Findings from a qualitative analysis on the provision of healthy foods in recreation and sports settings. *Health Promot. Chronic Dis. Prev. Can.* **2018**, *38*, 18–22. [CrossRef] [PubMed]

9. Cornwell, T.B.; Koenigstorfer, J. Sponsors as meso-level actors in sport: Understanding individual decisions as foundational to sustainability in food and drink. In *Routledge Handbook of Sport and the Environment*; McCullough, B.P., Kellison, T.B., Eds.; Routledge: London, UK, 2017; pp. 161–175, ISBN 978-1138666153.

10. Sukalakamala, P.; Sukalakamala, S.; Young, P. An exploratory study of the concession preferences of generation Y consumers. *J. Foodserv. Bus. Res.* **2013**, *16*, 378–390. [CrossRef]

11. Lopez-Gonzalez, H.; Estévez, A.; Jiménez-Murcia, S.; Griffiths, M.D. Alcohol drinking and low nutritional value food eating behavior of sports bettors in gambling advertisements. *Int. J. Ment. Health Addict.* **2018**, *16*, 81–89. [CrossRef] [PubMed]

12. Wakefield, K.L.; Blodgett, J.G. The effect of the servicescape on customers' behavioral intentions in leisure service settings. *J. Serv. Mark.* **1996**, *10*, 45–61. [CrossRef]

13. Bale, J. Space, place and body culture: Yi-Fu Tuan and a geography of sport. *Geogr. Ann. Ser. B Hum. Geogr.* **1996**, *78*, 163–171. [CrossRef]

14. Ramshaw, G.; Gammon, S. On home ground? Twickenham stadium tours and the construction of sport heritage. *J. Herit. Tour.* **2010**, *5*, 87–102. [CrossRef]

15. Norman, J. *Football Still Americans' Favorite Sport to Watch*; Gallup: Washington, DC, USA, 2018.

16. Statista. *The Most Popular Spectator Sports Worldwide*; Statista: Hamburg, Germany, 2017.

17. Birch, L.L.; Fisher, J.O. Development of eating behaviors among children and adolescents. *Pediatrics* **1998**, *101*, 539–549. [PubMed]

18. Mennalla, J.A.; Pepino, M.Y.; Reed, D.R. Genetic and environmental determinants of bitter perception and sweet preferences. *Pediatrics* **2005**, *115*, e216–e222. [CrossRef] [PubMed]

19. Stroebe, W.; van Koningsbruggen, G.M.; Papies, E.K.; Aarts, H. Why most dieters fail but some succeed: A goal conflict model of eating behavior. *Psychol. Rev.* **2013**, *120*, 110–138. [CrossRef] [PubMed]

20. Piaget, J. Intellectual evolution from adolescence to adulthood. *Hum. Dev.* **1972**, *15*, 1–12. [CrossRef]

21. Speer, P.W.; Esposito, C. Family problems and children's competencies over the early elementary school years. *J. Prev. Interv. Community* **2000**, *20*, 69–83. [CrossRef]

22. Mayring, P. Qualitative content analysis. *Forum Qualitative Sozialforschung/Forum Qual. Soc. Res.* **2000**, *1*, 20.

23. Fishbach, A.; Zhang, Y. Together or apart: When goals and temptations complement versus compete. *J. Personal. Soc. Psychol.* **2008**, *94*, 547–559. [CrossRef] [PubMed]

24. Rohr, M.; Kamm, F.; Koenigstorfer, J.; Groeppel-Klein, A.; Wentura, D. The color red supports avoidance reactions to unhealthy food. *Exp. Psychol.* **2015**, *62*, 335–345. [CrossRef] [PubMed]

25. Fishbach, A.; Dhar, R. Goals as excuses or guides: The liberating effect of perceived goal progress on choice. *J. Consum. Res.* **2005**, *32*, 370–377. [CrossRef]

26. Dhar, R.; Simonson, I. Making complementary choices in consumption episodes: Highlighting versus balancing. *J. Mark. Res.* **1999**, *36*, 29–44. [CrossRef]

27. Kelly, B.; Baur, L.A.; Bauman, A.E.; King, L.; Chapman, K.; Smith, B.J. Views of children and parents on limiting unhealthy food, drink and alcohol sponsorship of elite and children's sports. *Public Health Nutr.* **2013**, *16*, 130–135. [CrossRef] [PubMed]

28. Kelly, B.; King, L.; Bauman, A.E.; Baur, L.A.; Macniven, R.; Chapman, K.; Smith, B.J. Identifying important and feasible policies and actions for health at community sports clubs: A consensus-generating approach. *J. Sci. Med. Sport* **2014**, *17*, 61–66. [CrossRef] [PubMed]

29. Olstad, D.L.; Downs, S.M.; Raine, K.D.; Berry, T.R.; McCargar, L.J. Improving children's nutrition environments: A survey of adoption and implementation of nutrition guidelines in recreational facilities. *BMC Public Health* **2011**, *11*, 423. [CrossRef] [PubMed]

30. Drygas, W.; Ruszkowska, J.; Philpott, M.; Björkström, O.; Parker, M.; Ireland, R.; Roncarolo, F.; Tenconi, M. Good practices and health policy analysis in European sports stadia: Results from the 'Healthy Stadia' project. *Health Promot. Int.* **2013**, *28*, 157–165. [CrossRef] [PubMed]

![nutrients logo] *nutrients*

MDPI

Article

Nutrient Intake and Physical Exercise Significantly Impact Physical Performance, Body Composition, Blood Lipids, Oxidative Stress, and Inflammation in Male Rats

Richard J. Bloomer [1,*], John Henry M. Schriefer [1], Trint A. Gunnels [1], Sang-Rok Lee [2], Helen J. Sable [3], Marie van der Merwe [1], Randal K. Buddington [4] and Karyl K. Buddington [5]

[1] School of Health Studies, University of Memphis, Memphis, 106 Roane Fieldhouse, TN 38152, USA; Jschriefer44@yahoo.com (J.H.M.S.); trintagunnels@gmail.com (T.A.G.); mvndrmrw@memphis.edu (M.v.d.M.)

[2] Department of Kinesiology & Dance, New Mexico State University, Las Cruces, NM 88003, USA; srlee@nmsu.edu

[3] Department of Psychology, University of Memphis, Memphis, TN 38152, USA; hjsable@memphis.edu

[4] College of Nursing, University of Tennessee Health Science Center, Memphis, TN 38152, USA; rbudding@uthsc.edu

[5] Department of Biological Sciences, University of Memphis, Memphis, TN 38152, USA; kbudding@memphis.edu

* Correspondence: rbloomer@memphis.edu; Tel.: 901-678-5638; Fax: 901-678-3591.

Received: 31 July 2018; Accepted: 15 August 2018; Published: 17 August 2018

Abstract: Background: Humans consuming a purified vegan diet known as the "Daniel Fast" realize favorable changes in blood lipids, oxidative stress, and inflammatory biomarkers, with subjective reports of improved physical capacity. Objective: We sought to determine if this purified vegan diet was synergistic with exercise in male rats. Methods: Long–Evans rats ($n = 56$) were assigned to be exercise trained (+E) by running on a treadmill three days per week at a moderate intensity or to act as sedentary controls with normal activity. After the baseline physical performance was evaluated by recording run time to exhaustion, half of the animals in each group were fed ad libitum for three months a purified diet formulated to mimic the Daniel Fast (DF) or a Western Diet (WD). Physical performance was evaluated again at the end of month 3, and body composition was assessed using dual-energy x-ray absorptiometry. Blood was collected for measurements of lipids, oxidative stress, and inflammatory biomarkers. Results: Physical performance at the end of month 3 was higher compared to baseline for both exercise groups ($p < 0.05$), with a greater percent increase in the DF + E group (99%) than in the WD + E group (51%). Body fat was lower in DF than in WD groups at the end of month 3 ($p < 0.05$). Blood triglycerides, cholesterol, malondialdehyde, and advanced oxidation protein products were significantly lower in the DF groups than in the WD groups ($p < 0.05$). No significant differences were noted in cytokines levels between the groups ($p > 0.05$), although IL-1β and IL-10 were elevated three-fold and two-fold in the rats fed the WD compared to the DF rats, respectively. Conclusions: Compared to a WD, a purified diet that mimics the vegan Daniel Fast provides significant anthropometric and metabolic benefits to rats, while possibly acting synergistically with exercise training to improve physical performance. These findings highlight the importance of macronutrient composition and quality in the presence of ad libitum food intake.

Keywords: dietary restriction; macronutrients; physical exercise; free radicals; cytokines

1. Introduction

Both dietary restriction/modification and caloric restriction have been studied extensively for their ability to favorably alter body composition in humans and animals [1]. One form of dietary restriction is veganism, a plan that eliminates all animal products and improves health outcomes [2]. Unlike caloric restriction, which typically calls for a 10–30% reduction in daily dietary energy needs [1], veganism allows for ad libitum food intake.

The Daniel Fast (DF), a biblically inspired partial fast, is similar to veganism but much more stringent. Specifically, the DF is a form of dietary restriction that allows for ad libitum food intake but places firm restrictions on the types of food that are allowed [1,3–5], with choices primarily limited to fruits, vegetables, whole grains, legumes, nuts, seeds, and plant-based oils. No alcohol, sweeteners, or refined foods are allowed, resulting in carbohydrate sources that are complex with low glycemic indices. By default, the DF has an abundance of dietary fiber and plant-derived fatty acids and relatively high concentrations of antioxidants.

Our previous human studies involving the DF revealed health-specific benefits including, but not limited to, reductions in body mass and body fat, systolic and diastolic blood pressure, total and LDL-cholesterol, blood oxidative stress biomarkers, and C-reactive protein [3,4,6]. The DF has also been shown to increase antioxidant capacity as well as the ratio of blood nitrate/nitrite (a biomarker of nitric oxide), which may have implications for improving physical performance with enhanced hemodynamic responses [7,8]. While we received multiple anecdotal reports of improved vitality, vigor, and mood from research participants in these studies [3,4,6], objective physical performance measures were not investigated in a controlled study. It is possible that chronic consumption of the DF with and without regular exercise—which is well known to aid in body mass/fat control and to improve functional capacity [9–11]—may yield favorable changes in body mass/fat and physical performance.

The present study sought to determine the independent and combined influences of dietary composition (under conditions of ad libitum intake) and moderate exercise training on functional capacity, body composition, blood lipids, oxidative stress, and inflammation in male rats. The decision to use an animal model was to ensure the control of key variables known to impact outcome measures (e.g., sleep, stress, type and volume of food, volume and intensity of exercise). We hypothesized that exercise training would improve physical performance as well as other outcome measures and that the benefits of exercise would be of greater magnitude in animals fed the DF.

2. Materials and Methods

2.1. Overview of the Experimental Design

Male Long–Evans rats (*n* = 56), 3–4 weeks of age, were purchased from Harlan Laboratories, Inc. (Indianapolis, IN, USA). Upon arrival, all rats were individually housed in standard shoebox caging in a climate-controlled room (21 °C), employing a standard 12:12 h light–dark cycle (lights on at 8:00 h). During a two-week acclimation period, they were transitioned from consuming a standard rat chow (Harlan 1018) to their assigned diets by gradually replacing the standard chow diet with an increasing proportion of the experimental diets. During this two-week period, the rats were also familiarized with the treadmill on three separate days (i.e., walking on the treadmill for 5 min at 15–20 m·min^{-1}), and the 12:12 h light–dark cycle was progressively shifted to lights on at 300 h and off at 1500 h. The light–dark cycle was shifted to allow exercise training and testing to occur during the latter part of the light phase. All housing and experimental procedures were approved by The University of Memphis Institutional Animal Care and Use Committee (approval #0734) and were in accordance with the 8th edition of the *Guide for the Care and Use of Laboratory Animals*. Throughout the study, the rats had ad libitum access to food and water.

2.2. Functional Capacity Assessment

The pre-Intervention (Baseline) total treadmill running time was determined for all animals after the two-week acclimation period following established procedures [12–14]. The animals began the test by running on a motorized treadmill (Exer-6M Treadmill, Columbus Instruments, Columbus, OH, USA) at a speed of 20 m·min^{-1} without incline for 15 min. The speed was increased by 5 m·min^{-1} every 15 min to a maximum speed of 35 m·min^{-1}. A mild electrical shock (frequency current at 3.0 hz at 1.60 mA with a voltage of a 115) was provided when the animals could not maintain the set pace. Fatigue was considered to occur when a rat started to lower its hindquarters and raise its snout, resulting in a significantly altered gait, to the point of not being able to remain on the treadmill. When this degree of fatigue was noted, and the animal had difficulty remaining on the treadmill belt (regardless of the delivery of the electrical shock), the animal was taken off the treadmill, and the run time was recorded to the nearest second. Again, testing was performed during the latter part of the light phase, when the rats are the most active [15]. All rats repeated the same treadmill test at the end of the intervention period.

2.3. Anthropometric Assessments

Body mass was measured daily. At the end of the three-month intervention, the body composition of all animals, anesthetized with isoflurane, was evaluated by using dual-energy x-ray absorptiometry (DXA) (Discovery QDR series, Hologic Inc., Bedford, MA, USA). The reliability of the DXA exam was evaluated using three rats not participating in the study that were scanned a total of seven times each. The measured variance of the percent body fat for these animals was 0.34, 0.60, and 0.89. Despite the low variance, all experimental animals were scanned twice. If the first two scans provided percent body fat values that varied by more than 1.5%, a third scan was performed, the two scans that were closest were averaged, and the mean value of these two scans was included in the data analysis. These data were used to calculate lean body mass, fat mass, and percent body fat.

2.4. Dietary and Exercise Interventions

The rats were randomly assigned to one of four intervention groups: Western Diet with exercise (WD + E; $n = 14$); Western Diet without exercise (WD; $n = 14$); Daniel Fast with exercise (DF + E; $n = 14$); Daniel Fast without exercise (DF; $n = 14$). The diets were purchased from Research Diets, Inc. (New Brunswick, NJ) and provided in pellet form. The WD was formulated to mimic a typical human WD, containing 17% protein, 43% carbohydrates, and 40% fat—A large portion of which was saturated (milk fat) (product: D12079B). The DF included 15% protein, 60% carbohydrates, and 25% fat (product: D13092801). The macronutrient sources and quantities of the DF were based on the dietary intakes of human subjects in our prior DF studies [1,3,4]. The DF diet fed to the rats therefore mimics what our human subjects consumed in terms of macronutrient sources and composition, fiber, and micronutrients (i.e., antioxidants). The specific nutrient compositions of the WD and DF are provided in Table 1.

The rats consumed their assigned diet (WD or DF) for three months, beginning after the two-week acclimation period. Food was provided ad libitum and was not measured each day. We simply monitored the animals' body mass over time. Our omission of daily food recording may be considered a limitation of this work. Equal numbers of rats in each diet group were randomly assigned either exercise (+E) or no exercise. The animals in the no-exercise groups were placed on the treadmill three days per week for a period of 5 min, while it was turned off. The animals in the exercise groups performed moderate-intensity endurance exercise on a motorized treadmill three days per week (i.e., Monday, Wednesday, Friday) for the three-month intervention period. The speed and duration was progressively increased. Specifically, the animals began training at 20 m·min^{-1} for 15 min·day^{-1} (week 1), progressed to 25 m·min^{-1} for 30 min·day^{-1} (week 2), and then to 25 m·min^{-1} for 35 min·day^{-1} (weeks 3–12). This progressive increase in intensity and duration of exercise is typical for rodent training studies [16].

Table 1. Dietary composition of the Western Diet and Daniel Fast.

Nutrient	Western Diet		Daniel Fast	
	gm%	kcal%	g%	kcal%
Protein	20	17	15	15
Carbohydrate	50	43	58	59
Fat	21	40	11	25
Fiber	5	0	13	1
Total		100		100
kcal/g	4.7		3.9	
Casein	195	780	0	0
Soy Protein	0	0	170	680
DL-Methionine	3	12	3	12
Corn Starch	50	200	0	0
Corn Starch-Hi Maize 260 (70% Amylose and 30% Amylopectin)	0	0	533.5	2134
Maltodextrin 10	100	400	150	600
Sucrose	341	1364	0	0
Cellulose, BW200	50	0	100	0
Inulin	0	0	50	50
Milk Fat, Anhydrous	200	1800	0	0
Corn Oil	10	90	0	0
Flaxseed Oil	0	0	130	1170
Ethoxyquin	0.04	0	0.04	0
Mineral Mix S1001	35	0	35	0
Calcium Carbonate	4	0	4	0
Vitamin Mix V1001	10	40	10	40
Choline Carbonate	2	0	2	0
Ascorbic Acid Phosphate, 33% active	0	0	0.41	0
Cholesterol	1.5	0	0	0
Total	1001.54	4686	1187.95	4686
Saturated g/kg	122.6		7.8	
Monunsaturated g/kg	60.2		19.7	
Polyunsaturated g/kg	13.5		77.7	
Cholesterol mg/kg	2048		0	
Saturated %Fat	62.4		7.4	
Monunsaturated %Fat	30.7		18.7	
Polyunsaturated %Fat	6.9		73.9	
Ascorbic Acid mg/kg	0		114	

2.5. Blood Collection and Analysis

At the end of the three-month intervention, one half of the animals in each group were euthanized, with the remaining animals being retained for a separate long-term study. For blood collection, the rats were euthanized via CO_2 inhalation, the abdominal cavity was opened, and the blood was collected from the inferior vena cava, using a syringe with a 22-gauge needle, and placed in vacutainer tubes containing EDTA. Plasma was separated, and multiple aliquots were stored at $-70\ ^{\circ}C$ for the analysis of plasma lipids and biomarkers of oxidative stress and inflammation. Plasma triglycerides (TAG) and total cholesterol were analyzed following standard enzymatic procedures, as described by the reagents' manufacturer (Thermo Electron Clinical Chemistry; product #: TR22421 and TR13421 for TAG and cholesterol, respectively). Malondialdehyde (MDA) was analyzed following the procedures of Jentzsch and colleagues [17], using reagents purchased from Northwest Life Science Specialties (Vancouver, WA; product #: NWK-MDA01). Advanced Oxidation Protein Products (AOPP) were measured using the methods described by the reagent's manufacturer (Cell Biolabs, Inc., San Diego, CA, USA; product #: STA-318). A customized Milliplex® Map Kit was purchased from the EMD Millipore Corporation (Billerica, MA, USA) for the measurement of IL-1α, IL-4, IL-1β, IL-6, IL-13, IL-10, IFNγ, and TNF-α.

The cytokine concentrations were determined using the MAGPIX® platform with xPONENT software (MilliporeSigma, Burlington, MA, USA). All samples were analyzed in duplicate. Since blood was only obtained at the time of death, no baseline values are available for the biochemical measures.

2.6. Statistical Analysis

The treadmill run time was analyzed using a 4 (group) × 2 (time) analysis of variance (ANOVA). The data obtained from the DXA scan (fat mass, lean mass, percent body fat) and all biochemical measures were analyzed using a one-way ANOVA. Tukey post-hoc tests were used to identify significant group differences. All analyses were performed using JMP statistical software (version 4.0.3; SAS Institute; Cary, NC, USA). Statistical significance was set at $p \leq 0.05$. All data are expressed as the mean ± SEM.

3. Results

One animal in the WD + E group died during week two of the intervention, approximately 30 min following the exercise training session. The necropsy revealed the abdomen was filled with blood, with a suspected aneurism. All remaining animals completed the three-month intervention.

3.1. Physical Performance Data

After three months, the treadmill run time to exhaustion (Table 2) displayed a significant group effect (WD + E & DF + E) > (WD & DF) ($p < 0.0001$) and a significant time effect (month 3 > baseline) ($p = 0.0005$), as well as a group x time interaction effect ($p < 0.0001$). The run time increased in both exercise groups, as demonstrated by a change in the percent improvement from baseline to month 3 in the WD + E (+51%) and the DF + E (+99%) groups. The increase for the DF + E group exceeded that for the WD + E group ($p = 0.02$).

Table 2. Treadmill run time (min) to exhaustion of male rats assigned to two different diets with and without exercise.

	Western Diet + Exercise (WD+E)	Western Diet (WD)	Daniel Fast + Exercise (DF+E)	Daniel Fast (DF)
Pre-Intervention (Baseline)	35.5 ± 3.5	28.7 ± 3.3	29.3 ± 2.8	33.6 ± 4.1
Post-Intervention **	48.3 ± 1.9 *,†	24.4 ± 1.5	52.9 ± 1.9 *,†	28.8 ± 1.1

Values are mean ± SEM. The data show: a group effect for treadmill run time to exhaustion ($p < 0.0001$); * WD + E & DF + E > W & DF; a time effect for treadmill run time to exhaustion ($p = 0.0005$); ** Month 3 > Baseline, a group-by-time interaction effect for treadmill run time to exhaustion ($p < 0.0001$). † Month 3 > Baseline for WD + E and DF + E.

3.2. Anthropometric Data

All the rats gained weight during the three-month study ($p < 0.0001$; Table 3). The gain in body weight displayed a group effect ($p < 0.0001$), with the 205% increase in the WD group exceeding ($p < 0.05$) those in the WD + E (177%), DF + E (148%), and DF (168%) groups. There was no difference between DF and WD + E rats for body weight gain ($p > 0.05$). A group effect was noted for mean fat mass ($p < 0.0001$), with DF rats having a lower mean fat mass than the WD groups; also, mean fat mass in WD + E rats was lower than in WD rats ($p < 0.05$). A group effect was noted for body fat percentage ($p < 0.0001$), with both DF groups displaying lower values than the WD groups; also, body fat percentage in WD + E rats was lower than in WD rats ($p < 0.05$). There was no significant difference in lean mass between the groups ($p = 0.14$).

Table 3. Anthropometric data of male rats assigned to two different diets with and without exercise.

	Western Diet + Exercise	Western Diet	Daniel Fast + Exercise	Daniel Fast
Body Mass (g) Pre-Intervention	186.5 ± 3.3	187.0 ± 4.5	192.6 ± 2.7	185 ± 4.8
Body Mass (g) Post-Intervention	516.8 ± 10.7	571.1 ± 14.7	478.7 ± 11.3	496.8 ± 13.5
Fat Mass (g) Post-Intervention	161.6 ± 8.0	195.5 ± 8.4	100.73 ± 7.4	124.45 ± 9.8
Lean Mass (g) Post-Intervention	366.0 ± 9.2	386.8 ± 6.7	391.4 ± 8.8	376.5 ± 7.8
% Fat Post-Intervention	30.6 ± 1.3	33.5 ± 1.0	20.3 ± 1.3	24.6 ± 1.4

Values are mean ± SEM. A group effect was noted for body mass ($p < 0.0001$). A time effect was noted for body mass ($p < 0.0001$). A group-by-time interaction effect was noted for body mass ($p < 0.0001$). A group effect was noted for fat mass ($p < 0.0001$). A group effect was noted for % fat ($p < 0.0001$). No other statistically significant effects were noted ($p > 0.05$).

3.3. Biochemical Data

Since only one half of the animals were euthanized at the end of month 3, only samples from these animals were available for biochemical analysis ($n = 7$ per group). A group effect was noted for TAG ($p < 0.0001$; Figure 1), cholesterol ($p < 0.0001$; Figure 1), MDA ($p = 0.03$; Figure 2), and AOPP ($p < 0.0001$; Figure 2). The values for TAG, cholesterol, and AOPP in the WD group were different from those in all other groups (Figures 1 and 2B); also, the values in the WD + E group were different from those in all the other groups ($p < 0.05$). For MDA, the WD group displayed a higher value than the DF + E group ($p < 0.05$).

The cytokine results are presented in Table 4. The concentrations of four of the examined cytokines were below the limit of detection and were not included. Due to variation between animals, the differences in IL-1β and IL-10 concentrations in the WD group compared with the DF rats did not reach statistical significance ($p > 0.05$). The concentrations of the remaining cytokines were similar in both groups.

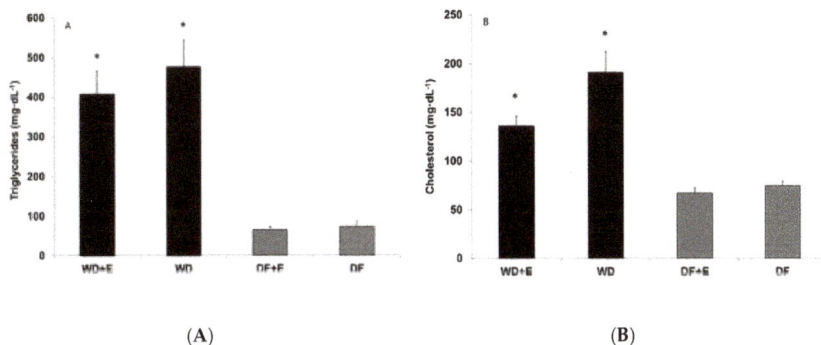

(A)   (B)

Figure 1. (**A**) Blood triglycerides (TAG) and (**B**) total cholesterol of male rats assigned to two different diets with and without exercise. Values are mean ± SEM. *A group effect was noted for TAG ($p < 0.0001$) and cholesterol ($p < 0.0001$). For TAG and cholesterol, the Western Diet (WD) group was different from all the other groups, and the Western Diet + Exercise (WD + E) group was different from all the other groups ($p < 0.05$).

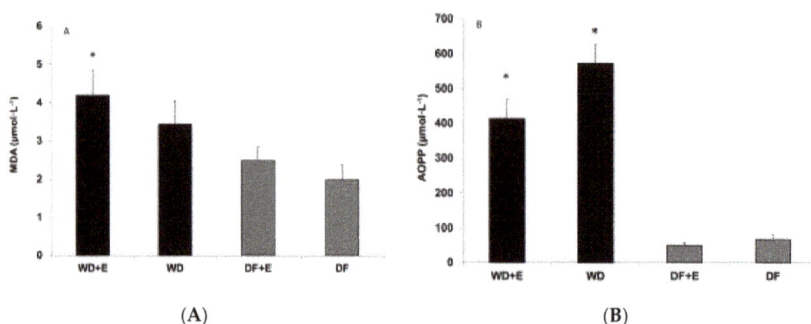

(A)   (B)

Figure 2. (**A**) Blood malondialdehyde (MDA) and (**B**) advanced oxidation protein products (AOPP) of male rats assigned to two different diets with and without exercise. Values are mean $\pm$ SEM. * A group effect was noted for MDA ($p = 0.03$) and AOPP ($p < 0.0001$). For AOPP, the Western Diet (WD) group resulted different from all the other groups; also the Western Diet + Exercise (WD + E) group resulted different than all the other groups ($p < 0.05$). For MDA, the Western Diet + Exercise (WD + E) group resulted different from the DF group ($p < 0.05$).

Table 4. Cytokine concentrations in male rats assigned to two different diets with and without exercise.

Variable	Western Diet + Exercise	Western Diet	Daniel Fast + Exercise	Daniel Fast
IL-4	37 ± 9.6	25.9 ± 7.8	36.3 ± 9.6	27.1 ± 9.6
IL-1β	174.5 ± 59.5	120.9 ± 55.1	31.4 ± 59.5	72.3 ± 65.2
IL-10	126.5 ± 36.7	95.5 ± 34	40.4 ± 34	63.4 ± 34
TNF-α	14.3 ± 5.1	13.8 ± 4.1	14.6 ± 3.6	12.6 ± 4.1

Values are mean $\pm$ SEM (pg$\cdot$mL^{-1}). No statistically significant differences were noted ($p > 0.05$).

4. Discussion

To our knowledge, this is the first study to investigate the impact of a purified diet that mimics the Daniel Fast, with and without exercise training, on physical performance, body composition, and blood-derived measures of health in rats. Our data clearly indicate that a purified vegan diet improves body composition, blood lipids, and measures of oxidative stress compared to a WD. In addition, exercise training for three months enhances physical performance but has little impact on blood lipids or oxidative stress when consuming the DF, with some noted improvement in blood lipids and AOPP when consuming the WD.

4.1. Physical Performance Findings

Several studies have previously reported that exercise training increases endurance physical performance of rats [16,18–20]. This led to the a priori expectation that the exercise training groups would exhibit better physical performance compared to the sedentary groups, which was confirmed. Our hypothesis that rats fed the DF would display a greater adaptive response to exercise than rats fed the WD was demonstrated by the respective 99% versus 51% increases in run times for the DF + E and WD + E groups. These findings are also in agreement with other studies indicating that a Western-style diet can blunt exercise-induced performance gain [21–24]. It is possible that a much higher volume or intensity of exercise may have yielded more robust findings for physical performance and associated variables, as the present study employed only a moderate intensity and volume of exercise.

Our prediction was that the WD with and without exercise would impair endurance performance because of its relatively high fat content, which over time may lead to increased fat mass, which has been linked with higher oxygen consumption and perceived exertion during exercise [25,26]. Moreover, high-fat diets have been implicated in impaired glucose metabolism by skeletal muscle [27], which is important for muscle oxidation during exercise [28]. Diets with high levels of saturated and monounsaturated fat and low levels of polyunsaturated fatty acids, such as the WD used in the present study, yield poor exercise performance in both animals and humans [22,29]. The DF diet provided low levels of saturated and monounsaturated fatty acids and high levels of polyunsaturated fatty acids and was associated with improved exercise performance.

In addition to the amount and form of dietary fat, the amount and type of carbohydrate may have contributed to our findings. High-carbohydrate diets increase time to exhaustion [30,31] by delaying the complete oxidation of muscle glycogen [30,32]. The complex low-glycemic carbohydrates in the DF have been reported to enhance physical performance [33,34]. Overall, it is conceivable that the differences in the amount and form of dietary fat and carbohydrate may have contributed to the divergence in physical performance for the WD and DF rats.

4.2. Anthropometric Findings

Despite ad libitum feeding in all groups, body mass was greater in the WD group than in all others. We anticipated that the exercise groups would have a lower overall body mass due to the increased caloric expenditure from the thrice-weekly exercise bouts [35]. However, DF animals, despite being allowed to freely feed throughout the three-month intervention period, showed only a slightly reduced body mass compared to animals in the WD groups; in contrast, they showed a remarkably lower body fat. Despite the dramatic difference in body fat between the WD and DF groups, only small differences were noted in lean body mass. These findings suggest that animal protein is not necessary for the development of lean body mass—an often expressed concern raised by many nutritionists. It should be noted that, despite the plant-based protein sources, all DF animals grew as expected over the course of the three-month period and, importantly, without accumulating the excessive amounts of fat measured in the WD groups.

4.3. Biochemical Findings

In agreement with our findings obtained from human subjects [3,4,6], blood lipids and oxidative stress biomarkers were lower in the DF groups than in the WD groups. In terms of plasma lipids, the six-fold and two- to three-fold higher TAG and cholesterol values in the WD rats compared to the DF rats are attributed to differences in the types and amounts of dietary carbohydrate (including fiber) and fat and are in line with the 20% reduction in total and LDL-cholesterol measured in human subjects after just 21 days of adherence to the Daniel Fast [3,6].

The differences between the WD and DF groups for both MDA and AOPP are indicative of differences in lipid peroxidation and protein oxidation. The approximately eight-fold decrease in AOPP in rats fed the DF exceeds the decrease in oxidative stress observed in human subjects following the Daniel Fast for 21 days [4]. This suggests that maintaining a DF regimen beyond 21 days can continue the decline in oxidative burden, but this hypothesis needs to be confirmed, including other measures of oxidative stress. While exercise training improved physical performance in both exercise groups, chronic ingestion of the WD appears to blunt the exercise-driven improvement. This observed response may be at least partially attributable to higher levels of oxidative stress. Increased oxidative stress alters protein function by damaging both contractile (primarily myosin, due to thiol group oxidation [36]) and enzymatic proteins, compromising the excitation–contraction coupling [37] and potentially slowing reaction rates, respectively. Moreover, the potential oxidation of mitochondrial enzymes required for energy production (e.g., succinate dehydrogenase, cytochrome oxidase [38]) could negatively impact endurance performance.

Although systemic inflammation is reduced in human subjects following the Daniel Fast for 21 days [3,6,39], this was not evident in the rats in the present study. The human studies suggest that, with larger samples sizes, the approximately three-fold and two-fold higher concentrations of IL-1β and IL-10, respectively, in the WD rats may have reached significance. This is important, as IL-1β is secreted by all nucleated cells, including macrophages, monocytes, B cells, fibroblasts, chondrocytes, and keratinocytes [40], and has been repeatedly labeled as one of the most significant factors in regulating both the local and the systemic onset of acute and chronic inflammation [41,42]. The cytokine patterns found in the different groups also confirms previous research indicating that physical inactivity coupled with a WD leads to higher inflammatory levels [43,44]. The higher IL-10 in the WD groups may have been due to a compensatory response to increased pro-inflammatory cytokines (e.g., IL-1β). More work is needed to investigate this possibility.

5. Conclusions

The present work is unique in investigating the effect of a purified diet, with and without exercise, on measures of functional capacity, body composition, and biochemical outcomes in male rats. The findings extend our prior work in human subjects partaking in the Daniel Fast dietary plan and emphasize that macronutrient composition, and not simply calorie intake, has an influence on the degree of adiposity, independent of exercise. Notably, compared to the glycemic- and saturated fatty acid-rich WD, the macronutrient mix (low levels of glycemic carbohydrate, high levels of polyunsaturated fat) of the DF, significantly improves body composition, plasma lipids, markers of oxidative stress, and physical performance as measured by treadmill run time to exhaustion. Future investigations using animal models and human subjects are needed to define the specific mechanisms responsible for the improved indicators of health and physical performance in response to the Daniel Fast dietary plan, with and without exercise.

Author Contributions: R.J.B. was responsible for designing the research study, performing biochemical analyses, performing statistical analyses, writing the paper, and had primary responsibility for the final content. J.H.M.S., T.A.G., S.-R.L. were responsible for conducting the research (animal training, animal testing, data collection, database management). M.v.d.M. was responsible for performing biochemical analyses and assistance with manuscript preparation. H.J.S. and K.K.B. were responsible for designing the research study and conducting the research. R.K.B. was responsible for designing the research study, conducting the research, and assistance with manuscript preparation. All authors read and approved of the manuscript.

Funding: This research was funded by the School of Health Studies at the University of Memphis.

Acknowledgments: Appreciation is extended to Donny Ray for assistance with animal handling during the necropsies, as well as to Matt Butawan for assistance with manuscript formatting.

Conflicts of Interest: No author declares a conflict of interest related to this work.

Abbreviations

AOPP	Advanced Oxidation Protein Products
DF	Daniel Fast
DXA	Dual-Energy X-ray Absorptiometry
+E	Exercise-Trained
MDA	Malondialdehyde
TAG	Triglycerides
WD	Western Diet

References

1. Trepanowski, J.F.; Canale, R.E.; Marshall, K.E.; Kabir, M.M.; Bloomer, R.J. Impact of caloric and dietary restriction regimens on markers of health and longevity in humans and animals: A summary of available findings. *Nutr. J.* **2011**, *10*, 107. [CrossRef] [PubMed]
2. Trepanowski, J.F.; Varady, K.A. Veganism is a viable alternative to conventional diet therapy for improving blood lipids and glycemic control. *Crit. Rev. Food Sci. Nutr.* **2015**, *55*, 2004–2013. [CrossRef] [PubMed]
3. Bloomer, R.J.; Kabir, M.M.; Canale, R.E.; Trepanowski, J.F.; Marshall, K.E.; Farney, T.M.; Hammond, K.G. Effect of a 21 day Daniel Fast on metabolic and cardiovascular disease risk factors in men and women. *Lipids Health Dis.* **2010**, *9*. [CrossRef] [PubMed]

4. Bloomer, R.J.; Kabir, M.M.; Trepanowski, J.F.; Canale, R.E.; Farney, T.M. A 21 day Daniel Fast improves selected biomarkers of antioxidant status and oxidative stress in men and women. *Nutr. Metab.* **2011**, *8*, 17. [CrossRef] [PubMed]

5. Trepanowski, J.F.; Bloomer, R.J. The impact of religious fasting on human health. *Nutr. J.* **2010**, *9*, 57. [CrossRef] [PubMed]

6. Trepanowski, J.F.; Kabir, M.M.; Alleman, R.J.; Bloomer, R.J. A 21-day Daniel fast with or without krill oil supplementation improves anthropometric parameters and the cardiometabolic profile in men and women. *Nutr. Metab.* **2012**, *9*, 82. [CrossRef] [PubMed]

7. Anderson, J.E. A role for nitric oxide in muscle repair: Nitric oxide-mediated activation of muscle satellite cells. *Mol. Biol. Cell* **2000**, *11*, 1859–1874. [CrossRef] [PubMed]

8. Powers, S.K.; Jackson, M.J. Exercise-induced oxidative stress: Cellular mechanisms and impact on muscle force production. *Physiol. Rev.* **2008**, *88*, 1243–1276. [CrossRef] [PubMed]

9. Burgomaster, K.A.; Hughes, S.C.; Heigenhauser, G.J.; Bradwell, S.N.; Gibala, M.J. Six sessions of sprint interval training increases muscle oxidative potential and cycle endurance capacity in humans. *J. Appl. Physiol.* **2005**, *98*, 1985–1990. [CrossRef] [PubMed]

10. Eliakim, A.; Moromisato, M.; Moromisato, D.Y.; Cooper, D.M. Functional and Muscle Size Response to 5 Days of Treadmill Training in Young Rats. *Pediatr. Exerc. Sci.* **1997**, *9*, 324–330. [CrossRef]

11. McClenton, L.S.; Brown, L.E.; Coburn, J.W.; Kersey, R.D. The effect of short-term VertiMax vs. *depth jump training on vertical jump performance. J. Strength Cond. Res.* **2008**, *22*, 321–325. [CrossRef] [PubMed]

12. Copp, S.W.; Davis, R.T.; Poole, D. C.; Musch, T.I. Reproducibility of endurance capacity and Vo$_{2peak}$ in male Sprague-Dawley rats. *J. Appl. Physiol.* **2009**, *106*, 1072–1078. [CrossRef] [PubMed]

13. Koch, L.G.; Meredith, T.A.; Fraker, T.D.; Metting, P.J.; Britton, S.L. Heritability of treadmill running endurance in rats. *Am. J. Physiol.* **1998**, *275*, R1455–R1460. [CrossRef] [PubMed]

14. Lightfoot, J.T.; Turner, M.J.; Knab, A.K.; Jedlicka, A.E.; Oshimura, T.; Marzec, J.; Gladwell, W.; Leamy, L.J.; Kleeberger, S.R. Quantitative trait loci associated with maximal exercise endurance in mice. *J. Appl. Physiol.* **2007**, *103*, 105–110. [CrossRef] [PubMed]

15. Evans, H.L. Rats' activity: Influence of light-dark cycle, food presentation and deprivation. *Physiol. Behav.* **1971**, *7*, 455–459. [CrossRef]

16. Huang, T.; Chang, F.; Lin, S.; Liu, S.; Hsieh, S.S.; Yang, R. Endurance treadmill running training benefits the biomaterial quality of bone in growing male Wistar rats. *J. Bone Miner. Metab.* **2008**, *26*, 350–357. [CrossRef] [PubMed]

17. Jentzsch, A.M.; Bachmann, H.; Fürst, P.; Biesalski, H.K. Improved analysis of malondialdehyde in human body fluids. *Free Radic. Biol. Med.* **1996**, *20*, 251–256. [CrossRef]

18. Dolinsky, V.W.; Jones, K.E.; Sidhu, R.S.; Haykowsky, M.; Czubryt, M.P.; Gordon, T.; Dyck, J.R. Improvements in skeletal muscle strength and cardiac function induced by resveratrol during exercise training contribute to enhanced exercise performance in rats. *J. Physiol.* **2012**, *590*, 2783–2799. [CrossRef] [PubMed]

19. Lee, J.S.; Bruce, C.R.; Spriet, L.L.; Hawley, J.A. Interaction of diet and training on endurance performance in rats. *Exp. Physiol.* **2001**, *86*, 499–508. [PubMed]

20. Mazzeo, R.S.; Horvath, S.M. Effects of training on weight, food intake, and body composition in aging rats. *Am. J. Clin. Nutr.* **1986**, *44*, 732–738. [CrossRef] [PubMed]

21. Helge, J.W.; Richter, E.A.; Kiens, B. Interaction of training and diet on metabolism and endurance during exercise in man. *J. Physiol.* **1996**, *492*, 293–306. [CrossRef] [PubMed]

22. Murray, A.J.; Knight, N.S.; Cochlin, L.E.; McAleese, S.; Deacon, R.M.; Rawlins, J.N.; Clarke, K. Deterioration of physical performance and cognitive function in rats with short-term high-fat feeding. *FASEB J.* **2009**, *23*, 4353–4360. [CrossRef] [PubMed]

23. Okano, G.; Sato, Y.; Murata, Y. Effect of elevated blood FFA levels on endurance performance after a single fat meal ingestion. *Med. Sci. Sports Exerc.* **1998**, *30*, 763–768. [CrossRef] [PubMed]

24. Starling, R.D.; Trappe, T.A.; Parcell, A.C.; Kerr, C.G.; Fink, W.J.; Costill, D.L. Effects of diet on muscle triglyceride and endurance performance. *J. Appl. Physiol.* **1997**, *82*, 1185–1189. [CrossRef] [PubMed]

25. Ekkekakis, P.; Lind, E. Exercise does not feel the same when you are overweight: The impact of self-selected and imposed intensity on affect and exertion. *Int. J. Obes.* **2006**, *30*, 652–660. [CrossRef] [PubMed]

26. Norman, A.C.; Drinkard, B.; McDuffie, J.R.; Ghorbani, S.; Yanoff, L.B.; Yanovski, J.A. Influence of excess adiposity on exercise fitness and performance in overweight children and adolescents. *Pediatrics* **2006**, *115*, e690–e696. [CrossRef] [PubMed]

27. Tanaka, S.; Hayashi, T.; Toyoda, T.; Hamada, T.; Shimizu, Y.; Hirata, M.; Ebihara, K.; Masuzaki, H.; Hosoda, K.; Fushiki, T. High-fat diet impairs the effects of a single bout of endurance exercise on glucose transport and insulin sensitivity in rat skeletal muscle. *Metabolism* **2007**, *56*, 1719–1728. [CrossRef] [PubMed]

28. Wahren, J.; Felig, P.; Ahlborg, G.; Jorfeldt, L. Glucose metabolism during leg exercise in man. *J. Clin. Investig.* **1971**, *50*, 2715–2725. [CrossRef] [PubMed]

29. Rowlands, D.S.; Hopkins, W.G. Effect of high-fat, high-carbohydrate, and high-protein meals on metabolism and performance during endurance cycling. *Int. J. Sport Nutr. Exerc. Metab.* **2002**, *12*, 318–335. [CrossRef] [PubMed]

30. Foskett, A.; Williams, C.; Boobis, L.; Tsintzas, K. Carbohydrate availability and muscle energy metabolism during intermittent running. *Med. Sci. Sports Exerc.* **2008**, *40*, 96–103. [CrossRef] [PubMed]

31. Helge, J.W.; Ayre, K.; Chaunchaiyakul, S.; Hulbert, A.J.; Kiens, B.; Storlien, L.H. Endurance in high-fat-fed rats: Effects of carbohydrate content and fatty acid profile. *J. Appl. Physiol.* **1998**, *85*, 1342–1348. [CrossRef] [PubMed]

32. Coyle, E.F.; Coggan, A.R.; Hemmert, M.K.; Ivy, J.L. Muscle glycogen utilization during prolonged strenuous exercise when fed carbohydrate. *J. Appl. Physiol.* **1986**, *61*, 165–172. [CrossRef] [PubMed]

33. DeMarco, H.; Sucher, K.; Cisar, C.; Butterfield, G. Pre-exercise carbohydrate meals: Application of glycemic index. *Med. Sci. Sports Exerc.* **1999**, *31*, 164–170. [CrossRef] [PubMed]

34. Thomas, D.; Brotherhood, J.; Brand, J. Carbohydrate feeding before exercise: Effect of glycemic index. *Int. J. Sports Med.* **1991**, *12*, 180–186. [CrossRef] [PubMed]

35. Ekelund, U.; Brage, S.; Besson, H.; Sharp, S.; Wareham, N.J. Time spent being sedentary and weight gain in healthy adults: Reverse or bidirectional causality? *Am. J. Clin. Nutr.* **2008**, *88*, 612–617. [CrossRef] [PubMed]

36. Liu, D.F.; Wang, D.; Stracher, A. The accessibility of the thiol groups on G- and F-actin of rabbit muscle. *Biochem. J.* **1990**, *266*, 453–459. [CrossRef] [PubMed]

37. Goldhaber, J.I.; Qayyum, M.S. Oxygen free radicals and excitation-contraction coupling. *Antiox. Redox Signal.* **2000**, *2*, 55–64. [CrossRef] [PubMed]

38. Haycock, J.W.; Jones, P.; Harris, J.B.; Mantle, D. Differential susceptibility of human skeletal muscle proteins to free radical induced oxidative damage: A histochemical, immunocytochemical and electron microscopical study in vitro. *Acta Neuropathol.* **1996**, *92*, 331–340. [CrossRef] [PubMed]

39. Alleman, R.J.; Harvey, I.C.; Farney, T.M.; Bloomer, R.J. Both a traditional and modified Daniel Fast improve the cardio-metabolic profile in men and women. *Lipids Health Dis.* **2013**, *12*, 114. [CrossRef] [PubMed]

40. Huang, C.; Huang, P.; Chen, C. Interleukin-1-beta, interleukin-10, and tumor necrosis factor-alpha in Chinese patients with ankylosing spondylitis. *Mid. Taiwan J. Med.* **2009**, *14*, 10–15.

41. Besedovsky, H.; del Rey, A.; Sorkin, E.; Dinarello, C.A. Immunoregulatory feedback between interleukin-1 and glucocorticoid hormones. *Science* **1986**, *233*, 652–654. [CrossRef] [PubMed]

42. Coppack, S.W. Pro-inflammatory cytokines and adipose tissue. *Proc. Nutr. Soc.* **2001**, *60*, 349–356. [CrossRef] [PubMed]

43. Lakhdar, N.; Denguezli, M.; Zaouali, M.; Zbidi, A.; Tabka, Z.; Bouassida, A. Diet and diet combined with chronic aerobic exercise decreases body fat mass and alters plasma and adipose tissue inflammatory markers in obese women. *Inflammation* **2013**, *36*, 1239–1247. [CrossRef] [PubMed]

44. Thompson, D.; Markovitch, D.; Betts, J.A.; Mazzatti, D.; Turner, J.; Tyrrell, R.M. Time course of changes in inflammatory markers during a 6-mo exercise intervention in sedentary middle-aged men: A randomized-controlled trial. *J. Appl. Physiol.* **2010**, *108*, 769–779. [CrossRef] [PubMed]

nutrients

MDPI

Article

Determinants of Behaviour Change in a Multi-Component Telemonitoring Intervention for Community-Dwelling Older Adults

Marije N. van Doorn-van Atten [1,*], Lisette C. P. G. M. de Groot [1], Jeanne H. M. de Vries [1] and Annemien Haveman-Nies [2]

[1] Division of Human Nutrition and Health, Wageningen University and Research, P.O. Box 17, 6700 AA Wageningen, The Netherlands; lisette.degroot@wur.nl (L.C.P.G.M.d.G.); jeanne.devries@wur.nl (J.H.M.d.V.)

[2] Strategic Communication Chair, Wageningen University and Research, P.O. Box 17, 6700 AA Wageningen, The Netherlands; annemien.haveman@wur.nl

* Correspondence: marije.vandoorn@wur.nl; Tel.: +31-(0)317-485062

Received: 5 July 2018; Accepted: 9 August 2018; Published: 10 August 2018

Abstract: Optimal diet quality and physical activity levels are essential for healthy ageing. This study evaluated the effects of a multi-component telemonitoring intervention on behavioural determinants of diet quality and physical activity in older adults, and assessed the mediating role of these determinants and two behaviour change techniques in the intervention's effects. A non-randomised controlled design was used including 214 participants (average age 80 years) who were allocated to the intervention or control group based on municipality. The six-month intervention consisted of self-measurements of nutritional outcomes and physical activity, education, and follow-up by a nurse. The control group received regular care. Measurements took place at baseline, after 4.5 months and at the end of the study. The intervention increased self-monitoring and improved knowledge and perceived behavioural control for physical activity. Increased self-monitoring mediated the intervention's effect on diet quality, fruit intake, and saturated fatty acids intake. Improved knowledge mediated the effect on protein intake. Concluding, this intervention led to improvements in behavioural determinants of diet quality and physical activity. The role of the hypothesised mediators was limited. Insight into these mechanisms of impact provides directions for future development of nutritional eHealth interventions for older adults, in which self-monitoring may be a promising behaviour change technique. More research is necessary into how behaviour change is established in telemonitoring interventions for older adults.

Keywords: older adults; diet quality; physical activity; telemonitoring; lifestyle intervention; mechanisms of impact; mediation analyses

---

1. Introduction

An increasing number of older adults lives longer and healthier. An optimal nutritional status contributes to healthy ageing. Conversely, ageing poses nutritional risks as deteriorations in health, cognitive, and physical functioning, as well as changes in social circumstances, may impair nutritional status [1]. In the Netherlands, 11 to 35% of community-dwelling older adults are undernourished and diet quality of community-dwelling older adults is suboptimal [2,3]. Furthermore, awareness concerning undernutrition is low among older adults [4,5] and nutrition knowledge and attitude seem to be poorer among older adults than among younger adults [6–8]. Good access to appropriate nutrition care, such as meal programs, nutrition education, nutritional monitoring, counselling, and therapy, contributes to an optimal nutritional status [9]. Additionally, physical activity (PA) levels of older adults are suboptimal, with about one third of the 70–79-year-old and about half of the adults aged 80 years and over failing to meet the WHO guidelines for PA [10]. Barriers are mentioned such as health status, fear, and lack of interest [11,12], with health status also acting as a facilitator (e.g., physical benefits of PA), together with enjoyment and social support [11]. Much is expected from eHealth as a way to improve nutrition and PA behaviour [13]. Advantages of eHealth include personalisation, scalability, accessibility, and reduced costs as compared to regular face-to-face care [13].

Reviews of eHealth interventions to improve nutrition behaviour in various settings show mixed results and mostly focus on younger populations [14–17]. eHealth interventions to improve nutritional outcomes in older adults are scarce. One pilot study focussed on providing computer-tailored dietary advice to older adults, in combination with improving physical activity and meaningful social roles. This appeared to be feasible, but effectiveness has yet to be affirmed in an RCT [18]. Another eHealth study focussed on nutritional counselling for older adults at increased cardiovascular risk, but effects on dietary intake were not evaluated [19]. The scarcity of nutritional eHealth interventions for older adults and mixed results of eHealth interventions to improve nutrition behaviour in a general population call for more research.

To explore the potential of nutritional eHealth interventions for older adults, it is not only necessary to know whether interventions are effective, but also how an intervention achieves its effects [14]. Ideally, interventions rely on a theoretical framework that specifies how an intervention results in effects on behavioural determinants and behaviour through behaviour change techniques (BCT's) [20]. Research shows that increased use of theory positively impacts effect sizes [21]. Testing a theoretical framework in order to verify the assumed relations deepens understanding of how interventions work and contributes to future intervention development. However, only a minority of nutritional eHealth studies that included a theoretical framework analysed the hypothesised mediators [14,15], and more insight is needed into what contributes to effective eHealth interventions to improve nutrition behaviour in older populations.

The PhysioDom Home Dietary Intake Monitoring (HDIM) study focused on telemonitoring of nutritional parameters and physical activity. This intervention resulted in improved compliance with the Dutch dietary guidelines for the intake of vegetables, fruit, dietary fibre, and protein, and to guidelines for PA [22]. Concerning the content of the PhysioDom HDIM intervention, the three most important BCT's were self-monitoring, goalsetting, and feedback, reflecting an application of control theory [20,23,24]. Effectiveness of self-monitoring has been confirmed in non-eHealth studies [25], but eHealth studies including self-monitoring to promote behaviour change show less optimistic results [21,26]. It has been shown that self-monitoring is more effective in combination with other BCT's such as goalsetting and tailored feedback [21,26]. According to the control theory, self-monitoring, goalsetting and feedback are key in behavioural self-management [24], which is relevant nowadays with the increasing focus on self-management of health and health-related behaviours [27].

All in all, we hypothesized that the intervention would result in an increased frequency of self-monitoring and goalsetting, and in improved perceived behavioural control, attitude, and knowledge, in turn improving diet quality and PA. In this article, we aimed to shed light on these hypothesised mechanisms of impact by studying changes in frequency of self-monitoring and goalsetting, by studying the effects on perceived behavioural control, attitude, and knowledge, and by studying the mediating role of self-monitoring, goalsetting, perceived behavioural control, attitude, and knowledge in the effects of PhysioDom HDIM on diet quality and PA.

2. Materials and Methods

2.1. Design

Measurements took place from April 2016 until June 2017, when the last participants finished the study. The study followed a non-randomised controlled design and had a duration of six months. Measurements took place at baseline (T0), after 4.5 months (T1), and after six months at the end of the study (T2). Telemonitoring measurements took place at the beginning of the study and three months after the start of the study, and only in the intervention group. The control group received regular care.

2.2. Ethics Approval and Consent to Participate

Written informed consent was obtained from all participants. The study was conducted in accordance with the Declaration of Helsinki, and all study procedures involving participants were approved by the Medical Ethical Committee of Wageningen University on the 18 February 2016, number NL53619.081.1.

2.3. Trial Registration

The study was registered at ClinicalTrials.gov (identifier NCT03240094), URL http://bit.ly/2zFTs3P.

2.4. Participants

Participants were recruited from February 2016 until September 2016 and were recruited from nine small to middle-sized municipalities in the Netherlands. Allocation of participants to the intervention or control group took place on municipality level to prevent contamination between the intervention and control group as local HCP's implemented the intervention. Five municipalities were non-randomly allocated to the intervention group and four other municipalities were allocated to the control group. Participants were recruited via letters from the two involved care organisations, via advertisements in local newspapers and public spaces, and invitation letters via post. Inclusion criteria were being 65 years or older and receiving home care and/or living in sheltered accommodation or a service flat. Exclusion criteria were cognitive impairment (Mini Mental State Examination (MMSE) <20), having cancer, receiving terminal care, being bedridden or bound to a wheelchair, or being unable to watch television. Persons who were interested to participate were visited by a researcher to receive more information, to have questions answered, to sign the informed consent, and to be screened on the exclusion criteria.

2.5. Intervention

The intervention consisted of telemonitoring measurements by participants, education concerning nutrition and PA, and follow-up by a nurse. These intervention components are further described below.

2.5.1. Telemonitoring Measurements

Firstly, participants performed self-measurements of body weight (weekly, with an A&D weighing scale, type UC-411PBT-C), steps (one week per month using a pedometer of A&D, type UW-101), and blood pressure (monthly or bi-monthly, only upon indication of a nurse). They also filled out questionnaires about their nutritional status, appetite, and diet quality using the Mini Nutritional Assessment Short-Form (MNA-SF) [28], Simplified Nutritional Appetite Questionnaire (SNAQ) [29], and Dutch Healthy Diet Food Frequency Questionnaire (DHD-FFQ) [30], respectively. Participants filled out these questionnaires at T0 during an interview with a researcher and three months later by means of a computer, tablet, or during a telephone interview with a researcher. Participants could view their telemonitoring results on a special television channel and could thus become aware of their nutritional status and changes in nutritional status. On this television channel, participants also received short text messages in which they were asked to write down their goals for diet quality (two times) and steps (daily, during one week per month).

2.5.2. Education

Secondly, participants received computer-tailored and non-tailored information about nutrition and physical activity. The computer-tailored information consisted of advice sent per post on how to improve compliance with ten Dutch food-based dietary guidelines and the Dutch guideline for physical activity. This advice was tailored to the participant's current compliance with the guidelines as measured by the DHD-FFQ. For example, advice concerning vegetable intake contained more accessible suggestions for participants with low compliance than for participants who were already compliant, for which suggestions were more focussed on maintaining this behaviour and diversity of vegetable intake. The non-tailored information consisted of three short television messages per week containing general information about nutrition and physical activity.

2.5.3. Follow-Up

Thirdly, a total of seven nurses was available to provide follow-up on the participants' self-measurements. On the project's website, they checked alerts that were activated in case of undernutrition, risk of undernutrition (based on MNA score, SNAQ score, weight loss of > five percent and/or body mass index (BMI) <20 kg/m^2), obesity (based on BMI >30 kg/m^2), or new blood pressure measurements. Nurses planned follow-up of these alerts with help of decision trees. In case of a good nutritional status, nurses kept monitoring without taking action. In case of risk of undernutrition, nurses contacted participants via telephone or a home visit. Nurses identified causes, provided suggestions on how to improve dietary intake, and gave a leaflet with advice to reverse the risk of undernutrition [31]. In case of undernutrition or obesity, nurses discussed with participants whether referral to a dietician or general practitioner was desirable.

2.6. Measurements

Measurements took place during a screening visit prior to the beginning of the study and at T0, T1, and T2. In the control group, the screening visit and T0 visit coincided. Data were collected by means of structured interviews at the participant's homes conducted by a trained researcher or research assistant. Furthermore, paper questionnaires were used to collect data.

2.6.1. Baseline Characteristics

Baseline characteristics were recorded during the screening visit and at T0 and included sex, age, body weight, height, number of morbidities, education level, civil status, living situation, country of birth, cognitive functioning as measured by the Mini Mental State Examination (MMSE) [32], physical functioning as measured by the Katz-15 [33], nutritional status as measured by the Mini Nutritional Assessment (MNA) [34], desire to lose weight, and type of received care. Body weight was measured without shoes and heavy clothes (scale of type UC-411PBT-C, A&D).

2.6.2. Frequency of Self-Monitoring and Goalsetting

Frequency of self-monitoring and goalsetting was measured at T0, T1, and T2 using a paper questionnaire with items derived from literature (Table 1). Self-monitoring was measured using four statements that were combined to form one scale for self-monitoring [35]. Goalsetting was measured using three statements that were combined to form one scale for goalsetting [36].

2.6.3. Behavioural Determinants

Behavioural determinants were measured at T0, T1, and T2 using a paper questionnaire with items derived from literature (Table 1). Perceived behavioural control (PBC) was measured using two items for self-efficacy and two items for controllability for both physical activity (PA) and healthy eating (HE) behaviour [37]. These items were combined into two scales for PA and HE. Attitude concerning HE and PA were each measured by six semantic differential items [38]. Items were combined to form scales of attitude concerning PA and HE. Crohnbach's alpha's for the abovementioned questionnaire items ranged from 0.67 to 0.80. Knowledge was measured using 11 statements concerning a healthy diet and physical activity that were answered with 'true', 'false', or 'I don't know'. A knowledge score (0–11) was generated based on the number of correct answers.

Table 1. Items to measure self-monitoring, goalsetting, perceived behavioural control, and attitude in the PhysioDom HDIM study.

Construct	Questionnaire Items	Answering Options (1–5)	Crohnbach's Alpha T0
Self-monitoring	1. How often in the past month have you kept track in your head of the amount of food you have eaten? 2. How often in the past month have you kept track in your head of the types of foods you have eaten during the course of the day? 3. How often in the past month have you kept track in your head of how physically active you have been during a week? 4. How often in the past month you have weighed yourself?	Never/a single time/a couple of times/every week/everyday	0.77
Goalsetting	1. How often in the past month did you set goals related to your weight? 2. How often in the past month did you set goals related to your eating habits? 3. How often in the past month did you set goals related to how much you exercise?	Never/a single time/a couple of times/every week/everyday	0.67
Perceived behavioural control healthy eating	1. I am confident that I can eat healthy in the coming month if I want to. 2. Whether I eat healthy in the coming month is entirely up to me. 3. Healthy eating in the coming month is for me... 4. How much control do you have over healthy eating in the coming month?	Totally disagree–totally agree Totally disagree–totally agree Difficult–easy No control–complete control	0.70
Perceived behavioural control physical activity	1. I am confident that I can be sufficiently physically active in the coming month if I want to. 2. Whether I am sufficiently physically active in the coming month is entirely up to me. 3. Sufficient physical activity in the coming month is for me... 4. How much control do you have over being sufficiently physically active in the coming month?	Totally disagree–totally agree Totally disagree–totally agree Difficult–easy No control–complete control	0.71
Attitude healthy eating	Healthy eating in the coming month is for me...	foolish–wise pleasant–unpleasant bad–good harmful–helpful unnecessary–necessary unenjoyable–enjoyable	0.70
Attitude physical activity	Physical activity in the coming month is for me...	foolish–wise pleasant–unpleasant bad–good harmful–helpful unnecessary–necessary boring–interesting	0.80

2.6.4. Compliance with Dutch Dietary Guidelines and Guidelines for Physical Activity

Compliance with Dutch dietary guidelines and guidelines for physical activity were evaluated using the DHD-FFQ, which was administered during a structured interview at T0 and T2. The DHD-FFQ contains 29 items that evaluate compliance with Dutch dietary guidelines and compliance with PA guidelines [30,39]. Additionally, for this study, compliance with guidelines for the intake of protein and vitamin D was evaluated as the DHD-FFQ contains questions on all relevant protein and vitamin D sources consumed by a Dutch elderly population [3,40]. This resulted in sub scores for compliance with guidelines for the intake of fruit ($\geq$200 g), vegetables ($\geq$150–200 g), dietary fibre ($\geq$14 g/4.2 MJ), fish (two times per week, from which at least one time fatty fish), saturated fatty acids (<10 en%), trans fatty acids (<1 en%), salt (<6 g), alcohol ($\leq$2 glasses for men,

≤1 glass for women), protein (≥70 grams for men, ≥55 grams for women), vitamin D (≥20 mcg), and physical activity (moderate physical activity for at least 30 min a day on at least five days a week). These scores ranged from 0–10, with higher scores indicating better compliance. A total score for diet quality was constructed by summing scores for vegetables, fruit, dietary fibre, fish, alcohol, saturated fatty acids, trans-fatty acids, and sodium. More information can be found elsewhere [30].

2.7. Statistics

The sample size was based on the primary outcome of the main study: nutritional status [34]. Aiming to detect a difference in MNA change of three, assuming a standard deviation of 6.1 [41], and taking into account a two-sided significance level of 0.05 and power of 80%, a sample size of 65 participants per group was required based on the formula $2 \times \frac{[(Z\alpha/2+Z\beta)^2 \times \sigma^2]}{\delta^2}$. We expected a drop-out rate of 30% at maximum, therefore we needed a sample size of at least 93 participants in each group.

Data were analysed using SPSS Statistics for Windows version 22 (IBM Corp., Armonk, NY, USA). Statistical analyses were carried out according to the intention-to-treat principle. Firstly, baseline characteristics of the intervention and control group were described using means (± standard deviations) or percentages. Differences between the groups were tested using independent *t*-tests, Mann-Whitney *U* tests in case of non-normality, or chi-square tests. Secondly, changes in self-monitoring and goalsetting and intervention effects on behavioural determinants were assessed using linear mixed models. Therefore, we first specified a model as large as possible including all main effects, possible interactions, and an unstructured covariance matrix. We then simplified the random part of the model by testing whether simpler covariance structures were allowed using the (REML) LR test, until a model was obtained that was as parsimonious as possible. Consequently, we simplified the fixed part of the model by including dummies for time points T1 and T2, treatment, the interaction terms of the time point dummies and treatment, age, sex, and also other covariates (e.g., BMI, education level, living situation, MNA, functional status, MMSE, municipality, diet, informal care) if they considerably (e.g., >10%) influenced the effect estimates. Thirdly, we performed mediation analyses to evaluate whether the effects of PhysioDom HDIM were mediated as hypothesised (Figure 1). The following outcomes were selected for mediation analyses as these were positively affected by the PhysioDom HDIM intervention: compliance with the Dutch dietary guidelines for the intake of fruit, vegetables, fibre, and protein, and compliance with Dutch guidelines for PA [22]. As mediation can also exist in absence of a significant intervention effect on the study's outcomes [42], also other components of the DHD-FFQ (alcohol, salt, saturated fatty acids, fish, and for this study vitamin D) and the total DHD-FFQ score were included as outcomes in the mediation analyses. To capture the longitudinal nature of the data, we used a multiple serial mediation model for each outcome and each hypothesised mediator in which the mediator at T1 and T2 was modelled in sequence (Figure 1). We regarded a parallel multiple mediator model less appropriate as the condition that no mediator causally influences another would probably not be fulfilled [42]. Using the PROCESS macro for SPSS version 2.16.3 we assessed whether indirect effects of the intervention on the selected outcomes through the hypothesised mediators were statistically significant [42]. Standard errors and confidence intervals of indirect effects were calculated using bootstrapping (10,000 samples). The analyses for diet quality and physical activity were adjusted for age, sex, and baseline values of the mediator.

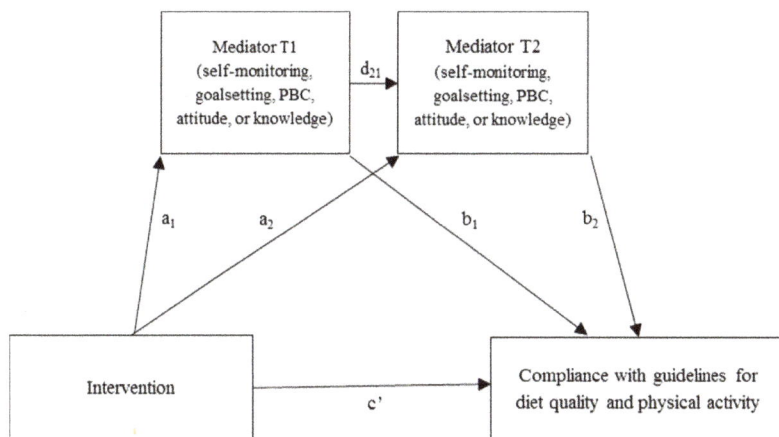

Figure 1. Hypothesised mediation pathways in the PhysioDom Home Dietary Intake Monitoring (HDIM) intervention. One model for each outcome and mediator.

3. Results

3.1. Baseline Characteristics

In total, 215 persons were screened, from which 97 were allocated to the intervention group and 107 to the control group. During the study, 21 intervention group participants and six control group participants were lost to follow-up. A flow chart with reasons for loss to follow-up can be found in Figure 2. Table 2 shows the baseline characteristics of the intervention and control group. Participants in the intervention group were slightly younger and had a higher BMI than participants in the control group. Participants in the intervention group lived less often alone and received more often informal care than control group participants.

Table 2. Baseline characteristics of participants of the PhysioDom HDIM study.

	Intervention Group (n = 97)		Control Group (n = 107)		p-Value [a]
	Mean	SD	Mean	SD	
Age (years)	78.4	7.2	81.0	7.9	0.02
BMI (kg/m^2)	29.2	4.5	27.7	5.4	0.04
Number of diagnoses	1.5	1.5	1.3	1.3	0.26
MMSE score	28.6	1.5	25.8	1.9	0.69
	Mdn	IQR	Mdn	IQR	
Katz-15 score	1.0	0–4	1.0	0–3	0.69
	Percentage		Percentage		
Sex (male)	35		23.4		0.09
Education level [b]					0.08
Low	17.5		10.3		
Moderate	55.7		49.5		
High	26.8		40.2		
Civil status					0.11
Married	42.3		27.1		
Single	7.2		13.1		
Divorced	7.2		10.3		
Widowed	43.3		49.5		

Table 2. *Cont.*

	Intervention Group (n = 97)		Control Group (n = 107)		p-Value [a]
	Mean	*SD*	**Mean**	*SD*	
Living alone	55.7		74.8		0.004
Born in the Netherlands	96.9		90.7		0.07
Desire to lose weight	52.7		39.4		0.07
Nutritional status					0.45
Normal nutritional status	79.2		83.8		
At risk of undernutrition	19.8		16.2		
Undernourished	1.0		0.0		
Type of care					
Domestic care	78.4		80.4		0.72
Personal care	32.0		29.9		0.75
Nursing care	9.3		2.8		0.05
Individual support	3.1		0.9		0.27
Informal care	32.0		11.2		<0.001

SD, Standard Deviation; BMI, Body Mass Index; MMSE, Mini-Mental State Examination. [a] Independent *t*-test, Mann-Whitney test, or chi-square test. [b] Low education level: primary school or less; intermediate level of education: secondary professional education or vocational school; High education level: higher vocational education, university.

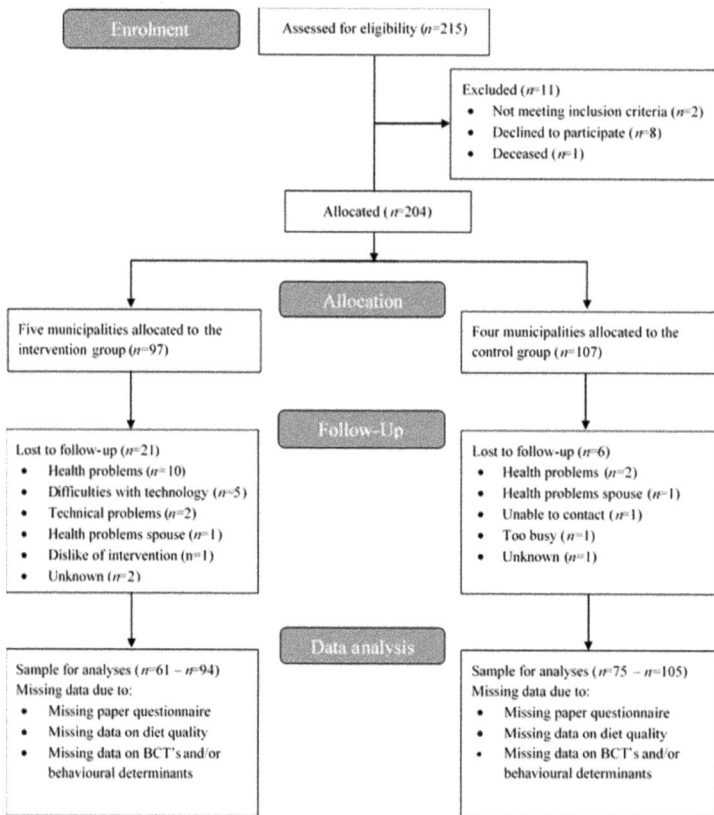

Figure 2. Flow diagram of participants of the PhysioDom HDIM study.

3.2. Changes in Self-Monitoring and Goalsetting and Effects on Behavioural Determinants

Table 3 shows changes in self-monitoring and goalsetting and shows the effects of the intervention on behavioural determinants. At baseline, there were no significant differences between the intervention and control group. During the intervention, several significant changes were observed. Firstly, the intervention group significantly increased scores for self-monitoring at T1 and T2, compared to the control group (T1: β = 0.49, 95% CI 0.19, 0.80; T2: β = 0.50, 95% CI 0.20, 0.80). Secondly, intervention participants perceived an increased behavioural control for physical activity at T2 compared to the control group (β = 0.26, 95% CI 0.08, 0.45). Thirdly, participants in the intervention group improved their knowledge at T1 and T2 compared to the control group, with the improvement at T2 being significant (β = 0.51, 95% CI 0.04, 0.99).

Table 3. Changes in self-monitoring and goalsetting and effects of the PhysioDom HDIM intervention on knowledge, perceived behavioural control, and attitude.

	Intervention Group						Control Group						Linear Mixed Models		
	T0		T1		T2		T0		T1		T2		β T1 (95% CI)	β T2 (95% CI)	N
	Mean	SD	Mean	SD	Mean	SD	Mean	SD	Mean	SD	Mean	SD			
Self-monitoring	2.9	1.2	3.5	0.9	3.3	1.1	3.1	1.2	3.1	1.2	3.0	1.3	0.49 (0.19, 0.80) **	0.50 (0.20, 0.80) **	199
Goalsetting	2.7	1.2	3.0	1.1	2.8	1.2	3.0	1.1	3.0	1.3	2.9	1.2	0.25 (−0.05, 0.55)	0.19 (−0.10, 0.48)	199
Knowledge [a]	7.3	2.1	8.2	1.9	8.3	1.8	7.5	2.0	7.5	2.1	7.6	2.2	0.51 (−0.09, 1.12)	0.51 (0.04, 0.99) *	198
PBC HE [b]	4.1	0.7	4.2	0.7	4.2	0.6	4.3	0.6	4.1	0.8	4.3	0.7	0.16 (−0.02, 0.33)	0.08 (−0.09, 0.25)	188
PBC PA	3.7	0.8	3.8	0.9	3.9	0.9	4.0	0.9	3.8	1.0	3.9	0.9	0.19 (−0.03, 0.41)	0.26 (0.08, 0.45) **	199
Attitude HE [c]	4.7	0.5	4.7	0.5	4.7	0.5	4.6	0.5	4.6	0.5	4.6	0.6	−0.01 (−0.18, 0.16)	0.00 (−0.17, 0.17)	188
Attitude PA [d]	4.5	0.6	4.4	0.8	4.5	0.7	4.5	0.7	4.4	0.8	4.5	0.7	−0.04 (−0.29, 0.22)	−0.05 (−0.26, 0.15)	190

SD, standard deviation; CI, confidence interval; PBC: perceived behavioural control; HE: healthy eating; PA: physical activity. All results are adjusted for age and sex. [a] Adjusted for age, sex, and Mini Mental State Examination score. [b] Adjusted for age, sex, BMI, living situation, nutritional status, and physical functioning. [c] Adjusted for age, sex, BMI, physical functioning and cognitive functioning. [d] Adjusted for age, sex, BMI, physical functioning. * $p < 0.05$; ** $p < 0.01$.

3.3. Effect Mediation

Four significant mediation pathways were found. Firstly, the effect of the intervention on compliance with the guidelines for the intake of fruit was mediated by increased self-monitoring behaviour at T1. Secondly, the effect of the intervention on compliance with the guidelines for the intake of protein was mediated by improvements in knowledge at T1 and T2 (Table 4). Thirdly, even though a significant effect of the intervention on the total DHD-FFQ score was lacking, we found significant mediation by self-monitoring at T1 (Appendix A). Likewise, increased self-monitoring mediated the intervention's effect on compliance with guidelines for the intake of saturated fat (Appendix A).

Table 4. Mediation of the intervention's effect on diet quality and physical activity.

	Indirect Effect 1 [a,b] (a₁ × b₁)		Indirect Effect 2 [a,c] (a₁ × d₂₁ × b₂)		Indirect Effect 3 [a,d] (a₂ × b₂)		
	β (SE)	95% CI	β (SE)	(95% CI)	β (SE)	(95% CI)	N
T0–T2 Fruit							
Self-monitoring	0.16 (0.10)	0.02, 0.45	0.02 (0.04)	−0.04, 0.13	0.04 (0.08)	−0.11, 0.24	141
Goalsetting	0.03 (0.08)	−0.08, 0.27	−0.00 (0.03)	−0.09, 0.03	−0.01 (0.04)	−0.15, 0.03	140
Knowledge	0.17 (0.14)	−0.01, 0.57	−0.02 (0.05)	−0.15, 0.05	−0.02 (0.06)	−0.21, 0.04	139
PBC HE	0.01 (0.05)	−0.05, 0.18	0.02 (0.03)	−0.01, 0.13	−0.02 (0.05)	−0.20, 0.04	136
Attitude HE	−0.00 (0.04)	−0.09, 0.07	0.00 (0.01)	−0.01, 0.03	−0.00 (0.05)	−0.12, 0.07	137
T0–T2 Vegetables							
Self-monitoring	−0.07 (0.11)	−0.11, 0.35	−0.06 (0.05)	−0.22, 0.01	−0.12 (0.11)	−0.39, 0.04	141
Goalsetting	0.01 (0.06)	−0.05, 0.22	−0.00 (0.03)	−0.07, 0.05	−0.00 (0.05)	−0.12, 0.08	140

Table 4. *Cont.*

	Indirect Effect 1 [a,b] ($a_1 \times b_1$)		Indirect Effect 2 [a,c] ($a_1 \times d_{21} \times b_2$)		Indirect Effect 3 [a,d] ($a_2 \times b_2$)		
	β (SE)	95% CI	β (SE)	(95% CI)	β (SE)	(95% CI)	N
Knowledge	−0.07 (0.09)	−0.32, 0.06	−0.01 (0.05)	−0.11, 0.08	−0.01 (0.06)	−0.18, 0.08	139
PBC HE	−0.01 (0.07)	−0.25, 0.06	−0.01 (0.03)	−0.13, 0.02	0.02 (0.05)	−0.04, 0.21	136
Attitude HE	−0.01 (0.07)	−0.23, 0.07	0.00 (0.01)	−0.01, 0.04	−0.00 (0.05)	−0.13, 0.08	137
T0–T2 Dietary fibre							
Self-monitoring	−0.07 (0.07)	−0.27, 0.03	0.03 (0.04)	−0.02. 0.14	0.07 (0.07)	−0.05, 0.26	141
Goalsetting	−0.00 (0.03)	−0.08, 0.05	0.01 (0.02)	−0.02, 0.09	0.02 (0.04)	−0.02, 0.15	140
Knowledge	0.02 (0.06)	−0.06, 0.22	0.01 (0.03)	−0.04, 0.10	0.1 (0.04)	−0.04, 0.14	139
PBC HE	0.05 (0.05)	−0.03, 0.20	−0.01 (0.02)	−0.08, 0.01	0.01 (0.03)	−0.02, 0.13	136
Attitude HE	0.00 (0.03)	−0.05, 0.06	0.00 (0.01)	−0.00, 0.03	−0.00 (0.02)	−0.09, 0.03	137
T0–T2 Protein							
Self-monitoring	−0.25 (0.19)	−0.76, 0.00	0.00 (0.08)	−0.16, 0.18	0.00 (0.17)	−0.37, 0.34	141
Goalsetting	−0.04 (0.12)	−0.42, 0.12	0.01 (0.06)	−0.05, 0.23	0.04 (0.09)	−0.05, 0.37	140
Knowledge	−0.07 (0.12)	−0.44, 0.07	0.12 (0.09)	0.006, 0.41	0.12 (0.13)	−0.04, 0.52	139
PBC HE	0.06 (0.09)	−0.04, 0.40	−0.02 (0.04)	−0.17, 0.02	0.02 (0.07)	−0.05, 0.28	136
Attitude HE	0.01 (0.07)	−0.08, 0.23	0.00 (0.01)	−0.01, 0.06	−0.01 (0.06)	−0.20, 0.08	137
T0–T2 Physical activity							
Self-monitoring	−0.02 (0.16)	−0.40, 0.29	−0.04 (0.08)	−0.27, 0.07	−0.10 (0.17)	−0.52, 0.18	141
Goalsetting	−0.02 (0.08)	−0.32, 0.07	−0.00 (0.05)	−0.51, 0.07	−0.01 (0.08)	−0.25, 0.10	140
Knowledge	−0.10 (0.13)	−0.49, 0.07	−0.04 (0.07)	−0.24, 0.05	−0.04 (0.09)	−0.35, 0.05	139
PBC PA	−0.04 (0.10)	−0.31, 0.10	0.07 (0.08)	−0.01, 0.34	0.16 (0.13)	−0.02, 0.51	137
Attitude PA	−0.003 (0.05)	−0.14, 0.07	0.001 (0.02)	−0.04, 0.06	0.008 (0.05)	−0.06, 0.17	133

SE: standard error; CI: confidence interval; PBC: perceived behavioural control; HE: healthy eating; PA: physical activity. All results were adjusted for age and sex. [a] Standard errors and confidence intervals for indirect effects were calculated with bootstrapping (10,000 samples). [b] Indirect effect of the intervention on the outcome Y through the mediator at T1. [c] Indirect effect of the intervention on the outcome Y through the mediators at T1 and T2 in serial. [d] Indirect effect of the intervention on the outcome Y through the mediator at T2.

4. Discussion

This study aimed to evaluate the effects of a multi-component telemonitoring intervention on behavioural determinants of nutrition and physical activity behaviour in older adults, and to evaluate the role of mediators in the effects on behaviour. The intervention resulted in improvements in self-monitoring, perceived behavioural control for physical activity, and knowledge. Furthermore, self-monitoring mediated the effect of the intervention on total diet quality score and compliance with the guidelines for the intake of fruit and saturated fatty acids. Knowledge mediated the effect of the intervention on compliance with the guidelines for the intake of protein.

Intervention group participants increased their self-monitoring, which mediated the effect of the intervention on total DHD-FFQ score, and the intake of fruit and saturated fatty acids. Scores for self-monitoring improved from 2.9 at T0 to 3.5 at T1 and 3.3 at T2, meaning that the frequency of self-monitoring increased from on average a few times per month to somewhere between a few times per month and weekly, suggesting that this rather small change is already sufficient to partly mediate the intervention's effect. Self-monitoring of diet, physical activity, and weight has mainly been used in weight loss programs with more frequent self-monitoring resulting in more weight loss as compared to less frequent self-monitoring [43]. Another study focussing on the effects of self-monitoring by means of mobile devices showed positive outcomes on dietary intake [44]. In an intervention study, the effect of daily tailored messaging on weight loss was mediated by self-monitoring of diet and physical activity [45]. In our study, self-monitoring mediated the effect of the intervention on total diet quality score and the intake of fruit and saturated fatty acids, but not the effect on other diet quality components. A possible explanation is that self-monitoring of diet was not very intensive as participants filled out the DHD-FFQ twice during the six-month intervention, as opposed to more frequent dietary self-monitoring encountered in other studies [44]. Apparently, also other mechanisms besides self-monitoring have caused the intervention to result in positive changes in diet quality and physical activity. Increasing the frequency of self-monitoring of dietary intake may strengthen the effect of this intervention.

The intervention resulted in increased perceived behavioural control for PA, but this did not mediate the intervention's effect on physical activity. This seems contradictory to the theory of planned behaviour that poses that PBC precedes behavioural intention and behaviour [46]. Furthermore, PBC is seen as a predictor of the translation of intention into behaviour [47,48]. With regard to older adults, PBC is considered as a consistent predictor of physical activity initiation and maintenance [49]. In contrast, an intervention study aiming to improve physical activity among participants with increased risk of type 2 diabetes shows that PBC did not predict physical activity or change in physical activity [50]. The authors argue that the TPB may be less accurate in explaining behaviour among clinical samples than among the often-used student samples, which could also explain the lack of mediation by PBC in this study [50]. All in all, other mechanisms besides the ones that we have measured may have been important in increasing physical activity levels of our study population, for example, awareness, enjoyment, or action planning [49].

The intervention had a positive effect on knowledge, and this mediated the intervention's effect on compliance with the guidelines for the intake of protein, but not the effects on compliance with other dietary guidelines. Knowledge score improved from 7.3 at T0 to 8.3 at T2, meaning that intervention group participants were able to answer one more knowledge statement correctly, from the eleven statements in total. This rather small effect size may explain why the intervention's effect was only limitedly mediated via knowledge. Nutrition education intervention studies among older adults have shown positive effects on knowledge [51–53], although in the study by Racine et al, this was not associated with better adherence to the DASH diet [51,52]. A review examining the relationship between nutrition knowledge and dietary intake showed positive but weak correlations [54]. The general idea is that nutritional knowledge is necessary but not sufficient for changing dietary habits, and that the association of knowledge with diet quality is complex and influenced by many other demographic and environmental factors [54]. Furthermore, the knowledge questionnaire used in this study assessed declarative knowledge, while procedural knowledge of nutrition (e.g., knowing how to read food labels or how to cook a healthy meal) might be more relevant for making healthy food choices [55]. Nevertheless, improved knowledge did mediate the intervention's effect on the compliance with dietary guidelines for protein intake. This is a relevant finding, as sufficient protein intake in older adults is necessary to counteract age-related loss of muscle mass [56]. Furthermore, older adults seem unaware of the importance of sufficient intake of protein [4]. This study suggests that increasing nutritional knowledge might be an effective and relatively easy way to improve protein intake in older adults.

The intervention did not result in significant changes in goalsetting, attitude, and perceived behavioural control for healthy eating. Several possible explanations could be given. The emphasis of the intervention was on self-monitoring of nutritional outcomes and PA. Participants received training and instructions to do these self-measurements and were reminded via a paper calendar and television messages, resulting in increased self-monitoring behaviour. Participants were also prompted to set goals for diet quality and PA, but only via the intervention manual and via television messages, which were not always read. This could explain the lack of significant effects on goal-setting. Secondly, attitude and perceived behavioural control for healthy eating were also targeted through television messages. Again, too little messages might have been read to have an impact on these behavioural determinants. Furthermore, television messages to target PBC for healthy eating were mainly focussed on the individual. However, PBC might also be affected by characteristics that are not easily targeted, such as impaired physical functioning, limited mobility, limited cooking skills, or more environmental determinants such as distance to a supermarket. All in all, to target goalsetting, attitude, and perceived behavioural control for healthy eating, a higher intervention dose might be necessary to result in change.

To our knowledge, this is the first study that aimed to unravel mechanisms of impact of an intervention that focussed on improving nutritional status in community-dwelling elderly through eHealth. Strengths of this study include the use of a theoretical framework and validated constructs to measure behavioural determinants. This study also has limitations that may have contributed to the limited significant findings from the mediation analyses. The population for analysis was smaller for the mediation analyses than for the mixed model analyses, as the method used for mediation analyses is less flexible concerning missing data. This could have resulted in a loss of power or have obscured mediation pathways. Secondly, using self-report measures of diet and physical activity instead of objective measures of behaviour may have led to weaker associations with the proposed behavioural determinants [57]. Furthermore, older adults might be less good in filling out TPB questionnaires than younger adults [50]. Thirdly, it is uncertain whether BCT's which have been proven successful in younger populations can be applied in older populations as well [58]. It may well be that some BCT's may be too complex for older adults with impaired physical functioning or in another way do not appeal to older adults, potentially explaining the limited results from the mediation analyses. Finally, it is uncertain whether effects on behavioural determinants and behaviour will sustain. Participants could keep the weighing scale and pedometer to keep track of their weight and daily steps. Employing BCT's enhances the chance that participants maintain their behaviour [13]. On the other hand, this study mainly focussed on individual determinants of health behaviour, while it is suggested that organisational and societal determinants are also important for achieving sustained change [59]. More research is necessary to assess the long-term effectiveness of nutritional eHealth interventions, and what exactly contributes to long-term impact.

Finally, this study showed that a multi-component telemonitoring intervention for community-dwelling older adults resulted in increased self-monitoring behaviour, and improved perceived behavioural control for physical activity and knowledge. The intervention's effect on total diet quality score, fruit intake, and saturated fatty acids intake was mediated by self-monitoring and the effect on protein intake was mediated by knowledge. Apparently, other unknown mediators have also played an important role in achieving the intervention's results on diet quality and physical activity. This calls for more research into which behaviour change techniques are effective in improving nutritional outcomes in older adults.

Author Contributions: Conceptualization, L.C.P.G.M.d.G., J.H.M.d.V. and A.H.-N.; Data curation, M.N.v.D.-v.A.; Formal analysis, M.N.v.D.-v.A.; Investigation, M.N.v.D.-v.A.; Methodology, M.N.v.D.-v.A., L.C.P.G.M.d.G., J.H.M.d.V. and A.H.-N.; Writing—original draft, M.N.v.D.-v.A.; Writing—review & editing, L.C.P.G.M.d.G., J.H.M.d.V. and A.H.-N.

Funding: This research was funded by [The European Union] grant number [CIP-ICT-PSP-2013-7].

Acknowledgments: The researchers thank all participants, health care professionals, and boards of Zorggroep Noord-West Veluwe and Opella.

Conflicts of Interest: The authors declare no conflict of interest. The funders had no role in the design of the study; in the collection, analyses, or interpretation of data; in the writing of the manuscript, and in the decision to publish the results.

Appendix A

Table A1. Mediation of the intervention's effect on total diet quality score and other diet quality components.

	Indirect Effect 1 [a,b] $(a_1 \times b_1)$		Indirect effect 2 [a,c] $(a_1 \times d_{21} \times b_2)$		Indirect Effect 3 [a,d] $(a_2 \times b_2)$		
	β (SE)	95% CI	β (SE)	(95% CI)	β (SE)	(95% CI)	N
T0–T2 Total DHD-FFQ score							
Self-monitoring	0.79 (0.50)	0.09, 2.11	0.07 (0.21)	−0.27, 0.65	0.16 (0.47)	−0.61, 1.35	141
Goalsetting	0.07 (0.30)	−0.24, 1.06	0.04 (0.16)	−0.15, 0.65	0.11 (0.26)	−0.15, 1.06	140
Knowledge	0.23 (0.37)	−0.28, 1.32	−0.17 (0.19)	−0.74, 0.07	−0.18 (0.27)	−1.10, 0.11	139
PBC HE	0.05 (0.23)	−0.28, 0.76	0.01 (0.11)	−0.20, 0.27	−0.01 (0.16)	−0.44, 0.28	136
Attitude HE	−0.03 (0.16)	−0.55, 0.19	−0.01 (0.06)	−0.19, 0.05	0.05 (0.24)	−0.30, 0.80	137

Table A1. *Cont.*

	Indirect Effect 1 [a,b] ($a_1 \times b_1$)		Indirect effect 2 [a,c] ($a_1 \times d_{21} \times b_2$)		Indirect Effect 3 [a,d] ($a_2 \times b_2$)		
	β (SE)	95% CI	β (SE)	(95% CI)	β (SE)	(95% CI)	N
T0–T2 Fish							
Self-monitoring	−0.07 (0.11)	−0.36, 0.10	0.01 (0.06)	−0.09, 0.14	0.02 (0.12)	−0.20, 0.30	141
Goalsetting	−0.01 (0.07)	−0.23, 0.07	−0.00 (0.03)	−0.08, 0.06	−0.00 (0.05)	−0.14, 0.09	140
Knowledge	0.09 (0.10)	−0.04, 0.38	−0.02 (0.06)	−0.17, 0.08	−0.02 (0.07)	−0.24, 0.08	139
PBC HE	0.05 (0.09)	−0.04, 0.35	−0.02 (0.04)	−0.18, 0.02	0.02 (0.06)	−0.04, 0.22	136
Attitude HE	0.00 (0.04)	−0.08, 0.09	−0.00 (0.01)	−0.04, 0.01	0.01 (0.05)	−0.06, 0.17	137
T0–T2 Saturated fatty acids							
Self-monitoring	0.31 (0.21)	0.02, 0.86	0.06 (0.09)	−0.05, 0.32	0.14 (0.18)	−0.14, 0.59	141
Goalsetting	0.05 (0.16)	−0.17, 0.51	−0.00 (0.06)	−0.18, 0.10	−0.01 (0.10)	−0.30, 0.15	140
Knowledge	0.03 (0.15)	−0.22, 0.40	−0.03 (0.08)	−0.24, 0.09	−0.03 (0.10)	−0.35, 0.09	139
PBC HE	−0.01 (0.10)	−0.29, 0.16	0.04 (0.06)	−0.03, 0.27	−0.04 (0.10)	−0.39, 0.08	136
Attitude HE	−0.03 (0.12)	−0.46, 0.12	−0.01 (0.03)	−0.12, 0.02	0.03 (0.12)	−0.15, 0.37	137
T0–T2 Salt							
Self-monitoring	0.05 (0.13)	−0.17, 0.37	0.02 (0.07)	−0.10, 0.20	0.04 (0.15)	−0.23, 0.40	141
Goalsetting	−0.00 (0.07)	−0.18, 0.12	0.01 (0.05)	−0.04, 0.19	0.03 (0.08)	−0.05, 0.34	140
Knowledge	0.00 (0.11)	−0.21, 0.27	−0.08 (0.07)	−0.32, 0.01	−0.08 (0.10)	−0.42, 0.03	139
PBC HE	−0.06 (0.09)	−0.37, 0.04	0.02 (0.04)	−0.02, 0.15	−0.02 (0.06)	−0.23, 0.04	136
Attitude HE	0.01 (0.05)	−0.05, 0.16	−0.00 (0.02)	−0.05, 0.02	0.01 (0.07)	−0.10, 0.20	137
T0–T2 Alcohol							
Self-monitoring	0.04 (0.04)	−0.02, 0.17	0.01 (0.04)	−.05, 0.11	0.02 (0.08)	−0.13, 0.20	141
Goalsetting	−0.00 (0.03)	−0.10, 0.05	0.00 (0.02)	−0.01, 0.06	0.01 (0.02)	−0.01, 0.09	140
Knowledge	−0.00 (0.03)	−0.08, 0.05	0.01 (0.03)	−0.03, 0.11	0.01 (0.04)	−0.03, 0.15	139
PBC HE	0.01 (0.02)	−0.01, 0.07	0.01 (0.02)	−0.01, 0.11	−0.01 (0.04)	−0.18, 0.03	136
Attitude HE	0.00 (0.03)	−0.02, 0.10	−0.00 (0.02)	−0.08, 0.01	0.01 (0.08)	−0.07, 0.31	137
T0–T2 Vitamin D							
Self-monitoring	0.00 (0.04)	−0.07, 0.08	−0.01 (0.03)	−0.06, 0.06	−0.01 (0.06)	−0.14, 0.11	141
Goalsetting	−0.00 (0.02)	−0.07, 0.03	−0.00 (0.01)	−0.05, 0.01	−0.01 (0.02)	−0.09, 0.01	140
Knowledge	0.01 (0.03)	−0.05, 0.08	0.03 (0.04)	−0.01, 0.16	0.04 (0.05)	−0.01, 0.24	139
PBC HE	0.02 (0.3)	−0.01, 0.12	−0.02 (0.03)	−0.12, 0.01	0.02 (0.05)	−0.03, 0.17	136
Attitude HE	0.00 (0.02)	−0.03, 0.08	−0.00 (0.00)	−0.02, 0.01	0.00 (0.02)	−0.03, 0.06	137

SE: standard error; CI: confidence interval; PBC: perceived behavioural control; HE: healthy eating, PA: physical activity. All results were adjusted for age and sex. [a] Standard errors and confidence intervals for indirect effects were calculated with bootstrapping (10,000 samples). [b] Indirect effect of the intervention on the outcome Y through the mediator at T1. [c] Indirect effect of the intervention on the outcome Y through the mediators at T1 and T2 in serial. [d] Indirect effect of the intervention on the outcome Y through the mediator at T2.

References

1. Morley, J.E. Undernutrition in Older Adults. *Fam. Pract.* **2012**, *29*, i89–i93. [CrossRef] [PubMed]
2. Schilp, J.; Kruizenga, H.M.; Wijnhoven, H.A.; Leistra, E.; Evers, A.M.; Van Binsbergen, J.J.; Deeg, D.J.; Visser, M. High Prevalence of Undernutrition in Dutch Community-Dwelling Older Individuals. *Nutrition* **2012**, *28*, 1151–1156. [CrossRef] [PubMed]
3. Ocke, M.C.; Buurma-Rethans, E.J.M.; De Boer, E.J.; Wilson-van den Hooven, C.; Etemad-Ghameslou, Z.; Drijvers, J.J.M.M.; Van Rossum, C.T.M. Diet of Community-Dwelling Older Adults: Dutch National Food Consumption Survey Older Adults 2010–2012. Available online: http://rivm.openrepository.com/rivm/handle/10029/305649 (accessed on 8 August 2018).
4. Ziylan, C.; Janssen, N.; De Roos, N.M.; De Groot, L.C.P.G.M. Undernutrition: Who Cares? Perspectives of Dietitians and Older Adults on Undernutrition. *BMC Nutr.* **2017**, *3*, 24.
5. Ziylan, C.; Haveman-Nies, A.; Van Dongen, E.J.I.; Kremer, S.; De Groot, C.P.G.M. Dutch Nutrition and Care Professionals' Experiences with Undernutrition Awareness, Monitoring, and Treatment among Community-Dwelling Older Adults: A. Qualitative Study. *BMC Nutr.* **2015**, *1*, 38. [CrossRef]
6. Parmenter, K.; Waller, J.; Wardle, J. Demographic Variation in Nutrition Knowledge in England. *Health Educ. Res.* **2000**, *15*, 163–174. [CrossRef] [PubMed]

7. Fischer, C.A.; Crockett, S.J.; Heller, K.E.; Skauge, L.H. Nutrition Knowledge, Attitudes, and Practices of Older and Younger Elderly in Rural Areas. *J. Am. Diet. Assoc.* **1991**, *91*, 1398–1401. [PubMed]
8. Stafleu, A.; Van Staveren, W.A.; De Graaf, C.; Burema, J.; Hautvast, J.G. Nutrition Knowledge and Attitudes Towards High-Fat Foods and Low-Fat Alternatives in Three Generations of Women. *Eur. J. Clin. Nutr.* **1996**, *50*, 33–41. [PubMed]
9. Kamp, B.J.; Wellman, N.S.; Russell, C. Position of the American Dietetic Association, American Society for Nutrition, and Society for Nutrition Education: Food and Nutrition Programs for Community-Residing Older Adults. *J. Nutr. Educ. Behav.* **2010**, *42*, 72–82. [CrossRef] [PubMed]
10. Bauman, A.; Merom, D.; Bull, F.C.; Buchner, D.M.; Fiatarone Singh, M.A. Updating the Evidence for Physical Activity: Summative Reviews of the Epidemiological Evidence, Prevalence, and Interventions to Promote "Active Aging". *Gerontologist* **2016**, *56*, S268–S280. [CrossRef] [PubMed]
11. Baert, V.; Gorus, E.; Mets, T.; Geerts, C.; Bautmans, I. Motivators and Barriers for Physical Activity in the Oldest Old: A Systematic Review. *Ageing Res. Rev.* **2011**, *10*, 464–474. [CrossRef] [PubMed]
12. Gellert, P.; Witham, M.D.; Crombie, I.K.; Donnan, P.T.; McMurdo, M.E.; Sniehotta, F.F. The Role of Perceived Barriers and Objectively Measured Physical Activity in Adults Aged 65-100. *Age Ageing* **2015**, *44*, 384–390. [CrossRef] [PubMed]
13. Celis-Morales, C.; Lara, J.; Mathers, J.C. Personalising Nutritional Guidance for More Effective Behaviour Change. *Proc. Nutr. Soc.* **2015**, *74*, 130–138. [CrossRef] [PubMed]
14. Harris, J.; Felix, L.; Miners, A.; Murray, E.; Michie, S.; Ferguson, E.; Free, C.; Lock, K.; Landon, J.; Edwards, P. Adaptive E-Learning to Improve Dietary Behaviour: A Systematic Review and Cost-Effectiveness Analysis. *Health Technol. Assess.* **2011**, *15*, 1–160. [CrossRef] [PubMed]
15. Norman, G.J.; Zabinski, M.F.; Adams, M.A.; Rosenberg, D.E.; Yaroch, A.L.; Atienza, A.A. A Review of Ehealth Interventions for Physical Activity and Dietary Behavior Change. *Am. J. Prev. Med.* **2007**, *33*, 336–345. [CrossRef] [PubMed]
16. Neville, L.M.; O'Hara, B.; Milat, A.J. Computer-Tailored Dietary Behaviour Change Interventions: A Systematic Review. *Health Educ. Res.* **2009**, *24*, 699–720. [CrossRef] [PubMed]
17. Kelly, J.T.; Reidlinger, D.P.; Hoffmann, T.C.; Campbell, K.L. Telehealth Methods to Deliver Dietary Interventions in Adults with Chronic Disease: A Systematic Review and Meta-Analysis. *Am. J. Clin. Nutr.* **2016**, *104*, 1693–1702. [CrossRef] [PubMed]
18. Lara, J.; O'Brien, N.; Godfrey, A.; Heaven, B.; Evans, E.H.; Lloyd, S.; Moffatt, S.; Moynihan, P.J.; Meyer, T.D.; Rochester, L.; et al. Pilot Randomised Controlled Trial of a Web-Based Intervention to Promote Healthy Eating, Physical Activity and Meaningful Social Connections Compared with Usual Care Control in People of Retirement Age Recruited from Workplaces. *PLoS ONE* **2016**, *11*, e0159703. [CrossRef] [PubMed]
19. Verheijden, M.; Bakx, J.C.; Akkermans, R.; Van den Hoogen, H.; Godwin, N.M.; Rosser, W.; Van Staveren, W.; Van Weel, C. Web-Based Targeted Nutrition Counselling and Social Support for Patients at Increased Cardiovascular Risk in General Practice: Randomized Controlled Trial. *J. Med. Internet Res.* **2004**, *6*, e44. [CrossRef] [PubMed]
20. Michie, S.; Richardson, M.; Johnston, M.; Abraham, C.; Francis, J.; Hardeman, W.; Eccles, M.P.; Cane, J.; Wood, C.E. The Behavior Change Technique Taxonomy (V1) of 93 Hierarchically Clustered Techniques: Building an International Consensus for the Reporting of Behavior Change Interventions. *Ann. Behav. Med.* **2013**, *46*, 81–95. [CrossRef] [PubMed]
21. Webb, T.L.; Joseph, J.; Yardley, L.; Michie, S. Using the Internet to Promote Health Behavior Change: A Systematic Review and Meta-Analysis of the Impact of Theoretical Basis, Use of Behavior Change Techniques, and Mode of Delivery on Efficacy. *J. Med. Int. Res.* **2010**, *12*, e4. [CrossRef] [PubMed]
22. Van Doorn-van Atten, M.N.; Haveman-Nies, A.; Van Bakel, M.M.; Ferry, M.; Franco, M.; De Groot, L.; De Vries, J.H.M. Effects of a Multi-Component Nutritional Telemonitoring Intervention on Nutritional Status, Diet Quality, Physical Functioning and Quality of Life of Community-Dwelling Older Adults. *Br. J. Nutr.* **2018**, *119*, 1185–1194. [CrossRef] [PubMed]
23. Abraham, C.; Michie, S. A Taxonomy of Behavior Change Techniques Used in Interventions. *Health Psychol.* **2008**, *27*, 379–387. [CrossRef] [PubMed]
24. Carver, C.S.; Scheier, M.F. Control Theory: A Useful Conceptual Framework for Personality-Social, Clinical, and Health Psychology. *Psychol. Bull.* **1982**, *92*, 111–135. [CrossRef] [PubMed]

25. Michie, S.; Abraham, C.; Whittington, C.; McAteer, J.; Gupta, S. Effective Techniques in Healthy Eating and Physical Activity Interventions: A Meta-Regression. *Health Psychol.* **2009**, *28*, 690–701. [CrossRef] [PubMed]

26. Morrison, L.G.; Yardley, L.; Powell, J.; Michie, S. What Design Features Are Used in Effective E-Health Interventions? A Review Using Techniques from Critical Interpretive Synthesis. *Telemed. J. E Health* **2012**, *18*, 137–144. [CrossRef] [PubMed]

27. Bodenheimer, T.; Lorig, K.; Holman, H.; Grumbach, K. Patient Self-Management of Chronic Disease in Primary Care. *JAMA* **2002**, *288*, 2469–2475. [CrossRef] [PubMed]

28. Kaiser, M.J.; Bauer, J.M.; Ramsch, C.; Uter, W.; Guigoz, Y.; Cederholm, T.; Thomas, D.R.; Anthony, P.; Charlton, K.E.; Maggio, M.; et al. Validation of the Mini Nutritional Assessment Short-Form (Mna-Sf): A Practical Tool for Identification of Nutritional Status. *J. Nutr Health Aging* **2009**, *13*, 782–788. [CrossRef] [PubMed]

29. Wilson, M.M.; Thomas, D.R.; Rubenstein, L.Z.; Chibnall, J.T.; Anderson, S.; Baxi, A.; Diebold, M.R.; Morley, J.E. Appetite Assessment: Simple Appetite Questionnaire Predicts Weight Loss in Community-Dwelling Adults and Nursing Home Residents. *Am. J. Clin. Nutr.* **2005**, *82*, 1074–1081. [CrossRef] [PubMed]

30. Van Lee, L.; Feskens, E.J.; Meijboom, S.; Hooft van Huysduynen, E.J.; Van't Veer, P.; De Vries, J.H.; Geelen, A. Evaluation of a Screener to Assess Diet Quality in the Netherlands. *Br. J. Nutr.* **2016**, *115*, 517–526. [CrossRef] [PubMed]

31. Ondervoeding. Eten Bij Ondervoeding En Bij Herstel Na Ziekte. Available online: https://webshop. voedingscentrum.nl/pdf/D752-20.pdf (accessed on 8 August 2018).

32. Folstein, M.F.; Folstein, S.E.; McHugh, P.R. "Mini-Mental State" A Practical Method for Grading the Cognitive State of Patients for the Clinician. *J. Psychiatr. Res.* **1975**, *12*, 189. [CrossRef]

33. Laan, W.; Zuithoff, N.P.; Drubbel, I.; Bleijenberg, N.; Numans, M.E.; De Wit, N.J.; Schuurmans, M.J. Validity and Reliability of the Katz-15 Scale to Measure Unfavorable Health Outcomes in Community-Dwelling Older People. *J. Nutr. Health Aging* **2014**, *18*, 848–854. [CrossRef] [PubMed]

34. Vellas, B.; Villars, H.; Abellan, G.; Soto, M.E.; Rolland, Y.; Guigoz, Y.; Morley, J.E.; Chumlea, W.; Salva, A.; Rubenstein, L.Z.; et al. Overview of the MNA—Its History and Challenges. *J. Nutr. Health Aging* **2006**, *10*, 456–463. [PubMed]

35. Nothwehr, F.; Dennis, L.; Wu, H. Measurement of Behavioral Objectives for Weight Management. *Health Educ. Behav.* **2007**, *34*, 793–809. [CrossRef] [PubMed]

36. Nothwehr, F.; Yang, J. Goal Setting Frequency and the Use of Behavioral Strategies Related to Diet and Physical Activity. *Health Educ. Res.* **2007**, *22*, 532–538. [CrossRef] [PubMed]

37. Francis, J.J.; Eccles, M.P.; Johnston, M.; Walker, A.; Grimshaw, J.; Foy, R.; Kaner, E.F.S.; Smith, L.; Bonetti, D. Constructing Questionnaires Based on the Theory of Planned Behaviour. Available online: http://openaccess. city.ac.uk/1735/1/TPB%20Manual%20FINAL%20May2004.pdf (accessed on 8 August 2018).

38. Blue, C.L.; Marrero, D.G. Psychometric Properties of the Healthful Eating Belief Scales for Persons at Risk of Diabetes. *J. Nutr. Educ. Behav.* **2006**, *38*, 134–142. [CrossRef] [PubMed]

39. *Richtlijnen Goede Voeding*; Gezondheidsraad: Den Haag, The Netherland, 2006.

40. Becker, W.; Lyhne, N.; Pedersen, A.N.; Aro, A.; Fogelholm, M.; Phorsdottir, I.; Alexander, J.; Anderssen, S.A.; Meltzer, H.M.; Pedersen, J.I. Nordic Nutrition Recommendations 2004-integrating nutrition and physical activity. *Scand. J. Nutr.* **2004**, *48*, 178–187. [CrossRef]

41. Nijs, K.A.; De Graaf, C.; Siebelink, E.; Blauw, Y.H.; Vanneste, V.; Kok, F.J.; Van Staveren, W.A. Effect of Family-Style Meals on Energy Intake and Risk of Malnutrition in Dutch Nursing Home Residents: A Randomized Controlled Trial. *J. Gerontol. A: Biol. Sci. Med. Sci.* **2006**, *61*, 935–942. [CrossRef]

42. Hayes, A.F. *Introduction to Mediation, Moderation, and Conditional Process Analysis: A Regression-Based Approach*; Guilford Publications: New York, NY, USA, 2013.

43. Burke, L.E.; Wang, J.; Sevick, M.A. Self-Monitoring in Weight Loss: A Systematic Review of the Literature. *J. Am. Diet. Assoc.* **2011**, *111*, 92–102. [CrossRef] [PubMed]

44. Lieffers, J.R.; Hanning, R.M. Dietary Assessment and Self-Monitoring with Nutrition Applications for Mobile Devices. *Can. J. Diet. Pract. Res.* **2012**, *73*, e253–e260. [CrossRef] [PubMed]

45. Wang, J.; Sereika, S.M.; Chasens, E.R.; Ewing, L.J.; Matthews, J.T.; Burke, L.E. Effect of Adherence to Self-Monitoring of Diet and Physical Activity on Weight Loss in a Technology-Supported Behavioral Intervention. *Patient Prefer. Adherence* **2012**, *6*, 221–226. [CrossRef] [PubMed]

46. Ajzen, I. The Theory of Planned Behavior. *Organ. Behav. Hum. Decis. Process.* **1991**, *50*, 179–211. [CrossRef]

47. Rhodes, R.E.; De Bruijn, G.J. What Predicts Intention-Behavior Discordance? A Review of the Action Control Framework. *Exerc. Sport Sci. Rev.* **2013**, *41*, 201–207. [CrossRef] [PubMed]
48. De Bruijn, G.-J. Exercise Habit Strength, Planning and the Theory of Planned Behaviour: An Action Control Approach. *Psychol. Sport Exerc.* **2011**, *12*, 106–114. [CrossRef]
49. Van Stralen, M.M.; De Vries, H.; Mudde, A.N.; Bolman, C.; Lechner, L. Determinants of Initiation and Maintenance of Physical Activity among Older Adults: A Literature Review. *Health Psychol. Rev.* **2009**, *3*, 147–207. [CrossRef]
50. Hardeman, W.; Kinmonth, A.L.; Michie, S.; Sutton, S. Theory of Planned Behaviour Cognitions Do Not Predict Self-Reported or Objective Physical Activity Levels or Change in the Proactive Trial. *Br. J. Health Psychol.* **2011**, *16*, 135–150. [CrossRef] [PubMed]
51. Southgate, K.M.; Keller, H.H.; Reimer, H.D. Determining Knowledge and Behaviour Change. After Nutrition Screening among Older Adults. *Can. J. Diet. Pract. Res.* **2010**, *71*, 128–133. [CrossRef] [PubMed]
52. Racine, E.; Troyer, J.L.; Warren-Findlow, J.; McAuley, W.J. The Effect of Medical Nutrition Therapy on Changes in Dietary Knowledge and Dash Diet Adherence in Older Adults with Cardiovascular Disease. *J. Nutr. Health Aging* **2011**, *15*, 868–876. [CrossRef] [PubMed]
53. Lyons, B.P. Nutrition Education Intervention with Community-Dwelling Older Adults: Research Challenges and Opportunities. *J. Community Health* **2014**, *39*, 810–818. [CrossRef] [PubMed]
54. Spronk, I.; Kullen, C.; Burdon, C.; O'Connor, H. Relationship between Nutrition Knowledge and Dietary Intake. *Br. J. Nutr.* **2014**, *111*, 1713–1726. [CrossRef] [PubMed]
55. Worsley, A. Nutrition Knowledge and Food Consumption: Can Nutrition Knowledge Change Food Behaviour? *Asia Pac. J. Clin. Nutr.* **2002**, *11*, 579–585. [CrossRef]
56. Paddon-Jones, D.; Rasmussen, B.B. Dietary Protein Recommendations and the Prevention of Sarcopenia: Protein, Amino Acid Metabolism and Therapy. *Curr. Opin. Clin. Nutr. Metab. Care* **2009**, *12*, 86–90. [CrossRef] [PubMed]
57. Armitage, C.J.; Conner, M. Efficacy of the Theory of Planned Behaviour: A Meta-Analytic Review. *Br. J. Soc. Psychol.* **2001**, *40*, 471–499. [CrossRef] [PubMed]
58. French, D.P.; Olander, E.K.; Chisholm, A.; Mc Sharry, J. Which Behaviour Change Techniques Are Most Effective at Increasing Older Adults' Self-Efficacy and Physical Activity Behaviour? A Systematic Review. *Ann. Behav. Med.* **2014**, *48*, 225–234. [CrossRef] [PubMed]
59. Grady, A.; Yoong, S.; Sutherland, R.; Lee, H.; Nathan, N.; Wolfenden, L. Improving the Public Health Impact of Ehealth and Mhealth Interventions. *Aust. N. Z. J. Public Health* **2018**, *42*, 118–119. [CrossRef] [PubMed]

nutrients

MDPI

Article

Promoting Healthy Diet, Physical Activity, and Life-Skills in High School Athletes: Results from the WAVE Ripples for Change Childhood Obesity Prevention Two-Year Intervention

Yu Meng [1,*], Melinda M. Manore [1], John M. Schuna Jr. [1], Megan M. Patton-Lopez [2], Adam Branscum [1] and Siew Sun Wong [3]

1   School of Biological and Population Health Sciences, College of Public Health and Human Sciences, Oregon State University, Corvallis, OR 97330, USA; melinda.manore@oregonstate.edu (M.M.M.); John.Schuna@oregonstate.edu (J.M.S.J.); Adam.Branscum@oregonstate.edu (A.B.)
2   Division of Health and Exercise Science, Western Oregon University, 240 Richard Woodcock Education Center, 345 Monmouth Ave N., Monmouth, OR 97361, USA; pattonlm@wou.edu
3   Family and Community Health, School of Biological and Population Health Sciences, College of Public Health and Human Sciences, Oregon State University, Corvallis, OR 97330, USA; siewsun.wong@oregonstate.edu
*   Correspondence: mengy@oregonstate.edu; Tel.: +1-435-881-9068

Received: 29 June 2018; Accepted: 22 July 2018; Published: 23 July 2018

Abstract: The purpose of this study was to compare changes in diet and daily physical activity (PA) in high school (HS) soccer players who participated in either a two-year obesity prevention intervention or comparison group, while controlling for sex, race/ethnicity, and socioeconomic status. Participants ($n = 388$; females = 58%; Latino = 38%; 15.3 ± 1.1 years, 38% National School Breakfast/Lunch Program) were assigned to either an intervention ($n = 278$; 9 schools) or comparison group ($n = 110$; 4 schools) based on geographical location. Pre/post intervention assessment of diet was done using Block Fat/Sugar/Fruit/Vegetable Screener, and daily steps was done using the Fitbit-Zip. Groups were compared over-time for mean changes (post-pre) in fruit/vegetables (FV), saturated fat (SF), added sugar, and PA (daily steps, moderate-to-vigorous PA) using analysis of covariance. The two-year intervention decreased mean added sugar intake (-12.1 g/day, CI (7.4, 16.8), $p = 0.02$); there were no differences in groups for FV or SF intake ($p = 0.89$). For both groups, PA was significantly higher in-soccer (9937 steps/day) vs. out-of-soccer season (8117 steps/day), emphasizing the contribution of organized sports to youth daily PA. At baseline, Latino youth had significantly higher added sugar intake (+14 g/day, $p < 0.01$) than non-Latinos. Targeting active youth in a diet/PA intervention improves diet, but out of soccer season youth need engagement to maintain PA (200).

Keywords: adolescent; free or reduced lunch; National School Lunch Program; Fitbit; soccer; low-income; Latino; added sugar; sport nutrition; sport

---

1. Introduction

Currently, 20.6% of United States (US) adolescents are obese [1]. Reducing obesity is best accomplished through prevention efforts, yet the practices and methods to implement this goal are still emerging. In 2017, 52.7% of public high school (HS) students participated in sports [2,3]. Since many youth begin playing sports at a young age and continue through HS, this time may represent a window of opportunity to begin building life-skills that emphasize daily physical activity (PA), and eating a healthy diet to fuel growth, performance and health. Thus, targeting youth involved in HS school sports for obesity prevention, may be an overlooked opportunity to engage these youth.

Diet and PA are two key lifestyle factors in preventing and reducing obesity risk. Developing these life-skills as part of the youth sport experience will benefit participants for a lifetime, long after school sports are over.

For US youth, soccer is one of the fastest growing sports and among one of the top five most popular games for male and female youth. Since 2000, over three million youth have played soccer annually [4]. Soccer is especially popular among Latino youth [5]. Although Latina girls have lower sport participation rates than their peers, soccer is the one sport they are most likely to play [5]. Based on data from the National Federation of State High School Associations, more than eight million adolescents participated in HS sports in 2016–2017 [3], yet there is limited nutrition education given to these youth related to diet and PA. We recently examined the sport nutrition knowledge, attitudes and beliefs of HS soccer players. Results show that sports nutrition knowledge is low (46%), especially in Latino youth (39%) and those participating in National School Breakfast/Lunch Program (NSLP) (41%) [6].

Children and adolescents with lower social economic status have higher obesity rates, poorer diets, and lower levels of PA [7–9]. Latino youth are especially at risk for obesity [10], with significantly higher obesity prevalence (25.8%) than non-Latino youth (11.0–22.0%) [1]. For Latino youth, obesity rates are high regardless of social economic status [8]. Khan et al. found higher body mass index (BMI, kg/m^2) levels among second-generation Mexican-American adults compared with those born in Mexico [11]. This increase in body weight may be due to the availability of cheap food that can be higher in added sugar and fat than healthier foods (e.g., fruit/vegetable (FV), whole grains, low-fat dairy/meats) [12]. Finally, Boutelle et al. [13] found that Latino parents with HS adolescents were more likely than White, Asian, and Black families to purchase fast-food for the family meals (>3 time/week). No study has examined the influences of parental practices on HS soccer players' diet and PA.

Promoting a healthy diet, daily PA, and life-skills through sport programs provides an opportunity to reach groups of diverse adolescents. Youth athletes see these programs as 'improving sport performance' and less as 'another classroom lesson'. The WAVE~Ripples for Change program was developed for HS soccer players to encourage PA outside of sport, and to teach nutrition and life-skills (e.g., meal planning, shopping on a budget, food preparation/cooking skills, and gardening) to support sustainable healthy eating and lifetime PA. Thus, the purpose of this study was to compare changes in diet and daily PA in HS soccer players who participated in either a two-year obesity prevention intervention or comparison group, while controlling for sex, race/ethnicity, and socioeconomic status. In order to examine the influences of parental practices on HS soccer player's diet and PA, youth's self-report of their parents' involvement in healthy eating (e.g., modeling, setting rules and providing healthy food at home), setting screen time limits, and modeling PA were compared to youth's self-reported FV, saturated fat, added sugar intakes, and measured PA using Fitbit-zip.

2. Materials and Methods

2.1. Program Overview and Experimental Design

The WAVE~Ripples for Change: Obesity Prevention in Active Youth (WAVE) program is a two-year integrated (research, education and extension) obesity prevention intervention targeting HS soccer players (aged 14–19 years). The intervention was age-specific and included health assessments, nutrition and diet questionnaires, face-to-face sports nutrition lessons, team-building workshops (TBWs) and experiential learning. The WAVE educational objectives were to encourage PA outside of sport, and teach sport nutrition and life-skills (e.g., meal planning, shopping on a budget, food preparation/cooking skills, and gardening) to support sustainable healthy eating and life-long PA among HS soccer players. Eligibility criteria included: (1) age 14–19 years; (2) enrolled in HS soccer; (3) living with a parent/caregiver; (4) no medical conditions preventing a normal diet; (5) internet access during the two-year study; and (6) proficiency in English. Figure 1 shows the experimental design of WAVE program for this study, see the full WAVE experimental design in Supplement

Figure S1. See Manore et al [6] for more details on baseline sports nutrition knowledge and Meng et al [14] for details on program process evaluation. This manuscript only presents the two-year intervention data for changes in diet (FV, added sugar, saturated fat) and PA (steps/day; minutes of moderate-to-vigorous PA (MVPA)). All study instruments, protocols, and consent procedures were approved by the Oregon State University Institutional Review Board (#6317).

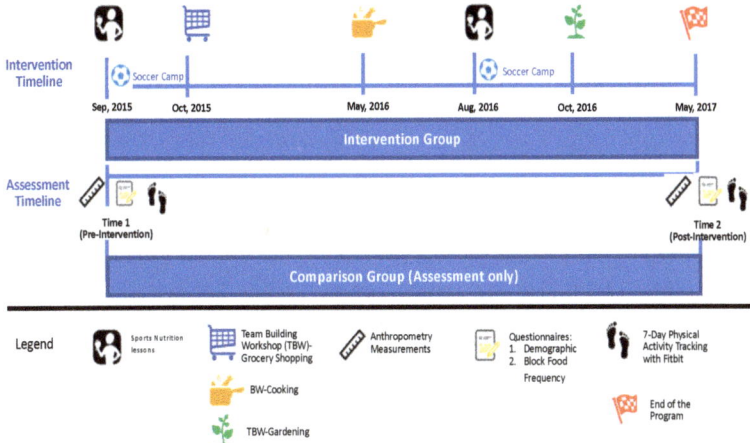

Figure 1. The WAVE program intervention experimental design (2015–2017) specific for diet and physical activity data.

2.2. Recruitment and Participants

We used a two-step recruitment process. First, soccer coaches and their schools were recruited through OSU 4-H Soccer, then soccer players were recruited through their coaches and parents recruited at soccer parent meetings. Soccer teams were then assigned (non-randomized) to either the intervention (n = 278; 9 schools,) or comparison (n = 110; 3 schools) groups based on geographical location.

The WAVE program recruited 864 HS soccer players and 72% were enrolled after submitting youth assent and parent consent forms [14]. For data analyses using only baseline (time 1) data, we included only those soccer players (n = 388; 14–18 years; 13 schools; 24 soccer teams) who completed a demographic/health history/soccer experience questionnaire, diet and anthropometry assessments, and PA using Fitbit-zip for at least two days. For data analyses examining changes over the two-year intervention, we included 202 participants who completed both assessments periods (time 1, time 2) and 97 participants who completed PA assessments at time 1 and time 2.

2.3. Intervention Delivery

The majority of the intervention (lessons, TBW, PA measurement) was delivered to teams during the fall soccer season and summer soccer camps. The WAVE HS sport nutrition curriculum was delivered to teams and their coaches by a registered dietitian nutritionist (RDN) trained in sport nutrition and experienced in collegiate/professional soccer. Prior to intervention, all lessons were pilot-tested and revised based on athletes and input from experts. During year 1 the sport nutrition topics covered were hydration and pre/during/post-exercise fueling, while year 2 focused on body composition/image; maintaining muscle and staying healthy, and eating-well while eating out. PA outside of soccer practice and out-of-soccer season were encouraged. Newsletters reinforced lessons and provided recipes or tips to meet sport nutrition needs. Life-skills trainings were taught via three TBWs focused on menu planning, grocery shopping, cooking, knife skills and safety, and gardening, and delivered to teams by Extension faculty and community partners. These educational approaches all included experiential learning opportunities (e.g., food demonstrations and tastings, cooking with recipes high in FV, shopping on a budget, planning a post-game team meal, garden harvesting). The sports nutrition lessons and TBW encouraged intakes of FV, whole grains, and low-fat proteins, while selecting less processed foods and added sugar. See Supplement Table S1 for more information of the WAVE sports nutrition lesson curriculum and newsletter topics.

2.4. Data Collection

Eligible participants enrolled in person or online. Questionnaires and assessments were completed at baseline (time 1) and post-intervention (time 2) (Figure 1). Based on the completion rate of intervention activities, each participant received gift cards (maximum = $65/year).

2.4.1. Demographics, Parental Practices, and Anthropometry

The baseline questionnaire collected demographic (sex, age, races/ethnicity), NSLP participation, soccer experience, and youth perceived parental practices. Youth indicated how many days/week (0, 1–3, 4–5, or 6–7 days) their parent participated in the following behaviors: (1) making healthy food available at home; (2) role modeling healthy eating; (3) setting rules/expectations on what youth eat; (4) setting rules/expectation around screen time; and (5) role modeling PA. Anthropometry (height, weight; Tanita scale TM-300A, Tanita Corp., Singapore) assessments were completed at time 1 and time 2. Researcher completed required anthropometry training sessions to assure protocol fidelity and test/retest reliability.

2.4.2. Dietary Assessment

A validated food intake screener (Block Fat/Sugar/Fruit/Vegetable Screener (Block-FSFV)) was used to measure youth's FV, saturated fat, and added sugar intakes [15]. The screener reliability ranging from 0.70 to 0.78 over a four-month test/retest period [15]. Participants reported the portion size and consumption frequency of consuming 35 foods over the past week. Summary data included estimated average daily intakes for FV (cup equivalents), saturated fat (g/day) and added sugar (g/day). Estimated daily energy intake was only used to identify over/under-reporting as defined by Boucher et al [16].

2.4.3. Physical Activity Assessment

At baseline (Time 1) and post-intervention (Time 2), all participants wore a Fitbit-Zip for 7-days. Fitbit-Zip correlates highly with the step data and MVPA data yielded by the ActiGraph among free-living adolescents [17]. This approach was pre-tested in the pilot study to assure viability and repeatability. Participants were given both oral and written instructions on how to use the Fitbit-zip. Daily texts/emails were sent to remind participants of wearing Fitbit. The comparison group's Fitbit-Zip screen was covered with duct tape to avoid the self-monitoring effect on PA. Fitbit-Zip minute-by-minute data were aggregated into average daily steps, while MVPA was identified as time spent at >100 steps/min/day. Substantial empirical evidence supports using the >100 steps/min threshold for defining ambulatory time spent at or above a moderate-intensity in adults [18]. Recent research, using regression-based calibration in older adolescents (15–17 years), shows that this threshold (>100 steps/min) appears appropriate for defining time spent at or above a moderate-intensity in this age group [19]. All participants used in the PA analyses had at least two valid days of Fitbit-Zip measured PA (8-h/day wearing time) and daily values for steps/day and MVPA (min/day) were averaged across valid days. The Fitbit-Zip has shown excellent reliability (ICC = 0.90) for measuring steps in the laboratory setting [20] and previous research among a large and diverse sample of children and adolescents has demonstrated that two days of pedometer monitoring provides an acceptable level of reliability (ICC > 0.85) when estimating weekly steps/day [21]. Although there remains no consensus wear time threshold for defining a "valid day" when working with minute-by-minute Fitbit-Zip step data, we chose to employ the 8-h/day criteria utilized by a previous validation study in adolescents [17].

2.5. Data Analysis

At baseline (Time 1) descriptive statistics were calculated. Analysis of variance (ANOVA) models were used to examine mean baseline differences in FV, saturated fat, added sugar intake, average daily steps, and MVPA (min/day) between race/ethnicity (Latino/non-Latino, white and others), sex (female/male), and NSLP (participants/non-participants); interactions were also examined. Pearson correlations were used to evaluate associations between diet and PA variables and youth self-reported parental practices.

A series of analysis of covariance (ANCOVA) models, with categorical independent variables for group assignment, race/ethnicity, sex, and NSLP, were used to evaluate mean changes (time 2-time 1) in PA and diet intakes (FV, saturated fat, added sugar) between groups (intervention/comparison). For each outcome variable, a separate ANCOVA model was fitted with change score (time 2-time 1) as the dependent variable and baseline value as a covariate. Analyses were conducted using SPSS (version 24, SPSS, IBM SPSS, Inc., Chicago, IL, USA) and R Core Team.

3. Results

Table 1 provides the demographic characteristics of participants who completed baseline assessments. Participants were 15.3 ± 1.1 years of age and were normal weight based on average body mass index (BMI) percentile [22]. More specifically, 1% of participants were below 5th percentile, 77% were between 5th–85th percentile, 14.9% were between 86th–95th percentile, and 7% were greater than the 95th percentile. Overall, 38.1% of the youth athletes participated in the NSLP, with most (78.4%) being Latino; 45% engage in ≥ 2 sports, and 42% reported ≥ 1 injuries in the past year that needed ≥ 1-week of rest.

Table 1. Baseline characteristics of participants who completed demographic questionnaire, diet and anthropometry assessments, and physical activity using Fitbit-zip (*n* = 388).

Baseline Characteristics	All (*n* = 388)	Intervention Group (*n* = 278)		Comparison Group (*n* = 110)	
		Female (*n* = 152)	Male (*n* = 126)	Female (*n* = 72)	Male (*n* = 3 8)
Age (year), mean (SD) [a]	15.3 (1.1)	15.1 (1.0)	15.3 (1.2)	15.4 (1.3)	15.9 (1.0)
Race, *n* (%)					
Latino	149 (38.4%)	43 (28.3%)	59 (46.8%)	34 (47.2%)	13 (34.2%)
Non-Latino [b]	239 (61.6%)	109 (71.7%)	67 (53.2%)	38 (52.8%)	25 (65.8%)
NSLP [c], *n* (%)	148 (38.1%)	43 (28.3%)	53 (42.1%)	35 (48.6%)	17 (44.7%)
Prepare meal for themselves, *n* (%)	229 (59.0%)	99 (65.1%)	66 (52.4%)	44 (61.1%)	20 (52.6%)
		Mean (SD) [a]			
BMI [d] percentile	62.8 (25.0)	62.3 (23.5)	56.1 (27.5)	73.7 (21.0)	62.7 (21.0)
Years play soccer (year)	7.6 (3.7)	8.2 (3.3)	8.0 (3.7)	5.8 (3.6)	6.6 (4.0)
Fruit and vegetable (FV) intakes (cup equivalent/day)	2.8 (1.6)	2.9 (1.6)	2.8 (1.6)	2.4 (1.5)	2.7 (1.6)
Saturated fat (g/day)	21.8 (10.4)	19.8 (9.5)	24.1 (11.6)	20.9 (9.2)	23.6 (10.0)
Added sugar (g/day)	47.9 (38.3)	38.2 (30.6)	53.4 (43.7)	56.5 (38.4)	51.6 (39.9)
Daily steps/day	9937 (3180)	9019 (2534)	10629 (3551)	10061(2997)	11080(3642)
Moderate-to-vigorous physical activity (MVPA) (min/day)	33.3 (15.5)	29.2 (12.2)	38.3 (18.0)	31.8 (14.0)	35.9 (15.7)

[a] Standard Deviation (SD); [b] Non-Latino includes participants who self-identified as White, American Indian/Alaska Native, Asian/Pacific Islander, or Black/African American; [c] NSLP = National School Lunch Program, participation indicates social economic status; [d] BMI = Body mass index (kg/m²), 5th percentile to less than the 85th percentile is considered normal or healthy weight [22].

3.1. Diet Assessment

For all participants at baseline (time 1), the average FV consumption was 2.8 cups/day (Table 1). There were no differences between sexes for FV and added sugar intakes, but females had significantly lower saturated fat intake than males (−4.4 g/day) (Table 2). Latinos had significantly higher added sugar intake than non-Latinos (+14.1 g/day), after adjusting for sex and NSLP. The two-year intervention resulted in participants significantly decreasing added sugar (−12g/day) and saturated fat (−2.7 g/day) intakes, with no changes in FV intake (Table 3). For the same time period, only saturated fat (−2.5 g/day) decreased in the comparison group. In comparing differences between groups over the two-year intervention, only added sugar intake was significantly lower in the intervention group (−10.4 g/day). This lower intake of sugar was attributed to a decrease in frequency (−11%) of consuming cake/cookie snack, ice cream, and ice-cream bars. There were no changes in sugar-sweetened beverage intake.

3.2. Physical Activity

For all participants (n = 388), baseline average steps/day were 9937 (males = 107.34 steps/day; females = 9353 steps/day) and MVPA was 33.3 min/day (Table 1). Neither the 2008 or 2018 Physical Activity Guidelines for Americans gives recommendation step guidelines for youth [23,24]. However, Adam et al. [25] translated the 2008 Physical Activity Guideline recommendation of 60 min/day of MVPA into step counts for youth 12–17 year (male ≥ 105,00 steps/day; female ≥ 9000 steps/day). Our participants met this step threshold recommendation. Overall, females were significantly less active (−1380 steps/day; −7.7 min MVPA/day) than males (Table 2). There were no differences in PA due to ethnicity or NSLP participation. During the two-year intervention, both groups had significantly higher PA in-season (time 1) compared to out-of-season (time 2; ~1500−2000 steps/day less) (Table 3). A similar response was observed in MVPA, with a higher level of MVPA (~17%) in-season vs. out-of-season. Over the two-year intervention, there were no significant differences between groups for PA.

To further explore how participating in soccer contributed to meeting the PA guidelines, we calculated the average minute-by-minute steps for participants (n = 97; 2-day Fitbit data) using a 24-h scale (see Figure 2). Between 2:30–6:00 p.m., participants averaged 18.6 and 8.8 steps/minute in-season and out-season, respectively. During this time, athletes accumulated an extra of 2060 steps/day in-soccer season, which explains the in-/out-soccer season difference of 1820 steps/day (pre/post-intervention differences).

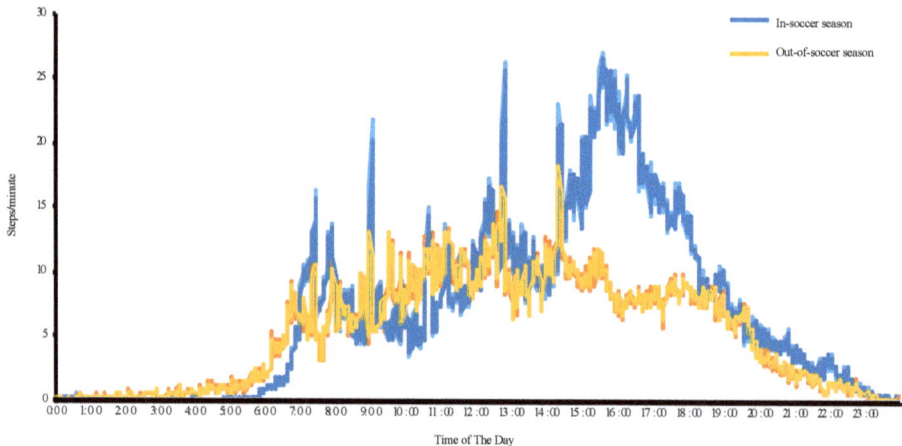

Figure 2. Daily pattern of steps-per-minute for all participants at baseline (in-soccer-season) and post-intervention (out-of-soccer season).

Table 2. Correlations between dietary and physical activity variable and sex, ethnicity and participation in NSLP at baseline (n = 388).

Variable	Fruits and Vegetables (cups/day)	Saturated Fat (g/day)	Added Sugar (g/day)	Averaged Daily Step Counts (steps/day)	Moderate-to-Vigorous Physical Activity (min/day)
	Mean difference (95% confident interval)				
Female [a]	−0.11 (−0.42, 0.21)	−4.41 (−6.17, −2.06) **	−6.91 (−14.45, 0.58)	−1380(−2013, −747) **	−7.67 (−10.73, −4.62) **
Latino [b]	−0.37 (−0.79, 0.05)	−1.73 (−4.46, 1.0)	14.08 (4.12, 24.03) **	−132 (−972, 706)	−0.07 (−4.11, 3.98)
NSLP [c]	−0.30 (−0.71, 0.12)	−1.48 (−4.21, 1.24)	7.52 (−2.43, 17.48)	156 (−682, 995)	1.36 (−2.68, 5.40)

* Group are significantly different (** $p < 0.01$). [a] Females are compared with males; [b] Latinos are compared with non-Latino (White, American Indian/Alaska Native, Asian/Pacific Islander, or Black/African American); [c] National School Breakfast/Lunch Program (NSLP) participants are compared with non-NSLP percipients.

Table 3. Pre/post-intervention changes in dietary intake and physical activity (PA) within and between groups over the WAVE two-year intervention (n = 202).

Variable	Sample Size (n)	Baseline (in-Season) Mean (SD [a])	Post-Intervention (Out-of-Season) Mean (SD)	Within Group Changes (Post-Pre) [b]	Intervention vs. Comparison Changes [b]	p-Value [c]
Fruit and vegetables (cup equivalent/day)	Intervention (n = 143)	2.7 (1.5)	2.6 (1.4)	−0.1 (−0.3, 0.1)	0.2 (−0.2, 0.5)	0.44
	Comparison (n = 59)	2.5 (1.5)	2.3 (1.2)	−0.2 (−0.6, 0.1)		
Saturated fat (g/day)	Intervention (n = 143)	21.2 (10.0)	18.7 (9.0)	−2.7 (−4.0, −1.4) *	−0.2 (−2.5, 2.2)	0.89
	Comparison (n = 59)	22.3 (9.4)	19.4 (9.0)	−2.5 (−4.5, −0.5) *		
Added sugar (g/day)	Intervention (n = 143)	43.6 (34.4)	33.7 (28.2)	−12.1 (−16.8, −7.4) *	−10.4 (−19.1, −1.6) *	0.02
	Comparison (n = 59)	57.6 (37.5)	50.6 (41.5)	−1.7 (−9.0, 5.5)		
Physical activity (average steps/day)	Intervention (n = 64)	9970 (3145)	7660 (3022)	−2058 (−2770, −1345) *	−473 (−1721, 775)	0.45
	Comparison (n = 33)	9849 (2977)	8752 (3186)	−1585 (−2587, −582) *		
Moderate-to-vigorous physical activity (MVPA) (min/day)	Intervention (n = 64)	34.3 (17.0)	24.3 (15.3)	−8.3 (−12.2, −4.5) *	−1.6 (−8.3, 5.2)	0.65
	Comparison (n = 33)	30.9 (13.2)	27.2 (16.8)	−6.8 (−12.2, −1.3) *		

* p value < 0.05; [a] = Standard Deviation (SD); [b] adjusted for the baseline value and wear time changes; [c] p-value for intervention vs. comparison.

3.3. Perceived Parental Support.

We also examined perceived parental support influences on youth's diet and PA at baseline for all the participants combined (Table 4). Youth had higher FV intake when parents modeled healthy eating habits. Youth had lower added sugar intake when parent modeled healthy eating habits, made healthy foods available at home, and led physically active lifestyles.

Table 4. Correlations between diet and physical activity and youth self-reported parental practice at baseline (*n* = 388).

Variable	Fruits and Vegetables (cups/day)	Saturated Fat (g/day)	Added Sugar (g/day)	Averaged Daily Step Counts (steps/day)	Moderate−to−Vigorous Physical Activity (MVPA) (min/day)
	Pearson's correlation coefficient				
Parents make healthy food available, day/week [a]	0.30 ***	0.09	−0.14 **	−0.02	−0.06
Parents role model healthy eating, day/week	0.23 ***	0.03	−0.11 *	−0.02	−0.04
Parents set rules on what youth eat, day/week	0.15 **	0.05	−0.10	0.03	0.02
Parents set rules on screen time, day/week	0.20 ***	0.03	−0.02	0.09	0.03
Parents role model PA, day/week	0.18 ***	0.02	−0.11 *	−0.01	−0.01

* Group are significantly different (* $p < 0.05$; ** $p < 0.01$; *** $p < 0.001$). [a] Youth self-reported the number of days in a week their parents have above parental practices, positive values indicant positive associations between days of parental practices and FV/saturated fat/added sugar/averaged daily steps/moderate-to-vigorous PA; negative value means the opposite.

4. Discussion

This is the first study to report changes in FV, saturated fat, and added sugar intakes, and PA in HS soccer players over a two-year intervention focused on teaching healthy eating behaviors, sport nutrition, lifetime PA, and life-skills for sport performance, health and obesity prevention. It is also the first study to evaluate whether diet and PA differ in HS athletes based on sex, race/ethnicity, social economic status, or youth self-reported parental practices.

4.1. Diet Assessment

Dietary assessments of FV, saturated fat, and added sugar intakes in HS soccer players are limited [26–28], and no published study has examined these factors in US Latino youth. The FV intakes in our participants (males = 2.8 cups/day; females = 2.7 cups/day) was higher than the mean FV intake reported in the 2013–2014 National Health and Nutrition Survey (NHANES) for adolescents age 12–18 years (1.9 cups/day) [29], but below the current Dietary Guideline for Americans (4.5–5 cups/day) (14–18 years) [30]. Three studies have examined differences in FV intake between youth sport participants and non-sport participants [31–33]. All found that sport participants had higher FV intake than non-sport participants, but no specific cup equivalents were reported. Others have examined the FV intake of youth athletes, including swimmers (1.6–2 cups/day) and rugby players (1.3 cups/day) [34,35], but no comparison group was included. Finally, Parnell et al [36] compared the FV intake of Canadian youth athletes to the Canadian diet recommendations for teens and found 40–54% met the FV guidelines of 7–8 servings per day. In contrast, only 13.9% of our participants met the Dietary Guidelines for Americans for FV intake [30].

For our participants, the average added sugar intake (males = 53 g/day; females = 44 g/day) was lower than that reported for 2013–2014 NHANES adolescents (81 g/day; 12–19 years) [29]. These differences may be due to methods used to assess added sugar intake. We used the Block-FSFV, which only includes 35 foods, while NHANES uses two 24-h recalls. Added sugar intake in our Latino soccer players was significantly higher (+14.1 g/day) than non-Latino soccer players. NHANES data (2003–2004) indicates that the major sources of added sugars of Latino youth were soda (28 g/day), fruit drinks (17 g/day), grain desserts (10 g/day), and candy (5 g/day) [37]. The high added sugar consumption in Latino youth has been attributed to acculturation and socioeconomic factors [38]. We, however, did not measure acculturation in our participants, so we cannot determine if acculturation affects added sugar intake. However, Manore et al. [6] found that Latino soccer players had lower sports nutrition knowledge (38.8%) compared to non-Latino players (48.4%), which may indicate lower general nutrition knowledge about added sugar intake.

Youth who self-reported that their parents frequently role modeled healthy eating habits and physically active lifestyles, and made healthy foods available at home had higher FV consumption, and lower added sugar intakes. These results support others who report that parental practices are important in shaping children's nutrition behavior [39,40]. Ranjit et al. [39] found that increasing access and availability to healthy foods could increase middle school aged adolescents' FV intake and decrease sugar-sweetened beverage intake. Battram et al. [40] also reported that parents were the primary gatekeepers and role models of children's sugar beverage consumption. To our knowledge our research is the first to examine the influence of perceived parental support on FV and added sugar intake among HS athletes.

4.2. Physical Activity

Cross-sectional research shows that youth who participate in sports have higher MVPA (60 min/day) than non-sport participants (43 min/day) [41]. Yet, no study has explored the importance of helping active youth maintain their PA 'out-of-sport-season' or during other breaks from their sport participation. We followed HS soccer players for two years and analyzed their steps/day and MVPA both in/out-of-soccer season. Our results showed that the PA level in-soccer season met the 2008 Physical Activity Guidelines for Americans [23], but dropped below recommendations by 1820 steps/day during out-of-soccer season. During soccer season, participants generally have two games (1.5-h/each) and three/four practices (2-h/each) per week. Wearing a Fitbit is not allowed during games, but players typically accumulate 3–7 miles/game based on their position [42]. If these additional steps were added to the in-season steps, PA would be even higher than reported here and the differences in in-/out-soccer season PA would be even greater. These results confirmed the importance of school sports in promoting PA and the need to maintain youth's PA when they are not engaged in organized sports. Although the WAVE project focused on active youth for obesity prevention, it is important to acknowledge that all youth need diet and PA life-skills. Thus, HS students who are not engaged in sports may need a different approach than used in this project to engage them in daily PA and learning life-skills around diet for obesity prevention.

We found no association between youth PA (steps and MVPA) and youth self-reported parental practices. Research examining the influence of parental modeling of PA on children's PA is equivocal, with some research being supportive [43,44], while others are not [45–47]. These data were all reported in young children (<12 years). No study has examined this association in active adolescents (≥14 years).

4.3. Strengths and Limitations

Our study had a number of strengths. First, we had a large sample (*n* = 388 at baseline; *n* = 202 completing pre/post assessments) of HS soccer players, with an average of 7.6-year playing experience; thus, they may be representative of many HS athletes who play at least one HS sport. Second, we had a diverse sample (58% female; 40% Latino; 40% NSLP participation), which allowed us to examine the association between diet and PA and sex, race/ethnicity, and social economic status. Third, the intervention lasted two years, compared to the 3–12 weeks interventions that are typically reported in the literature [48–50]. Finally, the WAVE intervention combining traditional face-to-face lessons and TBWs that included experiential learning that reinforced the lessons and allowed for practical application.

There were also limitations to this study. First, the Block-FSFV does not include Latino culturally-appropriate food or sport foods. The Block-FSFV was designed to focus on FV, saturated fat, and added sugar intake; thus, we could not calculate the percentage of energy from saturated fat and added sugar to compare to the 2015–2020 Dietary Guidelines for Americans recommendations. Second, dietary intake could have changed due to the time period (two years) we measured diet (pre-intervention: fall season 2015; post-intervention: spring season 2017). Typically, fall season is abundant in fruits and vegetables from the gardens and orchards, while spring is a time for celebration around graduation. However, there were no significant changes in FV intake across group or over time. Third, we did not measure the acculturation level, but all participants spoke English. Finally, the comparison group participants could have removed the duct-tape and monitored their daily steps, but we did not see differences between groups for steps in-/out-of-soccer season.

5. Conclusions

This is the first study to engage HS soccer players in a two-year obesity prevention program targeting sport nutrition education and healthy eating behaviors for growth, performance and health, and emphasizing the importance of daily PA and learning life-skills. Overall, the FV intakes were maintained over the intervention but were below recommendations, yet higher than those typically reported in the literature for youth and youth athletes. The intervention significantly decreased added sugar intake, by lowering the frequency of selecting cake/cookies and ice cream foods. Our lessons emphasized eating whole grains, FV, and lean protein sources and selecting food with less saturated fat and added sugar. Incorporating these eating behaviors into their lifestyle could lower added sugar intake. We found parents who modeled a healthy lifestyle, and made healthy foods available at home, had adolescents who reported lower added sugar intakes. Thus, parent involvement is important in shaping the healthy lifestyles of adolescents. Finally, we found PA recommendations were met during soccer season, but not out-of-season. Thus, it is important to engage active youth throughout the school year, helping them make daily PA a priority in their lifestyle. Based on our experience, self-monitoring alone is not enough to promote PA among active youth when they are not engaged in sport. Future studies should focus on maintaining PA in youth athletes when they are not engaged in sport, thus, helping them make the transition to being physically active adults.

Supplementary Materials: The following are available online at http://www.mdpi.com/2072-6643/10/7/947/s1, Figure S1: Expanded experimental design, Table S1: Sports nutrition lessons.

Author Contributions: All authors participated in all aspects of this study.

Funding: This project is funded by USDA National Institute of Food and Agriculture-Agriculture Food and Research Initiative (Award No. 2013-67001-20418).

Acknowledgments: We want to thank the many athletes, coaches, parents, schools, volunteers, and industry and community partners that helped make this possible.

Conflicts of Interest: All authors declare no conflict of interest.

References

1. Hales, C.M.; Carroll, M.D.; Fryar, C.D.; Ogden, C.L. Prevalence of Obesity among Adults and Youth: United States, 2015–2016. NCHS Data Brief; 2017. Available online: https://www.cdc.gov/nchs/products/databriefs/db288.htm (accessed on 26 November 2017).
2. National Center for Education Statistics. Back to School Statistics for 2017. Available online: https://nces.ed.gov/fastfacts/display.asp?id=372 (accessed on 13 July 2018).

3.   The National Federation of State High School Associations. Participation Statistics: 2016–2017 High School Athletics Participation Survey Results. Available online: http://www.nfhs.org/ParticipationStatistics/PDF/2016-17_Participation_Survey_Results.pdf (accessed on 26 June 2018).

4.   Colo Rapids Youth Soccer. US Youth Soccer Player Statistics. 2016. Available online: http://rapidsyouthsoccer.org/us-youth-soccer-player-statistics/ (accessed on 1 June 2018).

5.   Turner, R.W.; Perrin, E.M.; Coyne-Beasley, T.; Peterson, C.J.; Skinner, A.C. Reported sports participation, race, sex, ethnicity, and obesity in US adolescents from NHANES physical pctivity (PAQ_D). *Glob. Pediatr. Health* **2015**, *2*. [CrossRef]

6.   Manore, M.M.; Patton-Lopez, M.M.; Meng, Y.; Wong, S.S. Sport nutrition knowledge, behaviors and beliefs of high school soccer players. *Nutrients* **2017**, *9*, 350. [CrossRef] [PubMed]

7.   Janssen, I.; Boyce, W.F.; Simpson, K.; Pickett, W. Influence of individual- and area-level measures of socioeconomic status on obesity, unhealthy eating, and physical inactivity in Canadian adolescents. *Am. J. Clin. Nutr.* **2006**, *83*, 139–145. [CrossRef] [PubMed]

8.   Ogden, C.L.; Lamb, M.M.; Carroll, M.D.; Flegal, K.M. Obesity and socioeconomic status in children and adolescents: United States, 2005–2008. *NCHS Data Brief* **2010**, *51*, 1–8.

9.   Stalsberg, R.; Pedersen, A.V. Effects of socioeconomic status on the physical activity in adolescents: A systematic review of the evidence. *Scand. J. Med. Sci. Sports* **2010**, *20*, 368–383. [CrossRef] [PubMed]

10. Isasi, C.R.; Rastogi, D.; Molina, K. Health issues in Hispanic/Latino youth. *J. Lat. Psychol.* **2016**, *4*, 67–82. [CrossRef] [PubMed]

11. Khan, L.K.; Sobal, J.; Martorell, R. Acculturation, socioeconomic status, and obesity in Mexican Americans, Cuban Americans, and Puerto Ricans. *Int. J. Obes. Relat. Metab. Disord.* **1997**, *21*, 91–96. [CrossRef] [PubMed]

12. Drewnowski, A.; Darmon, N. Food choices and diet costs: an economic analysis. *J. Nutr.* **2005**, *135*, 900–904. [CrossRef] [PubMed]

13. Boutelle, K.N.; Fulkerson, J.A.; Neumark-Sztainer, D.; Story, M.; French, S.A. Fast food for family meals: Relationships with parent and adolescent food intake, home food availability and weight status. *Public Health Nutr.* **2007**, *10*, 16–23. [CrossRef] [PubMed]

14. Meng, Y.; Wong, S.S.; Manore, M.M.; Patton-López, M. WAVE~Ripples for change obesity two-year intervention in high school soccer players: process evaluation, best practices, and youth engagement. *Nutrients* **2018**, *10*, 711. [CrossRef] [PubMed]

15. Lalonde, I.; Graham, M.; Slovinec D'Angelo, M.; Beaton, L.; Brown, J.; Block, T. Validation of the Block Fat/Sugar/Fruit/Vegetable screener in a cardiac rehabilitation setting. *J. Cardiopulm. Rehabil. Prev.* **2008**, *28*, 340. [CrossRef]

16. Boucher, B.; Cotterchio, M.; Kreiger, N.; Nadalin, V.; Block, T.; Block, G. Validity and reliability of the Block98 food-frequency questionnaire in a sample of Canadian women. *Public Health Nutr.* **2006**, *9*, 84–93. [CrossRef] [PubMed]

17. Schneider, M.; Chau, L. Validation of the Fitbit Zip for monitoring physical activity among free-living adolescents. *BMC Res. Notes* **2016**, *9*, 448. [CrossRef] [PubMed]

18. Tudor-Locke, C.; Rowe, D.A. Using cadence to study free-living ambulatory behaviour. *Sports Med. Auckl. N. Z.* **2012**, *42*, 381–398. [CrossRef] [PubMed]

19. Tudor-Locke, C.; Schuna, J.M.; Han, H.; Aguiar, E.J.; Larrivee, S.; Hsia, D.S.; Ducharme, S.W.; Barreira, T.V.; Johnson, W.D. Cadence (steps/min) and intensity during ambulation in 6–20 year olds: The CADENCE-kids study. *Int. J. Behav. Nutr. Phys. Act.* **2018**, *15*, 20. [CrossRef] [PubMed]

20. Kooiman, T.J.M.; Dontje, M.L.; Sprenger, S.R.; Krijnen, W.P.; van der Schans, C.P.; de Groot, M. Reliability and validity of ten consumer activity trackers. *BMC Sports Sci. Med. Rehabil.* **2015**, *7*. [CrossRef] [PubMed]

21. Craig, C.L.; Tudor-Locke, C.; Cragg, S.; Cameron, C. Process and treatment of pedometer data collection for youth: the Canadian Physical Activity Levels among Youth study. *Med. Sci. Sports Exerc.* **2010**, *42*, 430–435. [CrossRef] [PubMed]

22. Centers for Disease Control and Prevention. Healthy Weight. Available online: https://www.cdc.gov/healthyweight/assessing/bmi/childrens_bmi/about_childrens_bmi.html (accessed on 25 January 2018).

23. U.S. Department of Health and Human Services: 2008 Physical Activity Guidelines for Americans. Available online: https://health.gov/paguidelines/guidelines/children.aspx (accessed on 9 March 2018).

24. U.S. Department of Health and Human Services. 2018 Physical Activity Guidelines Advisory Committee Submits Scientific Report. Available online: https://health.gov/news/blog-bayw/2018/03/2018-physical-activity-guidelines-advisory-committee-submits-scientific-report/ (accessed on 9 March 2018).

25. Adams, M.A.; Johnson, W.D.; Tudor-Locke, C. Steps/day translation of the moderate-to-vigorous physical activity guideline for children and adolescents. *Int. J. Behav. Nutr. Phys. Act.* **2013**, *10*, 49. [CrossRef] [PubMed]

26. Elizondo, R.H.T.; Bermudo, F.M.; Mendez, R.P.; Amorós, G.B.; Padilla, E.L.; de la Rosa, F.J.B. Nutritional intake and nutritional status in elite Mexican teenagers soccer players of different ages. *Nutr. Hosp.* **2015**, *32*, 1735–1743. [CrossRef]

27. Ruiz, F.; Irazusta, A.; Gil, S.; Irazusta, J.; Casis, L.; Gil, J. Nutritional intake in soccer players of different ages. *J. Sports Sci.* **2005**, *23*, 235–242. [CrossRef] [PubMed]

28. dos Santos, D.; da Silveira, J.Q.; Cesar, T.B. Nutritional intake and overall diet quality of female soccer players before the competition period. *Rev. Nutr.* **2016**, *29*, 555–565. [CrossRef]

29. Bowman, S.; Clemens, J.C.; Friday, J.E.; Lynch, K.L.; LaComb, R.P.; Moshfegh, A.J. *Food Patterns Equivalents Intakes by Americans: What We Eat in America, NHANES 2003–2004 and 2013–2014*; Food Surveys Research Group: Bethesda, MD, USA, 2017.

30. U.S. Department of Health and Human Services. 2015–2020 Dietary Guidelines for Americans, 8th ed. Available online: http://www.health.gov/dietaryguidelines/2015/guidelines/ (accessed on 7 March 2018).

31. Cavadini, C.; Decarli, B.; Grin, J.; Narring, F.; Michaud, P.-A. Food habits and sport activity during adolescence: Differences between athletic and non-athletic teenagers in Switzerland. *Eur. J. Clin. Nutr.* **2000**, *54*, S16–S20. [CrossRef] [PubMed]

32. Harrison, P.A.; Narayan, G. Differences in behavior, psychological factors, and environmental factors associated with participation in school sports and other activities in adolescence. *J. Sch. Health* **2003**, *73*, 113–120. [CrossRef] [PubMed]

33. Pate, R.R.; Heath, G.W.; Dowda, M.; Trost, S.G. Associations between physical activity and other health behaviors in a representative sample of US adolescents. *Am. J. Public Health* **1996**, *86*, 1577–1581. [CrossRef] [PubMed]

34. Collins, A.C.; Ward, K.D.; Mirza, B.; Slawson, D.L.; McClanahan, B.S.; Vukadinovich, C. Comparison of nutritional intake in US adolescent swimmers and non-athletes. *Health* **2012**, *4*, 873–880. [CrossRef] [PubMed]

35. Smith, D.R.; Jones, B.; Sutton, L.; King, R.F.G.J.; Duckworth, L.C. Dietary intakes of elite 14- to 19-year-old English academy rugby players during a pre-season training period. *Int. J. Sport Nutr. Exerc. Metab.* **2016**, *26*, 506–515. [CrossRef] [PubMed]

36. Parnell, J.A.; Wiens, K.P.; Erdman, K.A. Dietary intakes and supplement use in pre-adolescent and adolescent Canadian athletes. *Nutrients* **2016**, *8*, 526. [CrossRef] [PubMed]

37. Reedy, J.; Krebs-Smith, S.M. Dietary sources of energy, solid fats, and added sugars among children and adolescents in the United States. *J. Am. Diet. Assoc.* **2010**, *110*, 1477–1484. [CrossRef] [PubMed]

38. Pérez-Escamilla, R.; Putnik, P. The Role of acculturation in nutrition, lifestyle, and incidence of type 2 diabetes among Latinos. *J. Nutr.* **2007**, *137*, 860–870. [CrossRef] [PubMed]

39. Ranjit, N.; Evans, A.E.; Springer, A.E.; Hoelscher, D.M.; Kelder, S.H. Racial and ethnic differences in the home food environment explain disparities in dietary practices of middle school children in Texas. *J. Nutr. Educ. Behav.* **2015**, *47*, 53–60. [CrossRef] [PubMed]

40. Battram, D.S.; Piché, L.; Beynon, C.; Kurtz, J.; He, M. Sugar-sweetened beverages: Children's perceptions, factors of influence, and suggestions for reducing intake. *J. Nutr. Educ. Behav.* **2016**, *48*, 27–34.e1. [CrossRef] [PubMed]

41. Silva, G.; Andersen, L.B.; Aires, L.; Mota, J.; Oliveira, J.; Ribeiro, J.C. Associations between sports participation, levels of moderate to vigorous physical activity and cardiorespiratory fitness in childrenand adolescents. *J. Sports Sci.* **2013**, *31*, 1359–1367. [CrossRef] [PubMed]

42. Buchheit, M.; Mendez-Villanueva, A.; Simpson, B.M.; Bourdon, P.C. Match running performance and fitness in youth soccer. *Int. J. Sports Med.* **2010**, *31*, 818–825. [CrossRef] [PubMed]

43. Hutchens, A.; Lee, R.E. Parenting Practices and Children's Physical Activity: An Integrative Review. *J. Sch. Nurs. Off. Publ. Natl. Assoc. Sch. Nurses* **2018**, *34*, 68–85. [CrossRef] [PubMed]

44. Lloyd, A.B.; Lubans, D.R.; Plotnikoff, R.C.; Morgan, P.J. Paternal Lifestyle-Related Parenting Practices Mediate Changes in Children's Dietary and Physical Activity Behaviors: Findings From the Healthy Dads, Healthy Kids Community Randomized Controlled Trial. *J. Phys. Act. Health* **2015**, *12*, 1327–1335. [CrossRef] [PubMed]

45. Jago, R.; Davison, K.K.; Brockman, R.; Page, A.S.; Thompson, J.L.; Fox, K.R. Parenting styles, parenting practices, and physical activity in 10- to 11-year olds. *Prev. Med.* **2011**, *52*, 44–47. [CrossRef] [PubMed]

46. Chiarlitti, N.; Kolen, A. Parental influences and the relationship to their children's physical activity levels. *Int. J. Exerc. Sci.* **2017**, *10*, 205–212. [PubMed]

47. Erkelenz, N.; Kobel, S.; Kettner, S.; Drenowatz, C.; Steinacker, J.M.; Research Group "Join the Healthy Boat—Primary School". Parental activity as influence on children's BMI percentiles and physical activity. *J. Sports Sci. Med.* **2014**, *13*, 645–650. [PubMed]

48. Elliot, D.L.; Moe, E.L.; Goldberg, L.; DeFrancesco, C.A.; Durham, M.B.; Hix-Small, H. Definition and outcome of a curriculum to prevent disordered eating and body-shaping drug use. *J. Sch. Health* **2006**, *76*, 67–73. [CrossRef] [PubMed]

49. Laramée, C.; Drapeau, V.; Valois, P.; Goulet, C.; Jacob, R.; Provencher, V.; Lamarche, B. Evaluation of a theory-based intervention aimed at reducing intention to use restrictive dietary behaviors among adolescent female athletes. *J. Nutr. Educ. Behav.* **2017**, *49*, 497–504.e1. [CrossRef] [PubMed]

50. Nascimento, M.; Silva, D.; Ribeiro, S.; Nunes, M.; Almeida, M.; Mendes-Netto, R. Effect of a nutritional intervention in athlete's body composition, eating behaviour and nutritional knowledge: A comparison between adults and adolescents. *Nutrients* **2016**, *8*, 535. [CrossRef] [PubMed]

nutrients

MDPI

Article

Disparate Habitual Physical Activity and Dietary Intake Profiles of Elderly Men with Low and Elevated Systemic Inflammation

Dimitrios Draganidis [1], Athanasios Z. Jamurtas [1], Theodoros Stampoulis [2], Vasiliki C. Laschou [1], Chariklia K. Deli [1], Kalliopi Georgakouli [1], Konstantinos Papanikolaou [1], Athanasios Chatzinikolaou [2], Maria Michalopoulou [2], Constantinos Papadopoulos [3], Panagiotis Tsimeas [1], Niki Chondrogianni [4], Yiannis Koutedakis [1,5,6], Leonidas G. Karagounis [7,8] and Ioannis G. Fatouros [1,*]

[1]   School of Physical Education and Sport Science, University of Thessaly, Karies, 42100 Trikala, Greece; dimidraganidis@gmail.com (D.D.); ajamurt@pe.uth.gr (A.Z.J.); lavassia123@gmail.com (V.C.L.); delixar@pe.uth.gr (C.K.D.); kgeorgakouli@gmail.com (K.G.); guspapa93@gmail.com (K.P.); ptsimeas@gmail.com (P.T.); y.koutedakis@uth.gr (Y.K.)
[2]   School of Physical Education and Sports Science, Democritus University of Thrace, 69100 Komotini, Greece; stampoulistheodoros@gmail.com (T.S.); achatzin@phyed.duth.gr (A.C.); michal@phyed.duth.gr (M.M.)
[3]   First Department of Neurology, Aeginition Hospital, School of Medicine, National and Kapodistrian University, 11528 Athens, Greece; constantinospapadopoulos@yahoo.com
[4]   Institute of Biology, Medicinal Chemistry and Biotechnology, National Hellenic Research Foundation, 116 35 Athens, Greece; nikichon@eie.gr
[5]   Institute of Human Performance and Rehabilitation, Centre for Research and Technology—Thessaly (CERETETH), Karies, 42100 Trikala, Greece
[6]   Faculty of Education Health and Wellbeing, University of Wolverhampton, Walsall 14287, West Midlands, UK
[7]   Institute of Nutritional Science, Nestlé Research Centre, 1015 Lausanne, Switzerland; leonidas.karagounis@rdls.nestle.com
[8]   Experimental Myology and Integrative Physiology Cluster, Plymouth Marjon University, Plymouth PL6 8BH, UK
*   Correspondence: fatouros@otenet.gr; Tel.: +30-24-310-47-047

Received: 17 April 2018; Accepted: 1 May 2018; Published: 4 May 2018

Abstract: The development of chronic, low-grade systemic inflammation in the elderly (inflammaging) has been associated with increased incidence of chronic diseases, geriatric syndromes, and functional impairments. The aim of this study was to examine differences in habitual physical activity (PA), dietary intake patterns, and musculoskeletal performance among community-dwelling elderly men with low and elevated systemic inflammation. Nonsarcopenic older men free of chronic diseases were grouped as 'low' (LSI: $n = 17$; 68.2 ± 2.6 years; hs-CRP: <1 mg/L) or 'elevated' (ESI: $n = 17$; 68.7 ± 3.0 years; hs-CRP: >1 mg/L) systemic inflammation according to their serum levels of high-sensitivity CRP (hs-CRP). All participants were assessed for body composition via Dual Emission X-ray Absorptiometry (DEXA), physical performance using the Short Physical Performance Battery (SPPB) and handgrip strength, daily PA using accelerometry, and daily macro- and micronutrient intake. ESI was characterized by a 2-fold greater hs-CRP value than LSI ($p < 0.01$). The two groups were comparable in terms of body composition, but LSI displayed higher physical performance ($p < 0.05$), daily PA (step count/day and time at moderate-to-vigorous PA (MVPA) were greater by 30% and 42%, respectively, $p < 0.05$), and daily intake of the antioxidant vitamins A (6590.7 vs. 4701.8 IU/day, $p < 0.05$), C (120.0 vs. 77.3 mg/day, $p < 0.05$), and E (10.0 vs. 7.5 mg/day, $p < 0.05$) compared to ESI. Moreover, daily intake of vitamin A was inversely correlated with levels of hs-CRP ($r = -0.39$, $p = 0.035$). These results provide evidence that elderly men characterized by low levels of systemic inflammation are more physically active, spend more time in MVPA, and receive higher amounts of antioxidant vitamins compared to those with increased systemic inflammation.

Keywords: aging; chronic low-grade systemic inflammation; physical activity; nutrition; physical performance; chronic diseases

---

1. Introduction

Chronic exposure to antigens as well as to chemical, physical, and nutritional stressors that the immune system has to cope with, in combination with the dramatic increase in life expectancy, result in the overstimulation of the immune system with advancing age and the development of a chronic and persistent pro-inflammatory state [1,2]. This age-associated, low-grade, chronic inflammatory status has been termed as "inflammaging" [1] and is clinically assessed by measuring systemic concentrations of cytokines and acute-phase proteins, including interleukin-6 (IL-6), tumor necrosis factor-α (TNF-α), and C-reactive protein (CRP) [3]. Inflammaging represents a significant risk factor for age-related frailty, morbidity, and mortality [2,4] as many chronic diseases and geriatric syndromes such as cardiovascular diseases, atherosclerosis, metabolic syndrome, type 2 diabetes mellitus, neurodegenerative diseases, cancer, and chronic obstructive pulmonary disease have been associated with chronic inflammation [5–8]. Moreover, increased levels of IL-6, TNF-α, and CRP in the elderly have been associated with lower muscle mass and physical performance [9–11] as well as with increased risk for sarcopenia and osteoporosis [12–14]. Thus, the concept of inflammaging appears to be a key determinant of successful aging and longevity and as such a valuable tool to counteract age-related pathologies [2].

To date, inflammaging is defined as a complex and multifactorial process whose origin cannot be simply attributed to a specific number of factors/mechanisms, as a complete understanding of the extent to which different tissues, organs, and biological systems contribute to its pathophysiology is lacking [3,15]. However, both physical activity (PA) and nutrition are considered powerful lifestyle factors that may, cooperatively or independently, influence both healthy aging and lifespan in humans [16,17]. Specifically, being physically active substantially reduces the risk of developing cardiovascular [16,17] and metabolic diseases [16,18], obesity [16,19], frailty [16,20,21], sarcopenia [22], osteoporosis [17,23], cognitive impairment [24], and mental health disorders [17,25] in a dose-response manner [26,27]. Numerous studies reported that higher volume of habitual PA is related to lower levels of IL-6, CRP, and TNF-α in older adults [28–40]. Most of these studies, though, are based on self-reported PA estimations [28–33,36,37,40] that may result in increased risk of recall bias [41] and therefore do not provide an objective determination of different intensity levels (i.e., light, moderate, vigorous, or very vigorous PA). However, to our knowledge, four studies have utilized accelerometry to provide an objective assessment of PA [34,35,38,39]. In two of them, an inverse relationship between PA and disease-related (chronic obstructive pulmonary disease and obesity) systemic inflammation was revealed in middle-aged adults [34,35]. Similarly, two other studies reported that time spent in MVPA is negatively associated with markers of systemic inflammation in the healthy elderly [38,39]. Although these data clearly suggest that habitual PA is inversely associated with mediators of systemic inflammation in older adults, a direct comparison of objectively assessed PA, sedentary time, and PA-related energy expenditure among the elderly with low and increased systemic inflammation is still lacking.

Ideally, this comparison would be more conclusive by the concurrent examination of habitual PA/inactivity and dietary intake levels, since both factors may impact systemic inflammation. In fact, available data suggest that the role of nutrition and dietary pattern is pivotal for immune function and low-grade systemic inflammation [42–44]. Both macronutrient and micronutrient intake may interfere with immune responses, triggering either a pro-inflammatory or an anti-inflammatory effect [45]. Excessive consumption of glucose and saturated fatty acids (SFA) (particularly long-chain SFA) are reported to activate pro-inflammatory markers in insulin-sensitive tissues [45,46] and may result in systemic inflammation [15], while high phospholipid consumption, especially that of polyunsaturated fatty acids (PUFA) and monounsaturated fatty acids (MUFA), elicit antiinflammatory properties and reduce the risk of chronic inflammation and its associated chronic diseases [47]. On the other hand, consumption of either plant- or dairy-based protein or amino acids may offer antiinflammatory effects by reducing levels of inflammatory mediators [45,48]. Furthermore, adequate intake of antioxidants and trace elements, particularly vitamins A, C, E, and selenium, also enhances immunity and elicits a protective effect against chronic inflammatory conditions [44]. However, to our knowledge, the literature lacks evidence regarding differences in dietary habits among older healthy adults with low and high systemic inflammation.

Given the pivotal role of both PA and macronutrient/micronutrient intake in mediating immunity and chronic inflammatory responses, a direct comparison of them among older adults exhibiting low and elevated systemic inflammation may identify which parameters of these lifestyle factors function as discriminants of healthy aging and inflammaging. Therefore, the aim of the present study was to compare levels of objectively assessed habitual PA and dietary macronutrient/micronutrient intake, among otherwise healthy elderly men of low and increased systemic inflammation.

2. Materials and Methods

2.1. Experimental Design and Participants

A total of fifty community-dwelling elderly men aged 65–75 years were recruited from the surrounding area of Thessaly (Greece) through postings, newspaper, and media advertisements. All volunteers completed a health history questionnaire and were also examined by a physician. In order to be included in the study, volunteers had to initially meet all of the following inclusion/exclusion criteria: (a) nonsmokers; (b) independently living; (c) absence of chronic disease (i.e., cancer, metabolic, cardiovascular, neurological, pulmonary, or kidney disease); (d) absence of inflammatory disease (i.e., osteoarthritis, rheumatoid arthritis); (e) absence of type 2 diabetes, and (f) no recent or current use of antibiotics or other medication that could affect inflammatory status (i.e., corticosteroids). Subsequently, those who fulfilled these criteria underwent assessment of body height, body weight, body composition, handgrip strength, and physical performance (via the SPPB) testing to estimate their weight status and stage of sarcopenia according to the European Working Group on Sarcopenia in Older People (EWGSOP) [49]. Volunteers who were characterized as presarcopenic/sarcopenic were excluded from the study at this stage, since substantial loss of skeletal muscle mass is accompanied by significant performance decline [49], resulting in lower levels of habitual PA [50]. Volunteers who were classified as obese were also excluded since obesity is linked to metaflammation, an adipose-tissue-mediated chronic inflammatory state that differs in terms of pathophysiology from inflammaging [5,15]. Accordingly, thirty-four volunteers who fulfilled the eligibility criteria participated in the study. The determination of inflammatory status was based on two consecutive measurements of high-sensitivity CRP (hs-CRP) and participants were grouped as "low systemic inflammation" (LSI: hs-CRP < 1 mg/L) or "elevated systemic inflammation" (ESI: hs-CRP > 1 mg/L) according to a previous report [51]. Participants were then provided with accelerometers and food diaries to monitor their habitual PA and daily macronutrient/micronutrient intake, respectively, over a 7-day period. They were fully informed about the aim and the experimental procedures of the study, as well as about the benefits involved, before obtaining written consent. The Institutional Review Board of the University of Thessaly approved the study and all procedures were in accordance with the 1975 Declaration of Helsinki (as revised in 2000).

2.2. Body Composition

Standing body mass and height were measured on a beam balance with stadiometer (Beam Balance-Stadiometer, SECA, Vogel & Halke, Hamburg, Germany) with participants wearing light clothing and no shoes as described previously [52]. Body composition [including fat mass, fat-free mass (FFM), percent of fat, lean body mass (LBM)] was assessed by dual emission X-ray absorptiometry

(DXA, GE Healthcare, Lunar DPX NT, Diegem, Belgium) with participants in supine position as described before [53]. Appendicular lean mass (ALM) and skeletal muscle mass index (SMI) were calculated as the sum of muscle mass (kg) of the four limbs (based on DXA scan) and as ALM divided by height by meters squared (kg/m^2), respectively [49], while sarcopenia status was determined according to the criteria established by EWGSOP [49].

2.3. Physical Activity

Physical activity was monitored by using the accelerometers ActiGraph, GT3X+ (ActiGraph, Pensacola, FL, USA) over a 7-day period. Accelerometers were attached to elastic, adjustable belts and did not provide any feedback to the participants. Participants were taught how to wear the belt around the waist with the monitor placed on the right hip and they were asked to wear it throughout the day, except for bathing or swimming and sleep, for seven consecutive days. To be included in the analysis, participants had to have ≥four days with ≥10 wear hours/day (i.e., four valid days) [54]. Nonwear time was calculated using the algorithms developed by Choi et al. [55] for vector magnitude (VM) data and defined as periods of 90 consecutive minutes of zero counts per minute (cpm), including intervals with nonzero cpm that lasted up to 2 min and were followed by 30 consecutive minutes of zero cpm. Daily activity and sedentary time were estimated according to VM data and expressed as steps/day and time in sedentary (<199 cpm), light (200–2689 cpm), moderate (2690–6166 cpm), vigorous (6167–9642 cpm), and moderate-to-vigorous (≥2690 cpm) PA [56]. The manufacturer software ActiLife 6 was utilized to initialize accelerometers and download data using 60-s epoch length.

2.4. Dietary Assessment

Participants were taught by a registered dietitian how to estimate food servings and sizes of different food sources and how to complete food diaries. They were allowed to weigh out food servings, so that they could precisely report the amount of specific food portions, while they were also provided with colored photographs depicting different portion sizes that they could use to compare their food weights. Furthermore, complete instructions on how to describe portion sizes based on household measures or other standard units were also administered to our participants. Participants recorded their daily dietary intake for seven consecutive days, describing, in as much detail as possible all portions of food and drinks/water. For commercially available products, the name of the manufacturer, fat content (i.e., 1%. 2%, etc.), and other related information had to be noted. The Science Fit Diet 200 A (Science Technologies, Athens, Greece) dietary software was utilized to analyze diet recalls and data regarding total energy (kJ), protein (g/kg/day & g/day), leucine (g/day), branched chain amino acids (BCAA, g/day), carbohydrates (g/day), fat (g/day), vitamin A (IU/day), vitamin C (mg/day), vitamin E (mg/day), selenium (μg/day), polyunsaturated fatty acids (PUFA), and monounsaturated fatty acids (MUFA).

2.5. Systemic Inflammation

Blood samples were collected early in the morning between 07:00 and 09:00 am, after an overnight fasting. Participants were asked to avoid alcohol and abstain from intense physical activity for ≥48 h before blood sampling. Blood was drawn from an antecubital arm vein via a 10-gauge disposable needle equipped with a Vacutainer tube holder (Becton Dickinson) with participants seated. To separate serum, blood samples were allowed to clot at room temperature and then centrifuged (15,000× *g*, 15 min, 4 °C). The supernatant was dispensed in multiple aliquots (into Eppendorf tubes) and stored at −80 °C for later analysis of hs-CRP. Serum hs-CRP was quantitatively measured in duplicate using the C-Reactive Protein (Latex) High Sensitivity assay (CRP LX High Sensitive, Cobas®) on a Cobas Integra® 400 plus analyzer (Roche) with a detectable limit of 0.01 mg/dL and an inter-assay coefficient of one standard deviation (1 SD).

2.6. Statistical Analyses

All data are presented as means ± SD. The normality of data was examined using the Shapiro–Wilk test (n = 17/group). Because our data sets in most of our variables differed significantly from normal distribution, we rejected the hypothesis of normality and applied nonparametric tests. To test differences in body composition, daily PA-related parameters, and dietary macronutrient/micronutrient intake among the two groups (LSI vs. HSI) a Kruskal–Wallis test was applied. Pearson's correlation analysis was used to examine the relation of dietary antioxidant vitamins intake, number of steps, and time in MVPA per day with serum levels of hs-CRP. Correlation coefficients of $r < 0.2$, $0.2 < r < 0.7$ and $r > 0.7$ were defined as small, moderate, and high, respectively. Effect sizes (ES) and confidence intervals (CI) were also calculated for all dependent variables using the Hedge's g method corrected for bias. ES was interpreted as none, small, medium-sized, and large for values 0.00–0.19, 0.20–0.49, 0.50–0.79, and ≥0.8, respectively. The level of statistical significance was set at $p < 0.05$. Statistical analyses were performed using the SPSS 20.0 software (IBM SPSS Statistics). The G * Power program (G * Power 3.0.10) was utilized to perform power analysis. With our sample size of 17/group we obtained a statistical power greater than 0.80 at an α error of 0.05.

3. Results

Participants' characteristics are presented in Table 1. Participants were healthy and had no pathological levels of hs-CRP. The two groups, though, differed significantly in respect to hs-CRP values (ESI: 2.1 ± 0.8 vs. LSI: 0.7 ± 0.2 mg/dL, p = 0.00), with ESI displaying a 2-fold elevation in serum hs-CRP compared to LSI. Averaged BMI values in LSI and ESI were 27.3 ± 3.1 kg/m^2 and 27.9 ± 2.5 kg/m^2, respectively, which classifies them as nonobese according to the criteria established by the World Health Organization (WHO) [57]. Moreover, all participants were characterized as nonsarcopenic, since they exhibited SMI > 7.26 kg/m^2, handgrip strength > 30 kg, and physical performance score in SPPB > 8. No differences were detected in respect to BMI, fat mass, percent of fat, FFM, LBM, ALM, SMI, and handgrip strength among groups. However, significant differences were observed in physical performance, with LSI achieving a higher SPPB score compared to ESI (LSI: 11.9 ± 0.2 vs. ESI: 11.2 ± 1.0; χ^2 = 6.436, p = 0.016; ES = 0.90; 95% CI = −1.63, −0.17).

Table 1. Participants' characteristics.

Parameter	LSI (n = 17)	ESI (n = 17)
Age (years)	68.2 ± 2.6	68.7 ± 3.0
Body Height (m)	1.71 ± 0.07	1.73 ± 0.04
Body Weight (kg)	82.3 ± 8.5	85.2 ± 7.5
BMI (kg/m^2)	27.3 ± 3.1	27.9 ± 2.5
Fat Mass (kg)	24.1 ± 7.0	26.3 ± 4.1
Fat (%)	29.5 ± 6.6	31.8 ± 2.1
Fat-Free Mass (kg)	56.3 ± 4.6	58.4 ± 5.2
Lean Body Mass (kg)	53.3 ± 4.5	55.3 ± 5.1
ALM (kg)	23.2 ± 2.4	24.4 ± 2.1
SMI (kg/m^2)	8.12 ± 0.7	8.13 ± 0.6
Grip Strength (kg)	34.3 ± 5.5	36.7 ± 6.6
SPPB (score)	11.9 ± 0.2	11.2 ± 1.0 [1]
Sarcopenia Status	Non-Sarcopenic	Non-Sarcopenic
hs-CRP (mg/L)	0.7 ± 0.2	2.1 ± 0.8 [2]

Data are presented as mean ± SD. ALM: Appendicular Lean Mass; SMI: Skeletal Muscle Mass Index; SPPB: Short Physical Performance Battery; hs-CRP: High-Sensitivity CRP. [1] significant difference between groups, $p < 0.05$, [2] significant difference between groups, $p < 0.01$.

Results comparing sedentary time and PA among groups are shown in Figure 1. The two groups were comparable in sedentary time throughout the day (LSI: 378.2 ± 98.7 vs. ESI: 370.5 ± 95.9 min/day; χ^2 = 0.008, p = 0.927) and in the time they spent in light PA/day (LSI: 342.9 ± 93.1 vs. ESI: 331.7 ± 98.2 min/day; χ^2 = 0.357, p = 0.550), while a trend for significantly more time spent in moderate

PA/day by the LSI group was also observed (LSI: 59.5 ± 16.7 vs. ESI: 44.1 ± 18.2 min/day; $\chi^2 = 3.637$, $p = 0.057$). Interpretation of the level of moderate PA by group means examined in relation to the PA guidelines adopted by the WHO revealed that both groups met the recommendation for at least 150 min of moderate-intensity PA throughout the week.

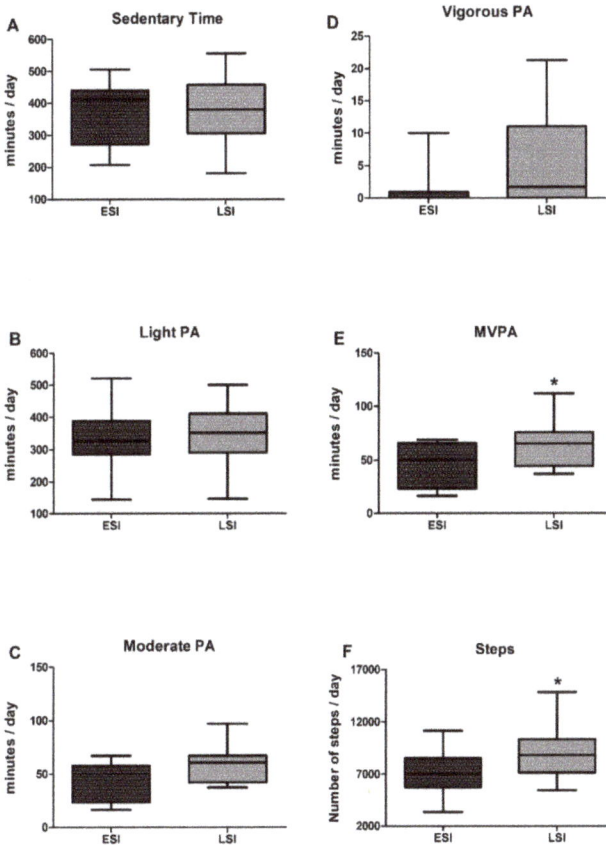

Figure 1. (A) Sedentary time, (B) time spent in light, (C) moderate, (D) vigorous, (E) moderate-to-vigorous (MVPA) PA, and (F) total step count throughout the day, in low (LSI) and elevated (ESI) systemic inflammation groups. Values are presented as mean ± SD. * denotes significant difference between groups at $p < 0.05$.

By performing an individual examination in both groups, we found that all participants in LSI and approximately 86% of participants in ESI met this criterion. Significant differences between LSI and ESI were observed in MVPA and daily step count, with LSI spending more time in MVPA throughout the day (LSI: 65.2 ± 21.5 vs. ESI: 45.9 ± 19.8 min/day; $\chi^2 = 3.997$, $p = 0.044$; ES = 0.91; 95% CI = −1.68, −0.13) and performing more steps (LSI: 9000.1 ± 2496 vs. ESI: 6968.3 ± 2075 steps/day; $\chi^2 = 4.087$, $p = 0.043$; ES = 0.86; 95% CI = −1.63, −0.08) than ESI, by 42% and 30%, respectively. The average step count/day for LSI was 9000.1 steps, which is close to the upper recommended limit for older adults (7100–10,000 steps/day) [58] while the ESI did not meet these recommendations, performing 6968.3 steps/day. Almost 86% of participants in the LSI group performed >7100 steps daily while slightly more than half (53%) of participants in the ESI group did so. A longitudinal analysis combining both groups revealed a trend for an inverse correlation between hs-CRP level and daily step count

($r = -0.37$, $p = 0.055$). Time in vigorous PA/day did not differ among groups (LSI: 5.3 ± 6.9 vs. ESI: 1.0 ± 2.6 min/day; $\chi^2 = 2.315$, $p = 0.128$), probably because of a high interindividual variability. Moreover, the two groups demonstrated similar PA-related energy expenditure throughout the day, as no differences observed in terms of kJ/day (LSI: 2554.3 ± 1033.5 vs. ESI: 2654.3 ± 1041.8 kJ/day, $p = 0.798$) and METs/day (LSI: 1.28 ± 0.1 vs. ESI: 1.23 ± 0.1 METs/day, $p = 0.203$) (Figure 2).

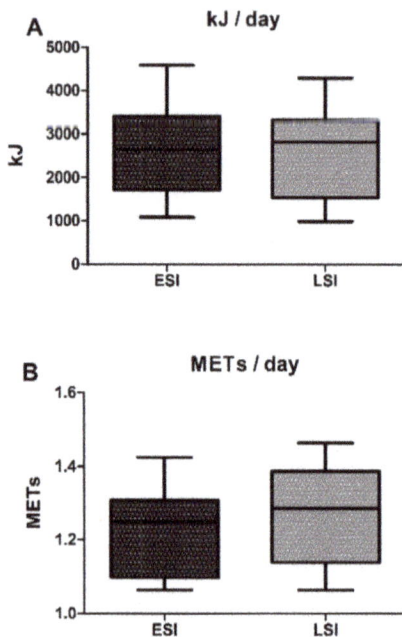

Figure 2. Daily PA-related energy expenditure expressed as (**A**) kJ and (**B**) METs in low (LSI) and elevated (ESI) systemic inflammation groups. Values are presented as mean $\pm$ SD.

LSI and ESI demonstrated similar total energy and macronutrient intake throughout the day (Table 2). The two groups had a daily energy intake of 6949.6–6794.8 kJ, constituted by 15–16% protein, 38% carbohydrate, and 42% fat. The mean protein intake in both groups was 0.8 g/kg body weight/day, which represents the recommended daily allowance (RDA) that meets 97.5% of the population [59]. However, approximately 46% of participants in both groups had a daily protein intake of 0.5–0.7 g/kg body weight/day. Separate analysis in leucine and BCAA intake revealed that both LSI and ESI received 0.6 g of leucine/kg body weight/day and 0.13–0.14 g of BCAAs/kg body weight/day, which meets the current recommendations for amino acid intake in adults [59]. The two groups, though, differed significantly in respect to daily antioxidant vitamin intake, with the LSI group receiving higher amounts of vitamin A (LSI: 6590.7 ± 2219 vs. ESI: 4701.8 ± 1552.6 IU/day; $\chi^2 = 5.616$, $p = 0.018$; ES = 0.95; 95% CI = 1.72, 0.18), vitamin C (LSI: 120.0 ± 55.5 vs. ESI: 77.3 ± 39.1 mg/day; $\chi^2 = 5.421$, $p = 0.020$; ES = 0.87; 95% CI = 1.63, 0.11), and vitamin E (LSI: 10.0 ± 2.9 vs. ESI: 7.5 ± 3.0 mg/day; $\chi^2 = 4.496$, $p = 0.034$; ES = 0.75; 95% CI = 1.50, 0.01) than ESI, by 37%, 59%, and 33%, respectively. Moreover, by performing a longitudinal analysis of both groups we observed that daily vitamin A intake was inversely correlated with levels of hs-CRP ($r = -0.39$, $p = 0.035$) (Figure 3). On the contrary, daily intake of selenium (LSI: 93.2 ± 29.8 vs. ESI: 96.1 ± 29.7 μg/day, $p = 0.793$), PUFA (LSI: 10.1 ± 2.4 vs. ESI: 8.9 ± 2.6 g/day, $p = 0.215$), and MUFA (LSI: 43.7 ± 10.8 vs. ESI: 37.9 ± 10.9 g/day, $p = 0.168$) was comparable in the two groups.

Table 2. Dietary macronutrient and micronutrient intake in LSI and ESI groups.

Parameter	LSI (*n* = 17)	ESI (*n* = 17)	*p* Value	χ^2
Total Energy (kJ/day)	6952.9 ± 1241.8	6797.8 ± 1136.8	0.771	0.085
Protein				
g/day	63.8 ± 20.3	66.9 ± 14.6	0.183	1.770
g/kg BM/day	0.8 ± 0.3	0.8 ± 0.2	0.817	0.054
% of total calories	15 ± 2.7	16 ± 3.0		
Leucine (g/day)	4.89 ± 1.7	5.13 ± 1.2	0.430	0.624
BCAAs (g/day)	11.38 ± 3.6	11.53 ± 2.4	0.533	0.389
Carbohydrates				
g/day	156.2 ± 37.6	154.9 ± 52.7	0.901	0.016
% of total calories	37.7 ± 6.9	37.5 ± 8.4		
Fat				
g/day	79.3 ± 12.5	73.7 ± 17.0	0.318	0.996
% of total calories	42.0 ± 4.0	41.7 ± 7.1		
PUFA (g/day)	10.1 ± 2.4	8.9 ± 2.6	0.275	1.191
MUFA (g/day)	43.7 ± 10.8	37.9 ± 10.9	0.359	0.840
Vitamin A (IU/day)	6590.7 ± 2219.6	4701.8 ± 1552.6 [1]	0.018	5.616
Vitamin C (mg/day)	120.0 ± 55.5	77.3 ± 39.1 [1]	0.020	5.421
Vitamin E (mg/day)	10.0 ± 2.9	7.5 ± 3.0 [1]	0.034	4.496
Selenium (µg/day)	93.2 ± 29.8	96.1 ± 29.7	0.589	0.292

Data are presented as mean ± SD. BM: Body mass; BCAA: Branched chain amino acids; PUFA: Polyunsaturated fatty acids; MUFA: Monounsaturated fatty acids. [1] Significant difference between groups.

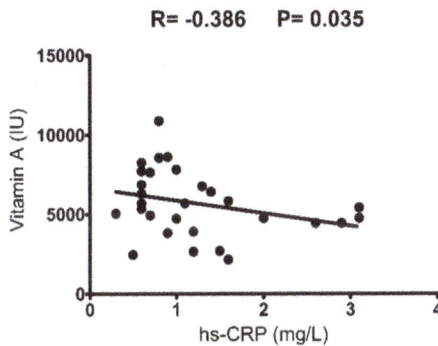

Figure 3. The relationship between serum hs-CRP level and daily dietary intake of Vitamin A.

4. Discussion

The present study is the first, to our knowledge, to compare the levels of habitual PA, sedentary time, and dietary intake between healthy elderly men with low and elevated low-grade systemic inflammation (inflammaging). Our findings suggest that older adults characterized by low levels of systemic inflammation perform more steps and spent more time in MVPA throughout the day and they receive higher amounts of dietary antioxidant vitamins (i.e., vitamins A, C, and E) on a daily basis compared to their counterparts with elevated systemic inflammation.

Participants were categorized as having either "low" or "elevated" low-grade systemic inflammation according to their serum levels of hs-CRP. This acute-phase protein is considered a valid and informative marker of inflammaging [60] and has been previously used as a single marker to identify levels of systemic inflammation in older adults [51]. The term inflammaging, first introduced by Franceschi and his colleagues [1], refers to the development of a chronic, low-grade inflammation phenotype with advancing age. However, the presence of obesity, either in young or older individuals, results in elevated systemic inflammation, which has been defined as metaflammation (metabolic inflammation) and is primarily mediated by the adipose tissue [5]. Although the underpinning mechanisms of inflammaging and metaflammation may be different, these two chronic inflammatory conditions may overlap [15]. Therefore, in an attempt to focus on inflammaging in this study, we included only nonobese elderly men (according to WHO criteria). Moreover, LSI and ESI groups were very homogeneous in terms of body composition, since they did not differ in body weight, fat mass, percent of fat, FFM, and LBM. All participants were also nonsarcopenic according to the criteria established by the EWGSOP [49], since the existence of sarcopenia could act as a covariate in our investigation, interfering with their ability to habitually perform PA [50].

Previous cross-sectional studies have investigated the association between habitual PA and inflammatory biomarkers in middle-aged and older adults [28–31,33–40]. However, only two utilized accelerometry to quantify not only the quantity but also the quality (intensity) of habitual PA in the otherwise healthy elderly with physiological and elevated chronic, low-grade systemic inflammation [38,39]. This study attempted to extend the current literature by providing insights concerning the differences in PA and dietary intake profile among elderly men with low and elevated low-grade systemic inflammation. The use of accelerometry to objectively assess the quantity and intensity of habitual PA is a strength of our study, as most of the previously cited studies [28–31,33,36,37,40] are based on questionnaires, self-reports, or interviews. The use of accelerometers over a 7-day period to assess PA and sedentary time has been reported to be a valid and reproducible methodological approach in the elderly [61].

Although sedentary time and time spent in light- and moderate-intensity activities throughout the day were similar between LSI and ESI, we noted that overall the LSI group performed more steps and spent more time in MVPA on a daily basis. This suggests that not only the volume of habitual PA but also the intensity in which daily physical activities are performed may interfere with the development of chronic, low-grade systemic inflammation in older individuals. Our findings further build on previous reports that higher volume of habitual PA is associated with lower levels of pro-inflammatory mediators in healthy elderly individuals [29,33,36] and COPD patients [34]. Moreover, this inverse association between PA and inflammation is suggested to be dose-dependent, so that the more physically active an individual is, the lower the chronic inflammatory milieu [29,31,40]. Although only a trend ($r = -0.37$, $p = 0.055$) for an inverse correlation between hs-CRP level and daily step number was observed in our study, possibly because of an interindividual variability in daily step counts of our participants (we used accelerometers whereas questionnaires were utilized by others), these findings collectively suggest that habitual PA may be associated with inflammaging in an inverse, dose-response pattern. Furthermore, it has been recently reported that the impact of PA on chronic low-grade inflammation is not only dose-dependent but also intensity-dependent, as moderate-to-vigorous activities induce greater improvements in the inflammatory profile of older adults while light- or moderate-intensity physical activities are accompanied by no changes in inflammatory mediators [62]. Indeed, Wahlin-Larsson et al. [39] found that in recreationally active elderly women, the time spent in MVPA is inversely associated with serum levels of CRP, a finding also reported in younger individuals [63]. The mechanism/s through which PA reduces or prevents low-grade systemic inflammation in the elderly remains to be elucidated. Observational, cross-sectional studies are not designed to identify the mechanisms that underline the effects of systematic PA on chronic inflammation and as such, more intervention studies are needed [41,62]. Based on the fact that inflammaging is tightly regulated by the balance between pro- and anti-inflammatory mediators [64], a possible mechanism could be that PA, and especially MVPA, suppresses the production of pro-inflammatory cytokines and molecules that trigger the inflammatory milieu, and enhances the production of anti-inflammatory mediators [41,62,65]. Moreover, the process of inflammaging may be further affected by the age-associated increase in the production of reactive oxygen and nitrogen species (RONS) that lead to redox balance disturbances and subsequent activation of the redox-sensitive NF-κB signaling pathway that stimulates the expression of numerous pro-inflammatory mediators such as TNF-α, IL-6, IL-1β, and CRP [48,66]. As such, a vicious cycle of RONS and pro-inflammatory molecule production is propagated, driving a chronic systemic pro-inflammatory phenotype [48,67]. Regular participation in moderate-to-vigorous intensity exercise has been shown to attenuate both basal and exercise-induced levels of oxidative damage, enhance the antioxidant capacity, and improve the DNA repair machinery in healthy, elderly individuals [68,69]. Thus, it can be proposed that systematic MVPA may prevent the development of inflammaging by lowering the production of RONS and levels of oxidative damage in the elderly.

LSI and ESI also differed significantly in terms of physical performance. More specifically, LSI exhibited higher performance in the SPPB test compared to ESI and this observation is in line with previous findings reporting that older adults with elevated systemic inflammation demonstrate lower physical performance [70,71]. Although the underlying mechanism leading from chronic inflammation to functional decline has not been clarified yet, it has been reported that systemic inflammation may impact physical performance by decreasing skeletal muscle mass [14,48]. However, in this study, the two groups demonstrated similar LBM, ALM, and SMI, indicating that the observed difference in physical performance was not muscle-mass-dependent. A previous report, though, by Wahlin-Larsson and colleagues [39] provided evidence that increased systemic inflammation influences muscle regeneration by decreasing the proliferation rate of myoblasts. In addition, increased inflammation and cytokine production may also reduce the quiescent satellite cells pool and attenuate their differentiation capacity [14]. Therefore, it can be assumed that elevated systemic inflammation may contribute to physical performance deterioration by attenuating the regeneration potential of the aged skeletal muscle.

We also utilized 7-day recalls to perform a thorough screening of the dietary intake in the LSI and ESI groups, focusing on macronutrients and micronutrients that have been shown to elicit either a pro- or an anti-inflammatory effect, and could be therefore characterized as 'key modifiers' in the process of inflammaging. LSI and ESI demonstrated similar energy and macronutrient intake, consuming 6794.8–6949.6 kJ/day composed of 15–16% protein, 38% carbohydrates, and 42% fat. Our group recently conducted a literature review suggesting that protein intake, especially that of whey protein and soy or isoflavone-enriched soy protein, may indirectly offer antioxidative and anti-inflammatory benefits beyond its ability to stimulate skeletal muscle protein synthesis [48]. Also, Zhou et al. [72] performed a meta-analysis on the effects of whey protein supplementation on levels of CRP, concluding that increased whey protein intake may induce favorable effects on individuals with elevated baseline CRP levels. However, in this study, we noted that daily protein intake was similar between LSI and ESI, with both groups receiving on average ~0.8 g/kg BM/day, which is in line with WHO RDA for protein [59]. BCAA and leucine intake were also compared among groups to provide a qualitative determination of daily protein intake. Although leucine is classified as a BCAA, we decided to present it separately because its role may differ from that of the other BCAAs, especially in the elderly where a higher amount of leucine should be consumed through diet to efficiently stimulate muscle protein synthesis and preserve muscle loss [73,74]. In our present work, we observed that LSI and ESI had a similar daily intake of BCAAs and leucine, meeting the recommendations for amino acid intake in adults [59]. Daily carbohydrate intake was also similar among groups (154–156 g/day), indicating that it does not play a prominent role in the development of inflammaging. Previous reports have noted that only increased consumption of high glycemic index carbohydrates may be associated with increased levels of inflammation [75]. Unfortunately, the determination of glycemic index and glycemic load in our participants' daily diets was not feasible.

Similarly, no differences were observed in total fat consumption among groups, with LSI and ESI receiving 79 and 74 g/day, respectively, which corresponds in both groups to 42% of daily energy intake. Although previous reports have indicated that increased fat consumption is associated with elevated systemic markers of inflammation [75,76], this was not the case here. High fat diets, and primarily SFA, have been reported to induce substantial alterations in the gut microbial flora (i.e., increases gut mucosa permeability, epithelial brier disruption) that result in enhanced translocation of lipolysaccharide (LPS) in the circulation, thus promoting the development of low-grade systemic inflammation [76,77]. However, it should be highlighted here that not all SFA demonstrate equal properties and consumption of specific SFA (i.e., C14:0, C15:0, C17:0, CLA, and trans-palmitoleic) has been associated with positive effects on cardiovascular health [78]. On the other hand, increased intake of MUFA and/or PUFA has been proposed to counteract the pro-inflammatory cascade by reducing the translocation of LPS in the circulation [76] and suppressing the eicosanoid and PAF inflammatory pathways [47]. Indeed, many studies have revealed an inverse association between higher intake of dietary PUFA and/or MUFA and levels of pro-inflammatory mediators such as hs-CRP and IL-6 [75]. In this study, although no statistically meaningful differences were observed in dietary MUFA and PUFA intake between groups, LSI displayed a higher intake of MUFA and PUFA, by 15% and 13.5%, respectively, compared to ESI.

Interestingly, we noted significant differences between LSI and ESI in terms of antioxidant vitamin intake. More specifically, daily dietary intake of vitamins A, C, and E in LSI was higher by 37%, 59%, and 33%, respectively, as compared to ESI. These vitamins play a major role in immune function, so that adequate intake enhances innate, cell-mediated, and humoral antibody immunity while deficiency promotes the opposite effects [44,79]. With aging, the production of reactive oxygen and nitrogen species and that of pro-inflammatory cytokines rises significantly, propagating a vicious cycle of oxidative stress and inflammation that promotes a chronic low-grade inflammatory state [48,67]. Vitamin A has been shown to promote a T-helper type 2 immune response by reducing the expression of pro-inflammatory cytokines (i.e., interferon-γ, TNF-α and IL-12) and adipocytokines (i.e., leptin) [44,79] while it may also inhibit the activation of the redox-sensitive nuclear factor-kappa B (NF-κB) [44,79], a principal mediator of the bidirectional interaction between oxidative stress and inflammation [48]. Moreover, the pivotal role of vitamin A in chronic inflammation is further supported by the fact that a deficit in vitamin A intake is associated with a pronounced pro-inflammatory state and inability to cope with pathogens, as well as with reduced phagocytic capacity of macrophages [44]. Vitamin C also reduces the production of pro-inflammatory cytokines through inhibition of the transcription factor NF-κB [44]. The anti-inflammatory effect of this micronutrient is further supported by a previous investigation where vitamin C intake was inversely associated with levels of CRP and tissue plasminogen activator (t-PA) antigen in elderly men [80]. Furthermore, vitamin C acts as a potent antioxidant, protecting cells from ROS-mediated oxidative damage, while it may also boost the synthesis of other antioxidants such as vitamin E [44]. Likewise, vitamin E is able to confer protection against oxidative stress by increasing the concentration of endogenous antioxidant enzymes, such as SOD, CAT, and GPX, and it also prevents oxidative damage in the cell membrane [44,81]. Evidence based on human studies indicates that vitamin E supplementation in older adults improves immune function [44] and is associated with a lower concentration of pro-inflammatory mediators [82]. Collectively, these data corroborate the higher antioxidant vitamin intake observed in LSI in the present study, indicating that vitamins A, C, and E may contribute to the control of low-grade systemic inflammation in the elderly. By contrast, no differences were observed in selenium intake between LSI and ESI, although selenium is also considered a micronutrient that may efficiently influence both innate and acquired immune function and may enhance the antioxidative defense system [44].

5. Conclusions

We found that elderly men with low levels of systemic inflammation are characterized by higher quality and quantity of habitual PA and ingested higher amounts of antioxidant vitamins A, C, and E through normal diet when compared to those with increased systemic inflammation. To the best of our knowledge, this is the first study to directly compare elderly men of low and increased low-grade systemic inflammation in respect to habitual PA and dietary profile. PA and antioxidant vitamin intake appear to be discriminant factors of inflammaging and healthy aging. Future research should further explore the cause and effect as well as the dose-response relationship between PA and/or antioxidant vitamins and inflammaging.

Author Contributions: D.D., I.G.F., A.Z.J., T.S., and L.G.K. conceived and designed the experiments; V.C.L., C.K.D., and N.C. performed biological assays; A.Z.J., and C.P. performed biological tissue sampling and medical monitoring; T.S., and K.G. performed dietary analyses; K.P., A.C. and P.T. performed physical performance measurements; M.M. collected and performed physical activity analyses; D.D., and T.S. analyzed the data; N.C., Y.K. and A.Z.J. contributed reagents/materials/analysis tools; D.D., L.G.K. and I.G.F. wrote the paper; all authors reviewed the manuscript.

Acknowledgments: This study was supported by the General Secretariat for Research and Technology (GSRT) and the Hellenic Foundation for Research and Innovation (HFRI). The authors are grateful to all participants for their contribution and commitment to this study.

Conflicts of Interest: The authors declare no conflict of interest.

References

1. Franceschi, C.; Bonafe, M.; Valensin, S.; Olivieri, F.; De Luca, M.; Ottaviani, E.; De Benedictis, G. Inflamm-aging. An evolutionary perspective on immunosenescence. *Ann. N. Y. Acad. Sci.* **2000**, *908*, 244–254. [CrossRef] [PubMed]

2. Franceschi, C. Inflammaging as a Major Characteristic of Old People: Can It Be Prevented or Cured? *Nutr. Rev.* **2007**, *65*, S173–S176. [CrossRef] [PubMed]

3. Calcada, D.; Vianello, D.; Giampieri, E.; Sala, C.; Castellani, G.; de Graaf, A.; Kremer, B.; van Ommen, B.; Feskens, E.; Santoro, A.; et al. The role of low-grade inflammation and metabolic flexibility in aging and nutritional modulation thereof: A systems biology approach. *Mech. Ageing Dev.* **2014**, *136–137*, 138–147. [CrossRef] [PubMed]

4. Hubbard, R.E.; O'Mahony, M.S.; Savva, G.M.; Calver, B.L.; Woodhouse, K.W. Inflammation and frailty measures in older people. *J. Cell. Mol. Med.* **2009**, *13*, 3103–3109. [CrossRef] [PubMed]

5. Franceschi, C.; Garagnani, P.; Vitale, G.; Capri, M.; Salvioli, S. Inflammaging and 'Garb-aging'. *Trends Endocrinol. Metab.* **2016**. [CrossRef] [PubMed]

6. Roxburgh, C.S.; McMillan, D.C. Role of systemic inflammatory response in predicting survival in patients with primary operable cancer. *Future Oncol.* **2010**, *6*, 149–163. [CrossRef] [PubMed]

7. Singh, T.; Newman, A.B. Inflammatory markers in population studies of aging. *Ageing Res. Rev.* **2011**, *10*, 319–329. [CrossRef] [PubMed]

8. De Martinis, M.; Franceschi, C.; Monti, D.; Ginaldi, L. Inflamm-ageing and lifelong antigenic load as major determinants of ageing rate and longevity. *FEBS Lett.* **2005**, *579*, 2035–2039. [CrossRef] [PubMed]

9. Schaap, L.A.; Pluijm, S.M.; Deeg, D.J.; Harris, T.B.; Kritchevsky, S.B.; Newman, A.B.; Colbert, L.H.; Pahor, M.; Rubin, S.M.; Tylavsky, F.A.; et al. Higher inflammatory marker levels in older persons: Associations with 5-year change in muscle mass and muscle strength. *J. Gerontol. Ser. A Biol. Sci. Med. Sci.* **2009**, *64*, 1183–1189. [CrossRef] [PubMed]

10. Schaap, L.A.; Pluijm, S.M.; Deeg, D.J.; Visser, M. Inflammatory markers and loss of muscle mass (sarcopenia) and strength. *Am. J. Med.* **2006**, *119*, 526.e9–526.e17. [CrossRef] [PubMed]

11. Visser, M.; Pahor, M.; Taaffe, D.R.; Goodpaster, B.H.; Simonsick, E.M.; Newman, A.B.; Nevitt, M.; Harris, T.B. Relationship of interleukin-6 and tumor necrosis factor-alpha with muscle mass and muscle strength in elderly men and women: The Health ABC Study. *J. Gerontol. Ser. A Biol. Sci. Med. Sci.* **2002**, *57*, M326–M332. [CrossRef]

12. Michaud, M.; Balardy, L.; Moulis, G.; Gaudin, C.; Peyrot, C.; Vellas, B.; Cesari, M.; Nourhashemi, F. Proinflammatory cytokines, aging, and age-related diseases. *J. Am. Med. Dir. Assoc.* **2013**, *14*, 877–882. [CrossRef] [PubMed]

13. Payette, H.; Roubenoff, R.; Jacques, P.F.; Dinarello, C.A.; Wilson, P.W.; Abad, L.W.; Harris, T. Insulin-like growth factor-1 and interleukin 6 predict sarcopenia in very old community-living men and women: The Framingham Heart Study. *J. Am. Geriatr. Soc.* **2003**, *51*, 1237–1243. [CrossRef] [PubMed]

14. Dalle, S.; Rossmeislova, L.; Koppo, K. The Role of Inflammation in Age-Related Sarcopenia. *Front. Physiol.* **2017**, *8*, 1045. [CrossRef] [PubMed]

15. Cevenini, E.; Monti, D.; Franceschi, C. Inflamm-ageing. *Curr. Opin. Clin. Nutr. Metab. Care* **2013**, *16*, 14–20. [CrossRef] [PubMed]

16. McPhee, J.S.; French, D.P.; Jackson, D.; Nazroo, J.; Pendleton, N.; Degens, H. Physical activity in older age: Perspectives for healthy ageing and frailty. *Biogerontology* **2016**, *17*, 567–580. [CrossRef] [PubMed]

17. Hammar, M.; Ostgren, C.J. Healthy aging and age-adjusted nutrition and physical fitness. *Best Pract. Res. Clin. Obstet. Gynaecol.* **2013**, *27*, 741–752. [CrossRef] [PubMed]

18. Bueno, D.R.; Marucci, M.F.N.; Rosa, C.; Fernandes, R.A.; de Oliveira Duarte, Y.A.; Lebao, M.L. Objectively Measured Physical Activity and Healthcare Expenditures Related to Arterial Hypertension and Diabetes Mellitus in Older Adults: SABE Study. *J. Aging Phys. Act.* **2017**, *25*, 553–558. [CrossRef] [PubMed]

19. Cooper, R.; Huang, L.; Hardy, R.; Crainiceanu, A.; Harris, T.; Schrack, J.A.; Crainiceanu, C.; Kuh, D. Obesity History and Daily Patterns of Physical Activity at Age 60–64 Years: Findings from the MRC National Survey of Health and Development. *J. Gerontol. Ser A Biol. Sci. Med. Sci.* **2017**, *72*, 1424–1430. [CrossRef] [PubMed]

20. Huisingh-Scheetz, M.; Wroblewski, K.; Kocherginsky, M.; Huang, E.; William, D.; Waite, L.; Schumm, L.P. Physical Activity and Frailty among Older Adults in the U.S. Based on Hourly Accelerometry Data. *J. Gerontol. Ser. A Biol. Sci. Med. Sci.* **2017**. [CrossRef] [PubMed]

21. Buchner, D.M.; Rillamas-Sun, E.; Di, C.; LaMonte, M.J.; Marshall, S.W.; Hunt, J.; Zhang, Y.; Rosenberg, D.E.; Lee, I.M.; Evenson, K.R.; et al. Accelerometer-Measured Moderate to Vigorous Physical Activity and Incidence Rates of Falls in Older Women. *J. Am. Geriatr. Soc.* **2017**, *65*, 2480–2487. [CrossRef] [PubMed]

22. Moore, D.R. Keeping older muscle "young" through dietary protein and physical activity. *Adv. Nutr.* **2014**, *5*, 599S–607S. [CrossRef] [PubMed]

23. Chastin, S.F.; Mandrichenko, O.; Helbostadt, J.L.; Skelton, D.A. Associations between objectively-measured sedentary behaviour and physical activity with bone mineral density in adults and older adults, the NHANES study. *Bone* **2014**, *64*, 254–262. [CrossRef] [PubMed]

24. Rolland, Y.; Abellan van Kan, G.; Vellas, B. Healthy brain aging: Role of exercise and physical activity. *Clin. Geriatr. Med.* **2010**, *26*, 75–87. [CrossRef] [PubMed]

25. Yoshida, Y.; Iwasa, H.; Kumagai, S.; Suzuki, T.; Awata, S.; Yoshida, H. Longitudinal association between habitual physical activity and depressive symptoms in older people. *Psychiatry Clin. Neurosci.* **2015**, *69*, 686–692. [CrossRef] [PubMed]

26. Hamer, M.; Lavoie, K.L.; Bacon, S.L. Taking up physical activity in later life and healthy ageing: The English longitudinal study of ageing. *Br. J. Sports Med.* **2014**, *48*, 239–243. [CrossRef] [PubMed]

27. Fielding, R.A.; Guralnik, J.M.; King, A.C.; Pahor, M.; McDermott, M.M.; Tudor-Locke, C.; Manini, T.M.; Glynn, N.W.; Marsh, A.P.; Axtell, R.S.; et al. Dose of physical activity, physical functioning and disability risk in mobility-limited older adults: Results from the LIFE study randomized trial. *PLoS ONE* **2017**, *12*, e0182155. [CrossRef] [PubMed]

28. Abramson, J.L.; Vaccarino, V. Relationship between physical activity and inflammation among apparently healthy middle-aged and older US adults. *Arch. Intern. Med.* **2002**, *162*, 1286–1292. [CrossRef] [PubMed]

29. Colbert, L.H.; Visser, M.; Simonsick, E.M.; Tracy, R.P.; Newman, A.B.; Kritchevsky, S.B.; Pahor, M.; Taaffe, D.R.; Brach, J.; Rubin, S.; et al. Physical activity, exercise, and inflammatory markers in older adults: Findings from the Health, Aging and Body Composition Study. *J. Am. Geriatr. Soc.* **2004**, *52*, 1098–1104. [CrossRef] [PubMed]

30. Elosua, R.; Bartali, B.; Ordovas, J.M.; Corsi, A.M.; Lauretani, F.; Ferrucci, L. Association between physical activity, physical performance, and inflammatory biomarkers in an elderly population: The InCHIANTI study. *J. Gerontol. Ser. A Biol. Sci. Med. Sci.* **2005**, *60*, 760–767. [CrossRef]

31. Fischer, C.P.; Berntsen, A.; Perstrup, L.B.; Eskildsen, P.; Pedersen, B.K. Plasma levels of interleukin-6 and C-reactive protein are associated with physical inactivity independent of obesity. *Scand. J. Med. Sci. Sports* **2007**, *17*, 580–587. [CrossRef] [PubMed]

32. Hamer, M.; Molloy, G.J.; de Oliveira, C.; Demakakos, P. Leisure time physical activity, risk of depressive symptoms, and inflammatory mediators: The English Longitudinal Study of Ageing. *Psychoneuroendocrinology* **2009**, *34*, 1050–1055. [CrossRef] [PubMed]

33. Jankord, R.; Jemiolo, B. Influence of physical activity on serum IL-6 and IL-10 levels in healthy older men. *Med. Sci. Sports Exerc.* **2004**, *36*, 960–964. [CrossRef] [PubMed]

34. Moy, M.L.; Teylan, M.; Weston, N.A.; Gagnon, D.R.; Danilack, V.A.; Garshick, E. Daily step count is associated with plasma C-reactive protein and IL-6 in a US cohort with COPD. *Chest* **2014**, *145*, 542–550. [CrossRef] [PubMed]

35. Nicklas, B.J.; Beavers, D.P.; Mihalko, S.L.; Miller, G.D.; Loeser, R.F.; Messier, S.P. Relationship of Objectively-Measured Habitual Physical Activity to Chronic Inflammation and Fatigue in Middle-Aged and Older Adults. *J. Gerontol. Ser. A Biol. Sci. Med. Sci.* **2016**, *71*, 1437–1443. [CrossRef] [PubMed]

36. Reuben, D.B.; Judd-Hamilton, L.; Harris, T.B.; Seeman, T.E. The associations between physical activity and inflammatory markers in high-functioning older persons: MacArthur Studies of Successful Aging. *J. Am. Geriatr. Soc.* **2003**, *51*, 1125–1130. [CrossRef] [PubMed]

37. Taaffe, D.R.; Harris, T.B.; Ferrucci, L.; Rowe, J.; Seeman, T.E. Cross-sectional and prospective relationships of interleukin-6 and C-reactive protein with physical performance in elderly persons: MacArthur studies of successful aging. *J. Gerontol. Ser. A Biol. Sci. Med. Sci.* **2000**, *55*, M709–M715. [CrossRef]

38. Valentine, R.J.; Woods, J.A.; McAuley, E.; Dantzer, R.; Evans, E.M. The associations of adiposity, physical activity and inflammation with fatigue in older adults. *Brain Behav. Immun.* **2011**, *25*, 1482–1490. [CrossRef] [PubMed]

39. Wahlin-Larsson, B.; Carnac, G.; Kadi, F. The influence of systemic inflammation on skeletal muscle in physically active elderly women. *Age* **2014**, *36*, 9718. [CrossRef] [PubMed]

40. Wannamethee, S.G.; Lowe, G.D.; Whincup, P.H.; Rumley, A.; Walker, M.; Lennon, L. Physical activity and hemostatic and inflammatory variables in elderly men. *Circulation* **2002**, *105*, 1785–1790. [CrossRef] [PubMed]

41. Tir, A.M.D.; Labor, M.; Plavec, D. The effects of physical activity on chronic subclinical systemic inflammation. *Arch. Ind. Hyg. Toxicol.* **2017**, *68*, 276–286. [CrossRef] [PubMed]

42. Huang, C.J.; Zourdos, M.C.; Jo, E.; Ormsbee, M.J. Influence of physical activity and nutrition on obesity-related immune function. *Sci. World J.* **2013**, *2013*, 752071. [CrossRef] [PubMed]

43. Panickar, K.S.; Jewell, D.E. The beneficial role of anti-inflammatory dietary ingredients in attenuating markers of chronic low-grade inflammation in aging. *Horm. Mol. Biol. Clin. Investig.* **2015**, *23*, 59–70. [CrossRef] [PubMed]

44. Wintergerst, E.S.; Maggini, S.; Hornig, D.H. Contribution of selected vitamins and trace elements to immune function. *Ann. Nutr. Metab.* **2007**, *51*, 301–323. [CrossRef] [PubMed]

45. Da Silva, M.S.; Rudkowska, I. Dairy nutrients and their effect on inflammatory profile in molecular studies. *Mol. Nutr. Food Res.* **2015**, *59*, 1249–1263. [CrossRef] [PubMed]

46. Donath, M.Y.; Shoelson, S.E. Type 2 diabetes as an inflammatory disease. *Nat. Rev. Immunol.* **2011**, *11*, 98–107. [CrossRef] [PubMed]

47. Lordan, R.; Tsoupras, A.; Zabetakis, I. Phospholipids of Animal and Marine Origin: Structure, Function, and Anti-Inflammatory Properties. *Molecules* **2017**, *22*. [CrossRef]

48. Draganidis, D.; Karagounis, L.G.; Athanailidis, I.; Chatzinikolaou, A.; Jamurtas, A.Z.; Fatouros, I.G. Inflammaging and Skeletal Muscle: Can Protein Intake Make a Difference? *J. Nutr.* **2016**. [CrossRef] [PubMed]

49. Cruz-Jentoft, A.J.; Baeyens, J.P.; Bauer, J.M.; Boirie, Y.; Cederholm, T.; Landi, F.; Martin, F.C.; Michel, J.P.; Rolland, Y.; Schneider, S.M.; et al. Sarcopenia: European consensus on definition and diagnosis: Report of the European Working Group on Sarcopenia in Older People. *Age Ageing* **2010**, *39*, 412–423. [CrossRef] [PubMed]

50. Mijnarends, D.M.; Koster, A.; Schols, J.M.; Meijers, J.M.; Halfens, R.J.; Gudnason, V.; Eiriksdottir, G.; Siggeirsdottir, K.; Sigurdsson, S.; Jonsson, P.V.; et al. Physical activity and incidence of sarcopenia: The population-based AGES-Reykjavik Study. *Age Ageing* **2016**, *45*, 614–620. [CrossRef] [PubMed]

51. Labonte, M.E.; Cyr, A.; Abdullah, M.M.; Lepine, M.C.; Vohl, M.C.; Jones, P.; Couture, P.; Lamarche, B. Dairy product consumption has no impact on biomarkers of inflammation among men and women with low-grade systemic inflammation. *J. Nutr.* **2014**, *144*, 1760–1767. [CrossRef] [PubMed]

52. Fatouros, I.G.; Douroudos, I.; Panagoutsos, S.; Pasadakis, P.; Nikolaidis, M.G.; Chatzinikolaou, A.; Sovatzidis, A.; Michailidis, Y.; Jamurtas, A.Z.; Mandalidis, D.; et al. Effects of L-carnitine on oxidative stress responses in patients with renal disease. *Med. Sci. Sports Exerc.* **2010**, *42*, 1809–1818. [CrossRef] [PubMed]

53. Draganidis, D.; Chondrogianni, N.; Chatzinikolaou, A.; Terzis, G.; Karagounis, L.G.; Sovatzidis, A.; Avloniti, A.; Lefaki, M.; Protopapa, M.; Deli, C.K.; et al. Protein ingestion preserves proteasome activity during intense aseptic inflammation and facilitates skeletal muscle recovery in humans. *Br. J. Nutr.* **2017**, *118*, 189–200. [CrossRef] [PubMed]

54. Gorman, E.; Hanson, H.M.; Yang, P.H.; Khan, K.M.; Liu-Ambrose, T.; Ashe, M.C. Accelerometry analysis of physical activity and sedentary behavior in older adults: A systematic review and data analysis. *Eur. Rev. Aging Phys. Act.* **2014**, *11*, 35–49. [CrossRef] [PubMed]

55. Choi, L.; Ward, S.C.; Schnelle, J.F.; Buchowski, M.S. Assessment of wear/nonwear time classification algorithms for triaxial accelerometer. *Med. Sci. Sports Exerc.* **2012**, *44*, 2009–2016. [CrossRef] [PubMed]
56. Keadle, S.K.; Shiroma, E.J.; Freedson, P.S.; Lee, I.M. Impact of accelerometer data processing decisions on the sample size, wear time and physical activity level of a large cohort study. *BMC Public Health* **2014**, *14*, 1210. [CrossRef] [PubMed]
57. WHO (World Health Organization). Obesity: Preventing and Managing the Global Epidemic. Report of a WHO Consultation. Available online: http://apps.who.int/iris/handle/10665/42330 (accessed on 4 May 2018).
58. Tudor-Locke, C.; Craig, C.L.; Aoyagi, Y.; Bell, R.C.; Croteau, K.A.; De Bourdeaudhuij, I.; Ewald, B.; Gardner, A.W.; Hatano, Y.; Lutes, L.D.; et al. How many steps/day are enough? For older adults and special populations. *Int. J. Behav. Nutr. Phys. Act.* **2011**, *8*, 80. [CrossRef] [PubMed]
59. WHO (World Health Organization). Protein and Amino Acid Requirements in Human Nutrition. Available online: http://apps.who.int/iris/bitstream/handle/10665/43411/WHO_TRS_935_eng.pdf?sequence=1 (accessed on 4 May 2018).
60. Morrisette-Thomas, V.; Cohen, A.A.; Fulop, T.; Riesco, E.; Legault, V.; Li, Q.; Milot, E.; Dusseault-Belanger, F.; Ferrucci, L. Inflamm-aging does not simply reflect increases in pro-inflammatory markers. *Mech. Ageing Dev.* **2014**, *139*, 49–57. [CrossRef] [PubMed]
61. Keadle, S.K.; Shiroma, E.J.; Kamada, M.; Matthews, C.E.; Harris, T.B.; Lee, I.M. Reproducibility of Accelerometer-Assessed Physical Activity and Sedentary Time. *Am. J. Prev. Med.* **2017**. [CrossRef] [PubMed]
62. Nimmo, M.A.; Leggate, M.; Viana, J.L.; King, J.A. The effect of physical activity on mediators of inflammation. *Diabetes Obes. Metab.* **2013**, *15* (Suppl. 3), 51–60. [CrossRef] [PubMed]
63. Sabiston, C.M.; Castonguay, A.; Low, N.C.; Barnett, T.; Mathieu, M.E.; O'Loughlin, J.; Lambert, M. Vigorous physical activity and low-grade systemic inflammation in adolescent boys and girls. *Int. J. Pediatr. Obes.* **2010**, *5*, 509–515. [CrossRef] [PubMed]
64. Franceschi, C.; Capri, M.; Monti, D.; Giunta, S.; Olivieri, F.; Sevini, F.; Panourgia, M.P.; Invidia, L.; Celani, L.; Scurti, M.; et al. Inflammaging and anti-inflammaging: A systemic perspective on aging and longevity emerged from studies in humans. *Mech. Ageing Dev.* **2007**, *128*, 92–105. [CrossRef] [PubMed]
65. Petersen, A.M.; Pedersen, B.K. The anti-inflammatory effect of exercise. *J. Appl. Physiol.* **2005**, *98*, 1154–1162. [CrossRef] [PubMed]
66. Chung, H.Y.; Kim, H.J.; Kim, J.W.; Yu, B.P. The inflammation hypothesis of aging: Molecular modulation by calorie restriction. *Ann. N. Y. Acad. Sci.* **2001**, *928*, 327–335. [CrossRef] [PubMed]
67. Baylis, D.; Bartlett, D.B.; Patel, H.P.; Roberts, H.C. Understanding how we age: Insights into inflammaging. *Longev. Healthspan* **2013**, *2*, 8. [CrossRef] [PubMed]
68. Fatouros, I.G.; Jamurtas, A.Z.; Villiotou, V.; Pouliopoulou, S.; Fotinakis, P.; Taxildaris, K.; Deliconstantinos, G. Oxidative stress responses in older men during endurance training and detraining. *Med. Sci. Sports Exerc.* **2004**, *36*, 2065–2072. [CrossRef] [PubMed]
69. Radak, Z.; Bori, Z.; Koltai, E.; Fatouros, I.G.; Jamurtas, A.Z.; Douroudos, I.I.; Terzis, G.; Nikolaidis, M.G.; Chatzinikolaou, A.; Sovatzidis, A.; et al. Age-dependent changes in 8-oxoguanine-DNA glycosylase activity are modulated by adaptive responses to physical exercise in human skeletal muscle. *Free Radic. Biol. Med.* **2011**, *51*, 417–423. [CrossRef] [PubMed]
70. Calvani, R.; Marini, F.; Cesari, M.; Buford, T.W.; Manini, T.M.; Pahor, M.; Leeuwenburgh, C.; Bernabei, R.; Landi, F.; Marzetti, E. Systemic inflammation, body composition, and physical performance in old community-dwellers. *J. Cachexia Sarcopenia Muscle* **2016**. [CrossRef] [PubMed]
71. Cesari, M.; Penninx, B.W.; Pahor, M.; Lauretani, F.; Corsi, A.M.; Rhys Williams, G.; Guralnik, J.M.; Ferrucci, L. Inflammatory markers and physical performance in older persons: The InCHIANTI study. *J. Gerontol. Ser. A Biol. Sci. Med. Sci.* **2004**, *59*, 242–248. [CrossRef]
72. Zhou, L.M.; Xu, J.Y.; Rao, C.P.; Han, S.; Wan, Z.; Qin, L.Q. Effect of whey supplementation on circulating C-reactive protein: A meta-analysis of randomized controlled trials. *Nutrients* **2015**, *7*, 1131–1143. [CrossRef] [PubMed]
73. Drummond, M.J.; Rasmussen, B.B. Leucine-enriched nutrients and the regulation of mammalian target of rapamycin signalling and human skeletal muscle protein synthesis. *Curr. Opin. Clin. Nutr. Metab. Care* **2008**, *11*, 222–226. [CrossRef] [PubMed]

74. Kimball, S.R.; Jefferson, L.S. Signaling pathways and molecular mechanisms through which branched-chain amino acids mediate translational control of protein synthesis. *J. Nutr.* **2006**, *136*, 227s–231s. [CrossRef] [PubMed]

75. Galland, L. Diet and inflammation. *Nutr. Clin. Pract.* **2010**, *25*, 634–640. [CrossRef] [PubMed]

76. Fritsche, K.L. The science of fatty acids and inflammation. *Adv. Nutr.* **2015**, *6*, 293S–301S. [CrossRef] [PubMed]

77. Bleau, C.; Karelis, A.D.; St-Pierre, D.H.; Lamontagne, L. Crosstalk between intestinal microbiota, adipose tissue and skeletal muscle as an early event in systemic low-grade inflammation and the development of obesity and diabetes. *Diabetes Metab. Res. Rev.* **2015**, *31*, 545–561. [CrossRef] [PubMed]

78. Lordan, R.; Tsoupras, A.; Mitra, B.; Zabetakis, I. Dairy Fats and Cardiovascular Disease: Do We Really Need to be Concerned? *Foods* **2018**, *7*. [CrossRef] [PubMed]

79. Garcia, O.P. Effect of vitamin A deficiency on the immune response in obesity. *Proc. Nutr. Soc.* **2012**, *71*, 290–297. [CrossRef] [PubMed]

80. Wannamethee, S.G.; Lowe, G.D.; Rumley, A.; Bruckdorfer, K.R.; Whincup, P.H. Associations of vitamin C status, fruit and vegetable intakes, and markers of inflammation and hemostasis. *Am. J. Clin. Nutr.* **2006**, *83*, 567–574. [CrossRef] [PubMed]

81. Chung, E.; Mo, H.; Wang, S.; Zu, Y.; Elfakhani, M.; Rios, S.R.; Chyu, M.C.; Yang, R.S.; Shen, C.L. Potential roles of vitamin E in age-related changes in skeletal muscle health. *Nutr. Res.* **2018**, *49*, 23–36. [CrossRef] [PubMed]

82. Calder, P.C.; Bosco, N.; Bourdet-Sicard, R.; Capuron, L.; Delzenne, N.; Dore, J.; Franceschi, C.; Lehtinen, M.J.; Recker, T.; Salvioli, S.; et al. Health relevance of the modification of low grade inflammation in ageing (inflammageing) and the role of nutrition. *Ageing Res. Rev.* **2017**, *40*, 95–119. [CrossRef] [PubMed]

nutrients

MDPI

Review

Physical Activity and Nutrition: Two Promising Strategies for Telomere Maintenance?

Estelle Balan [1], Anabelle Decottignies [2] and Louise Deldicque [1,*]

1   Institute of Neuroscience, Université catholique de Louvain, Place Pierre de Coubertin 1 L8.10.01, 1348 Louvain-La-Neuve, Belgium; estelle.balan@uclouvain.be
2   De Duve Institute, Université catholique de Louvain, Avenue Hippocrate 75, 1200 Brussels, Belgium; anabelle.decottignies@uclouvain.be
*   Correspondence: louise.deldicque@uclouvain.be; Tel.: +32-10-474443

Received: 14 November 2018; Accepted: 5 December 2018; Published: 7 December 2018

Abstract: As the world demographic structure is getting older, highlighting strategies to counteract age-related diseases is a major public health concern. Telomeres are nucleoprotein structures that serve as guardians of genome stability by ensuring protection against both cell death and senescence. A hallmark of biological aging, telomere health is determined throughout the lifespan by a combination of both genetic and non-genetic influences. This review summarizes data from recently published studies looking at the effect of lifestyle variables such as nutrition and physical activity on telomere dynamics.

Keywords: aging; exercise; diet; telomerase; TERRA; telomere length; senescence

1. Introduction

The proportion of the world population aged 60 years and over is increasing rapidly and is projected to rise above 20% in 2050, which will exceed the number of children in the world [1,2]. Indeed, most countries are seeing their demographic structure getting older. The aging of the population has major implications socially and economically as aging is characterized by a progressive loss of physiological integrity, leading to impaired function and autonomy [3]. This functional decline is the greatest risk factor for conditions that limit health span, i.e., quality of life at old age, and for the majority of chronic diseases such as type 2 diabetes, Alzheimer's disease, and various cancers [4,5]. Notably, senescence has become the greatest risk factor for death in developed countries [6]. With the increasing longevity, the maintenance of health and autonomy at old age becomes crucial. Today more than ever, highlighting strategies to counteract age-related disorders is a major public health concern.

Geroscience is an area that aims to explain the biological mechanisms of aging. Aging research has experienced an unprecedented advance over recent years, particularly with the discovery that the rate of aging is controlled, at least to some extent, by genetic pathways and biochemical processes conserved in evolution such as genomic instability, telomere attrition, epigenetic alterations, loss of proteostasis, deregulated nutrient sensing, mitochondrial dysfunction, cellular senescence, stem cell exhaustion, and altered intercellular communication [3]. Recent findings have revealed the importance of the regulation of telomere length and integrity during the aging process [7], as well as potential interventions to improve the health span such as physical activity and healthy diet [8]. Telomere attrition is associated with decreased life expectancy and increased risk of chronic disease [9] and has been described as one of the most important biological hallmarks of aging due to a key role in cellular senescence [3]. During the past decade, telomeres have evolved from a simple capsule hiding the ends of chromosomes to complex nucleoprotein structures with an active role in the protection of the genome and in the regulation of cellular senescence [10,11]. While previous reviews specifically focused on the regulation of telomere length by either nutrition [12] or exercise [13,14], the present review will give a broader view looking at the impact of lifestyle variables on human telomere dynamics with emphasis on diet and physical activity.

2. Telomeres

Mammalian telomeres consist of repetitive DNA G- and C-rich sequences (5′-TTAGGG-3′/3′-CCCTAA-5′) with the 3′ end of the G-strand extending beyond the 5′ end [15]. The double-stranded telomeric DNA is bound by the six-subunit shelterin complex: telomeric repeat factor 1 (TRF1), telomere repeat factor 2 (TRF2) and protection of telomere 1 (POT1) directly recognize TTAGGG repeats and they are interconnected with TRF1- and TRF2-interacting nuclear protein 2 (TIN2), POT1 and TIN2-interacting protein (TPP1) and repressor/activator protein 1 (RAP1) [16]. The Shelterin complex facilitates the formation of a lariat-like structure with a T- and a D-loop, allowing the telomere end to be hidden (Figure 1). This conformation represses the DNA damage response (DDR) at telomeres, thereby preventing the activation of the ataxia telangectasia mutated (ATM) and RAD3-related (ATR) kinases that induce cell cycle arrest in response to DNA double-strand breaks and other types of DNA damage [10,17].

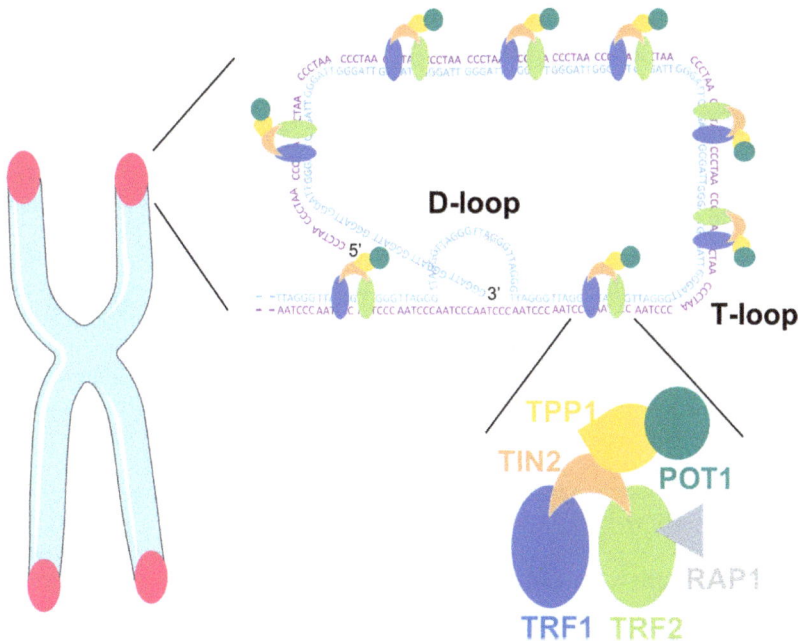

Figure 1. Telomeric DNA with the Shelterin complex facilitating the formation of D- and T-loop.

Many types of human cells lack telomerase, the enzyme responsible for telomere synthesis by adding nucleotides to the chromosome ends [18]. Telomerase consists of two core components; the catalytic subunit, telomerase reverse transcriptase (TERT) [19] and a RNA template (TERC) [20]. Hence, because of the "end-replication problem", i.e., DNA polymerase incapacity to maintain telomere length during cell divisions, somatic cells display gradual telomere shortening with age [21]. Critical loss of telomeric DNA or unprotected telomeres leads to insufficient chromosomes end protection and to the activation of the DNA damage response [17]. Damage at telomeres can also happen independently of cell division, notably in response to the accumulation of oxidative lesions, smoking behavior, or obesity [22]. While telomere shortening is considered as a protection against tumor development, loss of telomere function induces cellular senescence and impairs tissue turnover leading to the aging of the whole organism. The process of telomere attrition is not constant and differs between people [23,24], which can be explained by the impact of inflammation and oxidative stress on telomere shortening, which differs from one individual to the other [9]. Globally, telomere health is determined throughout the lifespan by a combination of both genetic and non-genetic factors.

3. Telomere Regulation by Nutrition

Lifestyle factors such as an unhealthy diet, physical inactivity, or smoking habits have been related to shorter leukocyte telomere length, a biomarker of the "biological age" of cells, as opposed to the "chronological age" [25]. Some studies have reported an association between diet [26–30] or consumption of specific foods [31] and leukocyte telomere length. To note, the rates of telomere shortening are similar in leukocytes and somatic cells, so that telomere length in leukocytes is now accepted to be representative of global telomere length in somatic cells [32].

3.1. Consumption of Specific Foods

Telomere length is positively associated with the consumption of legumes, nuts, seaweed, fruits, and 100% fruit juice, dairy products, and coffee, whereas it is inversely associated with consumption of alcohol, red meat, or processed meat [27,28,33,34]. Telomere attrition may represent a mechanism by which large sugar intake accelerates cardiometabolic disease [35]. Several studies suggest that reducing sugary beverage consumption could be associated with extended telomere length, independently of other characteristics such as age, sex, or body mass index [26–28]. Those results indicate that leukocyte telomere length maintenance may be sensitive to the metabolic effects of high sugar consumption over time [26]. Leung et al. examined the associations between the consumption of sugar-sweetened beverages (including soda, soft drinks, fruit-flavored drinks, sports drinks, and energy drinks), diet soda, fruit juice, and leukocyte telomere length in 5309 adults aged 20–65 years from the United States without any history of diabetes or cardiovascular disease [28]. After adjustment for sociodemographic and health-related characteristics, the consumption of sugar-sweetened beverages was associated with shorter telomeres, whereas the consumption of 100% fruit juice was associated with a higher telomere length. No significant association was observed between consumption of diet soda and telomere length [28]. As cross-sectional study may not be the most appropriate study design to assess telomere length, more recently, the same group conducted a longitudinal study to evaluate the associations between sugary foods and beverages and leukocyte telomere length in 65 overweight and obese pregnant women aged between 18 and 45 years. From ≤16 weeks gestation to 9 months postpartum, dietary intake was monitored using 24-h diet recalls and leukocyte telomere length was measured by real-time quantitative polymerase chain reaction (qPCR). From the baseline to 9 months post-partum, a low consumption of sugar-sweetened beverages was associated with longer leukocyte telomere length but no association was found between sugary foods and leukocyte telomere length [26].

People who regularly eat beans and whole grains are frequently spotlighted for increased longevity [36]. Boressen et al. (2016) tried to determine the feasibility of increasing navy beans or rice bran intake in colorectal cancer survivors to increase dietary fiber. The authors hypothesized that an increased amount of dietary fiber could positively regulate telomere length. Twenty-nine volunteers participated to a randomized-controlled trial with foods that included cooked navy beans powder (35 g/day), heat-stabilized rice bran (30 g/day), or no additional ingredient. The amount of navy beans powder or heat-stabilized rice bran consumed represented 4–9% of daily caloric intake. Over the intervention period of 4 weeks, no major gastrointestinal issues were reported and the dietary fiber amounts increased in the navy beans and rice bran groups at weeks 2 and 4 compared to baseline and the control group. At baseline, peripheral blood mononuclear cell (PBMC) telomere length was positively correlated with high density lipoprotein (HDL)-cholesterol and negatively correlated with lipopolysaccharide and age. Although a higher consumption of navy beans (35 g/day) or rice bran (30 g/day), known to contain fiber, iron, zinc, thiamin, niacin, vitamin B6, folate, and alpha-tocopherol, did not influence PBMC telomere length after the short intervention period of 4 weeks [31], the effect of a fiber-enriched diet on telomere length should be investigated in a healthy population over a longer period of time. This may be highly relevant in the context of colorectal cancer known to be associated with dysfunctional telomeres [37].

3.2. Diet Composition

While it is important to be aware of the effects of individual foods, it is even more critical to assess the role of cumulative nutrients contained in specific diets on telomere length, which better reflects reality. In 2015, Lee et al. compared the influence of the dietary pattern on leukocyte telomere length [27]. Dietary data were collected from a semi-quantitative food frequency questionnaire at baseline and leukocyte telomere length was assessed using qPCR 10 years later. A total of 1958 middle-aged and older Korean adults (40–69 years at baseline) were included in the study. The authors identified two major dietary patterns: "the prudent dietary pattern" was characterized by a high intake of whole grains, fish and seafood, legumes, vegetables, and seaweed, whereas the "western dietary pattern" included a high intake of refined grain, red meat or processed meat, and sweetened carbonated beverages. Using a multiple linear regression model adjusted for age, sex, body mass index, and other potential confounding variables, the "prudent dietary pattern" was found to be positively associated with leukocyte telomere length while an inverse trend was found in the "western dietary pattern". These results suggest that diet in the remote past, that is, 10 years earlier, may affect the degree of biological aging in middle-aged and older adults [27].

One of the best models of healthy eating is the Mediterranean diet which is characterized by a high intake of vegetables, legumes, nuts, fruits, and cereals (mainly unrefined); a moderate to high intake of fish; a low intake of saturated lipids but high intake of unsaturated lipids, particularly olive oil; a regular but moderate intake of alcohol, specifically wine [38]. This diet has been shown to prevent age-associated telomere shortening [29,30,39] and has been associated with reduced mortality risk in older people [40]. The possible mechanisms for the protective effect of the Mediterranean diet on telomeres will be discussed in the next section. In 4676 healthy women (42–70 years), the higher scores on the Mediterranean diet, evaluated by food frequency questionnaires, were associated with longer leukocyte telomere length [30]. In the same study, no association between prudent or western dietary patterns and telomere length was observed [30], while a prudent diet was previously found to be positively and a western diet negatively associated with leukocyte telomere length in 1958 middle-aged and older women and men [27]. Similarly, in 217 men and women aged 71–87 years, a greater adherence to a Mediterranean diet was associated with longer leukocyte telomere length and higher PBMC telomerase activity [29]. However, a recent study in 679 Australian men and women (57–68 years) found no association between diet quality and whole blood telomere length, including the Mediterranean diet. In this study, the authors assessed the dietary intake by using a 111-item food frequency questionnaire, which assessed self-reported intake of foods and beverages over the last 6 months, and the diet quality by three indices: the Dietary Guideline Index (DGI), the Recommended Food Score (RFS), and the Mediterranean Diet Score (MDS) [41]. Whole blood telomere length did not differ by age, smoking status, BMI, or physical activity but women had longer telomeres than men [41]. The discrepant results between studies could be explained by the use of different questionnaires to assess the diet quality and/or the populations studied. Longitudinal studies may be more suitable to determine the potential positive influence of diet on telomere health.

Of note, in animal models, calorie restriction has been shown to have a positive effect on telomere length [42] and to globally delay the onset of aging and age-related disease such as diabetes, cardiovascular diseases, various neurological disorders, cancer, and obesity [43,44], possibly via a reduction in oxidative stress [45,46]. In humans, the data are less convincing, probably because decreasing the caloric intake by a third or a half is very challenging in that population, certainly in the long-term.

Having presented which foods and diets were potentially beneficial for telomere health in general, the next section will attempt to summarize the mechanisms involved in those effects.

3.3. Mechanisms

Unhealthy dietary habits have been linked to an inflammatory state, contributing to progressive telomere attrition [47]. As unhealthy dietary habits increase the production of reactive oxygen species (ROS), it is possible that the impact on telomere erosion goes through an increased oxidation of telomeric DNA. Supporting this, is the observation that, because of their high content in guanine residues, telomeric sequences are highly prone to oxidation into 8-oxoG, at least in in vitro experiments [48]. When present at telomeres, 8-oxoG residues are likely decreasing the affinity of shelterin proteins for telomeric DNA and are, as well, disrupting the G-quadruplex structures of telomeres that play important roles at telomeres, like the regulation of telomerase activity [49]. Altogether, it is therefore possible that nutrients regulate telomere health by regulating oxidative stress and systemic inflammation [50]. Globally, it can be hypothesized that any antioxidant or anti-inflammatory diet could be protective for telomeres by slowing down telomeric shortening and delay the aging process. The intake of nutrients having antioxidant and anti-inflammatory properties, such as vitamin C or E, polyphenols, curcumin, or omega-3 fatty acid, has been associated with longer telomeres, at least in mouse [51].

The positive effects of the Mediterranean diet on telomeres may be due to its antioxidant and anti-inflammatory potential [52,53]. To understand whether the Mediterranean diet could prevent endothelial cellular senescence by regulating oxidative stress, the serum of 20 elderly subjects (age > 65 years; 10 men and 10 women) was collected before and after having randomly followed each of the 3 following diets for 4 weeks: a Mediterranean diet, a saturated fatty acid diet and a low fat and high carbohydrate diet [54]. Human endothelial cells incubated with the serum collected after ingestion of the Mediterranean diet produced lower intracellular ROS, unavoidable byproducts of aerobic metabolism, and the percentage of cells with telomere shortening was lower compared to baseline and the two other intervention diets. The authors postulated that those findings were possibly due to nutrients with antioxidant capacities included in the Mediterranean diet [54]. In 2015, a direct association was found between the pro-inflammatory capacity of the diet and telomere shortening in a population at high risk of cardiovascular disease. The diets with the higher pro-inflammatory scores were associated with a higher risk of having shorter telomeres and a two-fold risk of accelerated telomere shortening after a five-year follow-up period [47]. At a molecular level, exposure of human leukemic cells to the pro-inflammatory factor tumor necrosis factor alpha (TNFα) induced a senescence state, which was featured by prolonged growth arrest, increased beta-galactosidase activity, cyclin-dependent kinase inhibitor 1 (p21) activation, decreased telomerase activity, telomeric disturbances such as shortening, losses, and fusions, as well as additional chromosomal aberrations [55]. Those results indicate that TNFα alters telomere maintenance. Moreover, subjects with higher adherence to Mediterranean diet had lower plasmatic level of C-reactive protein (CRP), interleukin 6 (IL-6), TNFα, and nitrotyrosine, all markers of inflammation and/or oxidative stress [29]. As high levels of oxidative stress [56] and inflammation [57] are known to increase telomere attrition rate, the Mediterranean diet may protect telomere maintenance by downregulating both processes.

While a healthy diet may have an overall positive influence on telomeres, it seems that the benefit may be reduced in some individuals with specific genetic background [58]. For example, the *rs1800629* polymorphism at the *TNFα* gene has been shown to interact with the Mediterranean diet to modify triglyceride metabolism and inflammation status in patients suffering from the metabolic syndrome [58]. At baseline, the patients with the GG alleles had higher fasting and postprandial triglyceride and higher sensitivity C-reactive protein plasma levels than the patients with the GA or AA alleles. However, those differences between the polymorphisms observed at baseline disappeared after having followed a Mediterranean diet for 12 months, suggesting that the GG carriers were highly sensitive to this specific diet. Globally, understanding the role of gene–diet interactions may be an efficient strategy for personalized treatment of specific pathologies such as metabolic syndrome.

While some molecular mechanisms have already been highlighted, further research is needed to better understand how different diets and specific foods regulate biological aging in order to develop efficient nutritional strategies according to specific populations.

4. Telomere Regulation by Physical Activity

This section will deliberately present a positive view regarding the effects of physical activity on telomere dynamics, but it should be kept in mind that about half of the studies dealing with that topic found no association between physical activity and telomere length [13]. Obviously, further investigation will be needed to determine why the different findings are such discrepant from one study to the other. In addition, new analytical tools need to be developed to measure telomere length more accurately as well as new biomarkers for assessing biological aging [13].

4.1. Dose-Response

The beneficial effect of physical activity on telomere length has been reviewed and discussed by Denham et al. [59]. However, there is currently no clear consensus on the optimal exercise dose to exert the most beneficial response on telomere health. The effect of 9 different modes of physical activity, and thereby intensity levels, on leukocytes telomere length has been tested in US adults (20–84 years, N = 6503) [60]. The only mode of physical activity displaying an association with leukocyte telomere length was running, the most intense mode in that study. Another study used the data of a subgroup of the previously mentioned cohort (N = 5883) and found a strong positive association between the weekly amount of physical activity and telomere length in leukocytes [61]. However, a recent study indicated that moderate amounts of exercise are sufficient to protect telomere health, while higher amounts may not elicit additional benefits [62]. In 2010, telomere length was measured in skeletal muscle of 18 experienced middle-aged endurance runners versus 19 sedentary subjects [63]. No difference between groups was found. However, telomere length in the muscle of endurance athletes was inversely related to the number of years they spent running and the hours of spent training, which indicated that high level of chronic endurance could accelerate telomere attrition and thereby biological aging. More recently, leukocyte telomere length was determined in 61 young elite athletes and 64 healthy inactive controls [64]. Even with their high intensity and training volume, the young elite athlete had longer telomeres than their inactive peers. Finally, leukocyte telomere length was 11% higher in ultra-marathon runners compared to 56 healthy subjects, matched for age [65]. Altogether, these results suggest that high amounts of exercise may not reverse the beneficial impact of exercise on telomere length but further investigation is needed to see whether tissue-specific differences exist.

In humans, Diman et al. showed that a high intensity cycling exercise (75% VO₂ peak) boosted the transcription of skeletal muscle telomeres more than a moderate intensity exercise (50% VO₂ peak) of the same duration [66]. More details on the molecular mechanisms of this observation will be reported in a following section. In conclusion, due to the paucity of data, it remains unclear which of the intensity or the volume of each training session or the combination of both is crucial to induce the beneficial effects exercise has on telomere maintenance.

4.2. Physical Activity and Telomerase Activity

While physical activity has been associated with longer telomere length and protection against age-related telomere attrition [65,67–72], the mechanisms by which physical activity exerts its positive effects on telomeres are still largely unknown. As TERT, the catalytic subunit of the telomerase complex, is considered as the limiting factor for telomerase activity in human somatic cells, an increase in telomerase activity after exercise could promote telomere elongation. Chilton et al. were the first to look at the regulation of telomerase after one acute bout of exercise [73]. To that end, they investigated the acute exercise-induced response on telomeric-associated genes and microRNAs (miRNAs), i.e., small noncoding RNA molecules functioning in RNA silencing and post-transcriptionally regulating gene expression by base pairing with messenger RNA (mRNA). Blood samples were taken in 22 healthy young males before, immediately after, and 60 min after a 30-min bout of treadmill running at 80% VO$_2$ peak. In white blood cells, both TERT and sirtuin-6 (SIRT6) mRNA levels were increased immediately after exercise. Sixty minutes post-exercise, there was an upregulation of miR-186 and miR-96 expression, two miRNA controlling the expression of genes involved in telomere homeostasis [73]. In addition, telomeric repeat binding factor 2, interacting protein (TERF2IP) was identified as a potential binding target for miR-186 and miR-96 and demonstrated concomitant downregulation with the upregulation of those 2 miRNA at 60 min post-ex. TERF2IP is part of the shelterin complex and is recruited to telomeres via interaction with TRF2 [74]. TERF2IP deletion reduces telomere stability and increases telomere recombination [75]. However, TERF2IP/RAP1 has been found to be both a negative [76] and a positive regulator of telomere length [74]. Interestingly, TERF2IP/RAP1 is also known to play additional telomere-unrelated functions through the binding to extra-telomeric sites in the genome. Several regulatory functions have been attributed to the binding of TERF2IP/RAP1 outside telomeres, including the modulation of the nuclear factor-kappa B (NF-kB)-dependent pathway [77]. Whether the non-telomeric functions of TERF2IP/RAP1 play any role after exercise however warrants further investigation.

A very recent study tested whether an acute bout of exercise would induce a different response on telomerase activity in older vs. young individuals and whether this response would be gender-specific [78]. To test this hypothesis, age- and gender-related differences in telomerase and shelterin responses at 30, 60, and 90 min after a high intensity interval cycling exercise were determined in PBMC of 11 young (22 years) and 8 older (60 years) men and women. A larger increase in telomerase activity, as assessed by TERT mRNA levels, was found in the young compared to the older group after exercise [78]. The second main finding of that study was the higher TERT response to the acute endurance exercise in men compared to women, in whom the response was negligible, independently of age. Those results showed that aging is associated with reduced telomerase activation in response to high-intensity cycling exercise in men [78]. Another study showed that a 30-min treadmill running session was long enough to increase PBMC telomerase activity in 22 young healthy subjects including 11 women and 11 men [79]. Altogether, those recent studies confirm that the increasing telomerase activity after a single bout of exercise could be one of the mechanisms by which physical activity protects against aging [73,78,79].

Nevertheless, the increase in telomerase activity seems transient after acute exercise. The effect of a whole training program on telomerase activity and telomere length was investigated in 68 female and male caregivers, a population known to cope with chronic high stress, physical inactivity, and dealing with a high risk of disease [80]. Half of the subjects followed an endurance training program consisting in 40 min of aerobic exercise 3–5 times per week, while the other half remained inactive for 24 weeks. In aerobic trained caregivers, the leukocyte telomere length was lengthened after training while the telomere length was slightly shortened in the inactive group, as would be expected over a six month-period. However, no change in PBMC telomerase activity after the intervention was observed in either group [80]. Together with the findings from the acute exercise studies, it can be hypothesized that exercise-induced higher telomerase activity in PBMC may be a transient mechanism returning to basal level several hours after a single bout of exercise, though the exact kinetics still needs to be determined. In addition, telomerase is not active in all cell types, which implies that other mechanisms contribute to the exercise-induced beneficial effects on telomere length and integrity in those cells.

4.3. Physical Activity and Oxidative Stress

It is well established that moderate and regular physical activity is able to reduce the effect of aging by alleviating oxidative stress level [81]. Recently, an inverse relationship between the aerobic capacity and oxidative stress biomarkers in the blood was found in older Mexican adults [82]. Moreover, several studies indicate that oxidative stress accelerates telomere attrition [83–85].

Mechanistically, exercise transiently upregulates ROS production, which is counteracted by an antioxidant exercise-induced systemic adaptation response to protect the cells against oxidative damage [86,87]. This antioxidant response can be explained by the hormesis concept, namely that low levels of stress stimulate existing cellular and molecular pathways that improve the capacity of cells and organisms to withstand subsequent greater stress [88]. The antioxidant response leads to the activation of redox-sensitive transcription factors such as NF-*k*B, activator-protein 1 (AP-1) [89], and co-factors such as peroxisome proliferator-activated receptor gamma coactivator 1-alpha (PGC-1α) [81,90]. As a metabolic energy deprivation sensor, AMP-activated protein kinase (AMPK) is activated by exercise and triggers PGC-1α transcription and activation by allowing its nuclear translocation [91]. Once in the nucleus, PGC-1α induces the transcription of nuclear respiratory factor 1 (NRF1), an antioxidant factor. By activating the PGC-1α redox signaling pathway, exercise stimulates mitochondrial biogenesis and ameliorates the age-related decline in mitochondrial oxidative capacity [90].

4.4. Physical Activity and Regulation of TERRA

Mature muscle cells are one example of cells in which telomerase is not active and despite the absence of telomerase activity, physical activity has been shown to influence positively telomere length in skeletal muscle [92]. In the search of additional mechanisms, telomeric repeat containing RNAs, dubbed TERRA, have emerged as particularly interesting targets. For a long time, telomeres have been considered transcriptionally silent. Yet it turns out that telomeres are transcribed into TERRA molecules [93]. Located in the nucleus, TERRA are non-coding RNAs whose transcription is initiated from subtelomeric promoters. They consist of subtelomeric-derived sequences and G-rich telomeric repeats [93,94]. Once transcribed, TERRA remain partly associated with telomeres to play crucial functions, including telomere protection [95]. Diman et al. identified NFR1 as an important regulator of human telomere transcription in cultured cells. In addition to NRF1, PGC-1α as well as AMPK were found to be important molecular intermediates in the transcription of telomeres. As AMPK can be activated by high-intensity or long-lasting endurance exercise, it was tested in vivo whether an acute endurance exercise bout could upregulate telomere transcription in human skeletal muscles. Ten healthy young volunteers were submitted to a cycling endurance exercise of either low or high intensity and three muscle biopsies were taken before, directly after, and 2 h 30 min after exercise. Phosphorylation of acetyl-Coa carboxylase (ACC), a bona fide marker of AMPK activation, was induced after exercise, especially in the high intensity group. The same pattern of activation was found for the translocation of PGC-1α to the nucleus and for TERRA induction. As telomere transcription is activated by NRF1, an antioxidant factor, the upregulation of TERRA may be part of the antioxidant response that skeletal muscles set up to counteract exercise-induced ROS production [86]. Moreover, as they consist of a high content in guanine residues prone to oxidation, TERRA may possibly shield TTAGGG telomeric repeats from ROS [66]. Together, those results suggest that an acute bout of endurance exercise is sufficient to induce telomere transcription that, on a longer term, could possibly provide a mechanism for TERRA renewal and telomere protection in skeletal muscle.

5. Conclusions

Nowadays, the aging of the world population has major social and economic implications. Today more than ever, highlighting strategies to counteract age-related diseases is a major public health concern. In this review, we explored data from recently published studies looking at the influence of lifestyle variables such as nutrition and physical activity on one of the most important hallmarks of aging: the telomere.

Most studies indicate an important role of diet on the degree of biological aging. Indeed, a healthy diet characterized by a high intake of dietary fiber and unsaturated lipids exerts a protective role on telomere health, whereas high consumption of sugar and saturated lipids accelerates telomere attrition.

Nutrients **2018**, *10*, 1942

Those effects are likely to be globally mediated by oxidative stress and inflammation, as antioxidant and anti-inflammatory properties of nutrients are associated with longer telomeres. Physical activity may protect telomeres but more research is needed to establish a consensus on the optimal exercise dose (Figure 2). The beneficial effects of physical activity on telomeres could be driven by an increase in telomerase activity following an acute bout of exercise in PBMC, an alleviation of oxidative stress and a TERRA renewal in skeletal muscle. Further investigations are needed to study the other possible mechanisms contributing to the exercise-induced beneficial effects on telomere length and integrity.

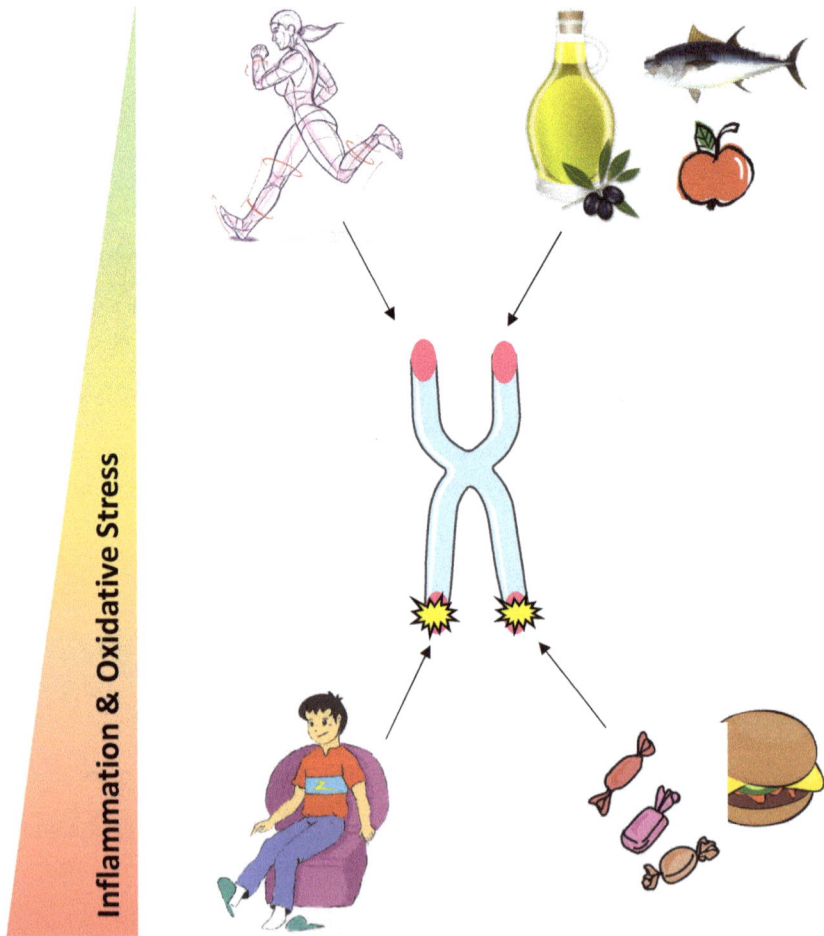

Figure 2. Potential influence of physical activity and nutrition on telomere health.

We propose that engaging in a healthy diet and regular physical activity could be both promising strategies to protect telomere maintenance and improve health span at old age. However, more research on the molecular based mechanisms is required.

Author Contributions: E.B. and L.D. wrote the first draft. A.D. corrected the first draft. All authors revised and approved the last version of the manuscript.

Funding: This work was funded by a Fonds Spécial de Recherche (FSR) from the Université catholique de Louvain.

Conflicts of Interest: The authors declare no conflict of interest.

References

1. United Nations. *World Population Ageing*; Economic and Social Affairs Population Division: New York, NY, USA, 2009.
2. Population Ageing and Sustainable Development. Available online: http://www.un.org/en/development/desa/population/publications/pdf/popfacts/PopFacts_2017-1.pdf (accessed on 8 August 2018).
3. Lopez-Otin, C.; Blasco, M.A.; Partridge, L.; Serrano, M.; Kroemer, G. The hallmarks of aging. *Cell* **2013**, *153*, 1194–1217. [CrossRef] [PubMed]
4. Omodei, D.; Fontana, L. Calorie restriction and prevention of age-associated chronic disease. *FEBS Lett.* **2011**, *585*, 1537–1542. [CrossRef] [PubMed]
5. Kennedy, B.K.; Berger, S.L.; Brunet, A.; Campisi, J.; Cuervo, A.M.; Epel, E.S.; Franceschi, C.; Lithgow, G.J.; Morimoto, R.I.; Pessin, J.E.; et al. Geroscience: Linking aging to chronic disease. *Cell* **2014**, *159*, 709–713. [CrossRef] [PubMed]
6. Kaeberlein, M.; Rabinovitch, P.S.; Martin, G.M. Healthy aging: The ultimate preventative medicine. *Science* **2015**, *350*, 1191–1193. [CrossRef] [PubMed]
7. Tzanetakou, I.P.; Nzietchueng, R.; Perrea, D.N.; Benetos, A. Telomeres and their role in aging and longevity. *Curr. Vasc. Pharmacol.* **2014**, *12*, 726–734. [CrossRef] [PubMed]
8. Mercken, E.M.; Carboneau, B.A.; Krzysik-Walker, S.M.; de Cabo, R. Of mice and men: The benefits of caloric restriction, exercise, and mimetics. *Ageing Res. Rev.* **2012**, *11*, 390–398. [CrossRef] [PubMed]
9. Sanders, J.L.; Newman, A.B. Telomere length in epidemiology: A biomarker of aging, age-related disease, both, or neither? *Epidemiol. Rev.* **2013**, *35*, 112–131. [CrossRef]
10. Arnoult, N.; Karlseder, J. Complex interactions between the DNA-damage response and mammalian telomeres. *Nat. Struct. Mol. Biol.* **2015**, *22*, 859–866. [CrossRef]
11. Karlseder, J.; Smogorzewska, A.; de Lange, T. Senescence induced by altered telomere state, not telomere loss. *Science* **2002**, *295*, 2446–2449. [CrossRef]
12. Vidacek, N.S.; Nanic, L.; Ravlic, S.; Sopta, M.; Geric, M.; Gajski, G.; Garaj-Vrhovac, V.; Rubelj, I. Telomeres, Nutrition, and Longevity: Can We Really Navigate Our Aging? *J. Gerontol. Ser. A Biol. Sci. Med. Sci.* **2017**, *73*, 39–47. [CrossRef]
13. Chilton, W.; O'Brien, B.; Charchar, F. Telomeres, Aging and Exercise: Guilty by Association? *Int. J. Mol. Sci.* **2017**, *18*, 2573. [CrossRef] [PubMed]
14. Arsenis, N.C.; You, T.; Ogawa, E.F.; Tinsley, G.M.; Zuo, L. Physical activity and telomere length: Impact of aging and potential mechanisms of action. *Oncotarget* **2017**, *8*, 45008–45019. [CrossRef] [PubMed]
15. Greider, C.W. Telomeres. *Curr. Opin. Cell Biol.* **1991**, *3*, 444–451. [CrossRef]
16. De Lange, T. Shelterin: The protein complex that shapes and safeguards human telomeres. *Genes Dev.* **2005**, *19*, 2100–2110. [CrossRef] [PubMed]
17. Maciejowski, J.; de Lange, T. Telomeres in cancer: Tumour suppression and genome instability. *Nat. Rev. Mol. Cell Biol.* **2017**, *18*, 175–186. [CrossRef] [PubMed]
18. Zhang, F.; Cheng, D.; Wang, S.; Zhu, J. Human Specific Regulation of the Telomerase Reverse Transcriptase Gene. *Genes* **2016**, *7*, 30. [CrossRef] [PubMed]
19. Nakamura, T.M.; Cech, T.R. Reversing time: Origin of telomerase. *Cell* **1998**, *92*, 587–590. [CrossRef]
20. Yu, G.L.; Bradley, J.D.; Attardi, L.D.; Blackburn, E.H. In vivo alteration of telomere sequences and senescence caused by mutated Tetrahymena telomerase RNAs. *Nature* **1990**, *344*, 126–132. [CrossRef]
21. Harley, C.B.; Futcher, A.B.; Greider, C.W. Telomeres shorten during ageing of human fibroblasts. *Nature* **1990**, *345*, 458–460. [CrossRef]
22. Muraki, K.; Nyhan, K.; Han, L.; Murnane, J.P. Mechanisms of telomere loss and their consequences for chromosome instability. *Front. Oncol.* **2012**, *2*, 135. [CrossRef]
23. Von Zglinicki, T. Oxidative stress shortens telomeres. *Trends Biochem. Sci.* **2002**, *27*, 339–344. [CrossRef]
24. Aviv, A.; Chen, W.; Gardner, J.P.; Kimura, M.; Brimacombe, M.; Cao, X.; Srinivasan, S.R.; Berenson, G.S. Leukocyte telomere dynamics: Longitudinal findings among young adults in the Bogalusa Heart Study. *Am. J. Epidemiol.* **2009**, *169*, 323–329. [CrossRef] [PubMed]
25. Eitan, E.; Hutchison, E.R.; Mattson, M.P. Telomere shortening in neurological disorders: An abundance of unanswered questions. *Trends Neurosci.* **2014**, *37*, 256–263. [CrossRef] [PubMed]

26. Leung, C.W.; Laraia, B.A.; Coleman-Phox, K.; Bush, N.R.; Lin, J.; Blackburn, E.H.; Adler, N.E.; Epel, E.S. Sugary beverage and food consumption, and leukocyte telomere length maintenance in pregnant women. *Eur. J. Clin. Nutr.* **2016**, *70*, 1086–1088. [CrossRef] [PubMed]

27. Lee, J.Y.; Jun, N.R.; Yoon, D.; Shin, C.; Baik, I. Association between dietary patterns in the remote past and telomere length. *Eur. J. Clin. Nutr.* **2015**, *69*, 1048–1052. [CrossRef] [PubMed]

28. Leung, C.W.; Laraia, B.A.; Needham, B.L.; Rehkopf, D.H.; Adler, N.E.; Lin, J.; Blackburn, E.H.; Epel, E.S. Soda and cell aging: Associations between sugar-sweetened beverage consumption and leukocyte telomere length in healthy adults from the National Health and Nutrition Examination Surveys. *Am. J. Public Health* **2014**, *104*, 2425–2431. [CrossRef] [PubMed]

29. Boccardi, V.; Esposito, A.; Rizzo, M.R.; Marfella, R.; Barbieri, M.; Paolisso, G. Mediterranean diet, telomere maintenance and health status among elderly. *PLoS ONE* **2013**, *8*, e62781. [CrossRef]

30. Crous-Bou, M.; Fung, T.T.; Prescott, J.; Julin, B.; Du, M.; Sun, Q.; Rexrode, K.M.; Hu, F.B.; De Vivo, I. Mediterranean diet and telomere length in Nurses' Health Study: Population based cohort study. *BMJ* **2014**, *349*, g6674. [CrossRef]

31. Borresen, E.C.; Brown, D.G.; Harbison, G.; Taylor, L.; Fairbanks, A.; O'Malia, J.; Bazan, M.; Rao, S.; Bailey, S.M.; Wdowik, M.; et al. A Randomized Controlled Trial to Increase Navy Bean or Rice Bran Consumption in Colorectal Cancer Survivors. *Nutr. Cancer* **2016**, *68*, 1269–1280. [CrossRef]

32. Daniali, L.; Benetos, A.; Susser, E.; Kark, J.D.; Labat, C.; Kimura, M.; Desai, K.; Granick, M.; Aviv, A. Telomeres shorten at equivalent rates in somatic tissues of adults. *Nat. Commun.* **2013**, *4*, 1597. [CrossRef]

33. Liu, J.J.; Crous-Bou, M.; Giovannucci, E.; De Vivo, I. Coffee Consumption Is Positively Associated with Longer Leukocyte Telomere Length in the Nurses' Health Study. *J. Nutr.* **2016**, *146*, 1373–1378. [CrossRef] [PubMed]

34. Pavanello, S.; Hoxha, M.; Dioni, L.; Bertazzi, P.A.; Snenghi, R.; Nalesso, A.; Ferrara, S.D.; Montisci, M.; Baccarelli, A. Shortened telomeres in individuals with abuse in alcohol consumption. *Int. J. Cancer* **2011**, *129*, 983–992. [CrossRef] [PubMed]

35. D'Mello, M.J.; Ross, S.A.; Briel, M.; Anand, S.S.; Gerstein, H.; Pare, G. Association between shortened leukocyte telomere length and cardiometabolic outcomes: Systematic review and meta-analysis. *Circ. Cardiovasc. Genet.* **2015**, *8*, 82–90. [CrossRef] [PubMed]

36. Hayat, I.; Ahmad, A.; Masud, T.; Ahmed, A.; Bashir, S. Nutritional and health perspectives of beans (*Phaseolus vulgaris* L.): An overview. *Crit. Rev. Food Sci. Nutr.* **2014**, *54*, 580–592. [CrossRef] [PubMed]

37. Tanaka, H.; Beam, M.J.; Caruana, K. The presence of telomere fusion in sporadic colon cancer independently of disease stage, TP53/KRAS mutation status, mean telomere length, and telomerase activity. *Neoplasia* **2014**, *16*, 814–823. [CrossRef] [PubMed]

38. Willett, W.C.; Sacks, F.; Trichopoulou, A.; Drescher, G.; Ferro-Luzzi, A.; Helsing, E.; Trichopoulos, D. Mediterranean diet pyramid: A cultural model for healthy eating. *Am. J. Clin. Nutr.* **1995**, *61*, 1402s–1406s. [CrossRef] [PubMed]

39. Garcia-Calzon, S.; Martinez-Gonzalez, M.A.; Razquin, C.; Corella, D.; Salas-Salvado, J.; Martinez, J.A.; Zalba, G.; Marti, A. Pro12Ala polymorphism of the PPARgamma2 gene interacts with a mediterranean diet to prevent telomere shortening in the PREDIMED-NAVARRA randomized trial. *Circ. Cardiovasc. Genet.* **2015**, *8*, 91–99. [CrossRef]

40. McNaughton, S.A.; Bates, C.J.; Mishra, G.D. Diet quality is associated with all-cause mortality in adults aged 65 years and older. *J. Nutr.* **2012**, *142*, 320–325. [CrossRef]

41. Milte, C.M.; Russell, A.P.; Ball, K.; Crawford, D.; Salmon, J.; McNaughton, S.A. Diet quality and telomere length in older Australian men and women. *Eur. J. Nutr.* **2018**, *57*, 363–372. [CrossRef]

42. Finkel, T. The metabolic regulation of aging. *Nat. Med.* **2015**, *21*, 1416–1423. [CrossRef]

43. Sun, L.; Sadighi Akha, A.A.; Miller, R.A.; Harper, J.M. Life-span extension in mice by preweaning food restriction and by methionine restriction in middle age. *J. Gerontol. Ser. A Biol. Sci. Med. Sci.* **2009**, *64*, 711–722. [CrossRef] [PubMed]

44. Mattison, J.A.; Colman, R.J.; Beasley, T.M.; Allison, D.B.; Kemnitz, J.W.; Roth, G.S.; Ingram, D.K.; Weindruch, R.; de Cabo, R.; Anderson, R.M. Caloric restriction improves health and survival of rhesus monkeys. *Nat. Commun.* **2017**, *8*, 14063. [CrossRef] [PubMed]

45. Saraswat, K.; Rizvi, S.I. Novel strategies for anti-aging drug discovery. *Expert Opin. Drug Discov.* **2017**, *12*, 955–966. [CrossRef] [PubMed]

46. Walsh, M.E.; Shi, Y.; Van Remmen, H. The effects of dietary restriction on oxidative stress in rodents. *Free Radic. Biol. Med.* **2014**, *66*, 88–99. [CrossRef] [PubMed]

47. Garcia-Calzon, S.; Zalba, G.; Ruiz-Canela, M.; Shivappa, N.; Hebert, J.R.; Martinez, J.A.; Fito, M.; Gomez-Gracia, E.; Martinez-Gonzalez, M.A.; Marti, A. Dietary inflammatory index and telomere length in subjects with a high cardiovascular disease risk from the PREDIMED-NAVARRA study: Cross-sectional and longitudinal analyses over 5 y. *Am. J. Clin. Nutr.* **2015**, *102*, 897–904. [CrossRef] [PubMed]

48. Oikawa, S.; Kawanishi, S. Site-specific DNA damage at GGG sequence by oxidative stress may accelerate telomere shortening. *FEBS Lett.* **1999**, *453*, 365–368. [CrossRef]

49. Ahmed, W.; Lingner, J. Impact of oxidative stress on telomere biology. *Differ. Res. Biol. Divers.* **2018**, *99*, 21–27. [CrossRef]

50. Aviv, A. Leukocyte telomere length: The telomere tale continues. *Am. J. Clin. Nutr.* **2009**, *89*, 1721–1722. [CrossRef]

51. Thomas, P.; Wang, Y.J.; Zhong, J.H.; Kosaraju, S.; O'Callaghan, N.J.; Zhou, X.F.; Fenech, M. Grape seed polyphenols and curcumin reduce genomic instability events in a transgenic mouse model for Alzheimer's disease. *Mutat. Res.* **2009**, *661*, 25–34. [CrossRef]

52. Garcia-Calzon, S.; Moleres, A.; Martinez-Gonzalez, M.A.; Martinez, J.A.; Zalba, G.; Marti, A. Dietary total antioxidant capacity is associated with leukocyte telomere length in a children and adolescent population. *Clin. Nutr.* **2015**, *34*, 694–699. [CrossRef]

53. Schwingshackl, L.; Hoffmann, G. Mediterranean dietary pattern, inflammation and endothelial function: A systematic review and meta-analysis of intervention trials. *Nutr. Metab. Cardiovasc. Dis.* **2014**, *24*, 929–939. [CrossRef] [PubMed]

54. Marin, C.; Delgado-Lista, J.; Ramirez, R.; Carracedo, J.; Caballero, J.; Perez-Martinez, P.; Gutierrez-Mariscal, F.M.; Garcia-Rios, A.; Delgado-Casado, N.; Cruz-Teno, C.; et al. Mediterranean diet reduces senescence-associated stress in endothelial cells. *Age* **2012**, *34*, 1309–1316. [CrossRef] [PubMed]

55. Beyne-Rauzy, O.; Recher, C.; Dastugue, N.; Demur, C.; Pottier, G.; Laurent, G.; Sabatier, L.; Mansat-De Mas, V. Tumor necrosis factor alpha induces senescence and chromosomal instability in human leukemic cells. *Oncogene* **2004**, *23*, 7507–7516. [CrossRef] [PubMed]

56. Kondo, T.; Hirose, M.; Kageyama, K. Roles of oxidative stress and redox regulation in atherosclerosis. *J. Atheroscler. Thrombosis* **2009**, *16*, 532–538. [CrossRef]

57. Masi, S.; Nightingale, C.M.; Day, I.N.; Guthrie, P.; Rumley, A.; Lowe, G.D.; von Zglinicki, T.; D'Aiuto, F.; Taddei, S.; Klein, N.; et al. Inflammation and not cardiovascular risk factors is associated with short leukocyte telomere length in 13- to 16-year-old adolescents. *Arterioscler. Thrombosis Vasc. Biol.* **2012**, *32*, 2029–2034. [CrossRef] [PubMed]

58. Gomez-Delgado, F.; Alcala-Diaz, J.F.; Garcia-Rios, A.; Delgado-Lista, J.; Ortiz-Morales, A.; Rangel-Zuniga, O.; Tinahones, F.J.; Gonzalez-Guardia, L.; Malagon, M.M.; Bellido-Munoz, E.; et al. Polymorphism at the TNF-alpha gene interacts with Mediterranean diet to influence triglyceride metabolism and inflammation status in metabolic syndrome patients: From the CORDIOPREV clinical trial. *Mol. Nutr. Food Res.* **2014**, *58*, 1519–1527. [CrossRef] [PubMed]

59. Denham, J.; O'Brien, B.J.; Charchar, F.J. Telomere Length Maintenance and Cardio-Metabolic Disease Prevention Through Exercise Training. *Sports Med.* **2016**, *46*, 1213–1237. [CrossRef]

60. Loprinzi, P.D.; Sng, E. Mode-specific physical activity and leukocyte telomere length among U.S. adults: Implications of running on cellular aging. *Prev. Med.* **2016**, *85*, 17–19. [CrossRef]

61. Tucker, L.A. Physical activity and telomere length in U.S. men and women: An NHANES investigation. *Prev. Med.* **2017**, *100*, 145–151. [CrossRef]

62. Denham, J.; O'Brien, B.J.; Prestes, P.R.; Brown, N.J.; Charchar, F.J. Increased expression of telomere-regulating genes in endurance athletes with long leukocyte telomeres. *J. Appl. Physiol.* **2016**, *120*, 148–158. [CrossRef]

63. Rae, D.E.; Vignaud, A.; Butler-Browne, G.S.; Thornell, L.E.; Sinclair-Smith, C.; Derman, E.W.; Lambert, M.I.; Collins, M. Skeletal muscle telomere length in healthy, experienced, endurance runners. *Eur. J. Appl. Physiol.* **2010**, *109*, 323–330. [CrossRef] [PubMed]

64. Muniesa, C.A.; Verde, Z.; Diaz-Urena, G.; Santiago, C.; Gutierrez, F.; Diaz, E.; Gomez-Gallego, F.; Pareja-Galeano, H.; Soares-Miranda, L.; Lucia, A. Telomere Length in Elite Athletes. *Int. J. Sports Physiol. Perform.* **2017**, *12*, 994–996. [CrossRef]

65. Denham, J.; Nelson, C.P.; O'Brien, B.J.; Nankervis, S.A.; Denniff, M.; Harvey, J.T.; Marques, F.Z.; Codd, V.; Zukowska-Szczechowska, E.; Samani, N.J.; et al. Longer leukocyte telomeres are associated with ultra-endurance exercise independent of cardiovascular risk factors. *PLoS ONE* **2013**, *8*, e69377. [CrossRef]

66. Diman, A.; Boros, J.; Poulain, F.; Rodriguez, J.; Purnelle, M.; Episkopou, H.; Bertrand, L.; Francaux, M.; Deldicque, L.; Decottignies, A. Nuclear respiratory factor 1 and endurance exercise promote human telomere transcription. *Sci. Adv.* **2016**, *2*, e1600031. [CrossRef] [PubMed]

67. Cherkas, L.F.; Hunkin, J.L.; Kato, B.S.; Richards, J.B.; Gardner, J.P.; Surdulescu, G.L.; Kimura, M.; Lu, X.; Spector, T.D.; Aviv, A. The association between physical activity in leisure time and leukocyte telomere length. *Arch. Intern. Med.* **2008**, *168*, 154–158. [CrossRef] [PubMed]

68. LaRocca, T.J.; Seals, D.R.; Pierce, G.L. Leukocyte telomere length is preserved with aging in endurance exercise-trained adults and related to maximal aerobic capacity. *Mech. Ageing Dev.* **2010**, *131*, 165–167. [CrossRef]

69. Ludlow, A.T.; Zimmerman, J.B.; Witkowski, S.; Hearn, J.W.; Hatfield, B.D.; Roth, S.M. Relationship between physical activity level, telomere length, and telomerase activity. *Med. Sci. Sports Exerc.* **2008**, *40*, 1764–1771. [CrossRef]

70. Puterman, E.; Lin, J.; Blackburn, E.; O'Donovan, A.; Adler, N.; Epel, E. The power of exercise: Buffering the effect of chronic stress on telomere length. *PLoS ONE* **2010**, *5*, e10837. [CrossRef]

71. Werner, C.; Furster, T.; Widmann, T.; Poss, J.; Roggia, C.; Hanhoun, M.; Scharhag, J.; Buchner, N.; Meyer, T.; Kindermann, W.; et al. Physical exercise prevents cellular senescence in circulating leukocytes and in the vessel wall. *Circulation* **2009**, *120*, 2438–2447. [CrossRef]

72. Sanft, T.; Usiskin, I.; Harrigan, M.; Cartmel, B.; Lu, L.; Li, F.Y.; Zhou, Y.; Chagpar, A.; Ferrucci, L.M.; Pusztai, L.; et al. Randomized controlled trial of weight loss versus usual care on telomere length in women with breast cancer: The lifestyle, exercise, and nutrition (LEAN) study. *Breast Cancer Res. Treat.* **2018**. [CrossRef]

73. Chilton, W.L.; Marques, F.Z.; West, J.; Kannourakis, G.; Berzins, S.P.; O'Brien, B.J.; Charchar, F.J. Acute exercise leads to regulation of telomere-associated genes and microRNA expression in immune cells. *PLoS ONE* **2014**, *9*, e92088. [CrossRef] [PubMed]

74. Li, B.; Oestreich, S.; de Lange, T. Identification of human Rap1: Implications for telomere evolution. *Cell* **2000**, *101*, 471–483. [CrossRef]

75. Sfeir, A.; Kabir, S.; van Overbeek, M.; Celli, G.B.; de Lange, T. Loss of Rap1 induces telomere recombination in the absence of NHEJ or a DNA damage signal. *Science* **2010**, *327*, 1657–1661. [CrossRef] [PubMed]

76. O'Connor, M.S.; Safari, A.; Liu, D.; Qin, J.; Songyang, Z. The human Rap1 protein complex and modulation of telomere length. *J. Biol. Chem.* **2004**, *279*, 28585–28591. [CrossRef] [PubMed]

77. Martinez, P.; Blasco, M.A. Telomeric and extra-telomeric roles for telomerase and the telomere-binding proteins. *Nat. Rev. Cancer* **2011**, *11*, 161–176. [CrossRef] [PubMed]

78. Cluckey, T.G.; Nieto, N.C.; Rodoni, B.M.; Traustadottir, T. Preliminary evidence that age and sex affect exercise-induced hTERT expression. *Exp. Gerontol.* **2017**, *96*, 7–11. [CrossRef] [PubMed]

79. Zietzer, A.; Buschmann, E.E.; Janke, D.; Li, L.; Brix, M.; Meyborg, H.; Stawowy, P.; Jungk, C.; Buschmann, I.; Hillmeister, P. Acute physical exercise and long-term individual shear rate therapy increase telomerase activity in human peripheral blood mononuclear cells. *Acta Physiol.* **2017**, *220*, 251–262. [CrossRef] [PubMed]

80. Puterman, E.; Weiss, J.; Lin, J.; Schilf, S.; Slusher, A.L.; Johansen, K.L.; Epel, E.S. Aerobic exercise lengthens telomeres and reduces stress in family caregivers: A randomized controlled trial—Curt Richter Award Paper 2018. *Psychoneuroendocrinology* **2018**, *98*, 245–252. [CrossRef]

81. Sallam, N.; Laher, I. Exercise Modulates Oxidative Stress and Inflammation in Aging and Cardiovascular Diseases. *Oxid. Med. Cell. Longev.* **2016**, *2016*, 7239639. [CrossRef]

82. Rosado-Perez, J.; Mendoza-Nunez, V.M. Relationship Between Aerobic Capacity with Oxidative Stress and Inflammation Biomarkers in the Blood of Older Mexican Urban-Dwelling Population. *Dose-Response* **2018**, *16*, 1559325818773000. [CrossRef]

83. Oeseburg, H.; de Boer, R.A.; van Gilst, W.H.; van der Harst, P. Telomere biology in healthy aging and disease. *Pflugers Arch.* **2010**, *459*, 259–268. [CrossRef] [PubMed]

84. Kurz, D.J.; Decary, S.; Hong, Y.; Trivier, E.; Akhmedov, A.; Erusalimsky, J.D. Chronic oxidative stress compromises telomere integrity and accelerates the onset of senescence in human endothelial cells. *J. Cell Sci.* **2004**, *117*, 2417–2426. [CrossRef] [PubMed]

85. Richter, T.; von Zglinicki, T. A continuous correlation between oxidative stress and telomere shortening in fibroblasts. *Exp. Gerontol.* **2007**, *42*, 1039–1042. [CrossRef] [PubMed]

86. Powers, S.K.; Ji, L.L.; Leeuwenburgh, C. Exercise training-induced alterations in skeletal muscle antioxidant capacity: A brief review. *Med. Sci. Sports Exerc.* **1999**, *31*, 987–997. [CrossRef]

87. Radak, Z.; Chung, H.Y.; Goto, S. Systemic adaptation to oxidative challenge induced by regular exercise. *Free Radic. Biol. Med.* **2008**, *44*, 153–159. [CrossRef] [PubMed]

88. Peake, J.M.; Markworth, J.F.; Nosaka, K.; Raastad, T.; Wadley, G.D.; Coffey, V.G. Modulating exercise-induced hormesis: Does less equal more? *J. Appl. Physiol.* **2015**, *119*, 172–189. [CrossRef] [PubMed]

89. Meyer, M.; Pahl, H.L.; Baeuerle, P.A. Regulation of the transcription factors NF-kappa B and AP-1 by redox changes. *Chem.-Biol. Interact.* **1994**, *91*, 91–100. [CrossRef]

90. Wu, Z.; Puigserver, P.; Andersson, U.; Zhang, C.; Adelmant, G.; Mootha, V., Troy, A., Cinti, S.; Lowell, B.; Scarpulla, R.C.; et al. Mechanisms controlling mitochondrial biogenesis and respiration through the thermogenic coactivator PGC-1. *Cell* **1999**, *98*, 115–124. [CrossRef]

91. Ji, L.L. Redox signaling in skeletal muscle: Role of aging and exercise. *Adv. Physiol. Educ.* **2015**, *39*, 352–359. [CrossRef]

92. Osthus, I.B.; Sgura, A.; Berardinelli, F.; Alsnes, I.V.; Bronstad, E.; Rehn, T.; Stobakk, P.K.; Hatle, H.; Wisloff, U.; Nauman, J. Telomere length and long-term endurance exercise: Does exercise training affect biological age? A pilot study. *PLoS ONE* **2012**, *7*, e52769. [CrossRef]

93. Azzalin, C.M.; Reichenbach, P.; Khoriauli, L.; Giulotto, E.; Lingner, J. Telomeric repeat containing RNA and RNA surveillance factors at mammalian chromosome ends. *Science* **2007**, *318*, 798–801. [CrossRef] [PubMed]

94. Schoeftner, S.; Blasco, M.A. Developmentally regulated transcription of mammalian telomeres by DNA-dependent RNA polymerase II. *Nat. Cell Biol.* **2008**, *10*, 228–236. [CrossRef] [PubMed]

95. Rippe, K.; Luke, B. TERRA and the state of the telomere. *Nat. Struct. Mol. Biol.* **2015**, *22*, 853–858. [CrossRef] [PubMed]

MDPI

St. Alban-Anlage 66

4052 Basel

Switzerland

Tel. +41 61 683 77 34

Fax +41 61 302 89 18

www.mdpi.com

Nutrients Editorial Office

E-mail: nutrients@mdpi.com

www.mdpi.com/journal/nutrients